ISLAM AND THE WEST

FRUITS OF KNOWLEDGE
ISLAM, SCIENCE AND ATHEISM

WAGIH H. MAKKY, PH.D.

authorHOUSE

AuthorHouse™
1663 Liberty Drive
Bloomington, IN 47403
www.authorhouse.com
Phone: 1 (800) 839-8640

© 2019 Wagih H. Makky, Ph.D. All rights reserved.

No part of this book may be reproduced, stored in a retrieval system, or transmitted by any means without the written permission of the author.

Published by AuthorHouse 03/28/2019

ISBN: 978-1-7283-0292-8 (sc)
ISBN: 978-1-7283-0291-1 (hc)
ISBN: 978-1-7283-0293-5 (e)

Library of Congress Control Number: 2019902525

Print information available on the last page.

Any people depicted in stock imagery provided by Getty Images are models, and such images are being used for illustrative purposes only.
Certain stock imagery © Getty Images.

This book is printed on acid-free paper.

Because of the dynamic nature of the Internet, any web addresses or links contained in this book may have changed since publication and may no longer be valid. The views expressed in this work are solely those of the author and do not necessarily reflect the views of the publisher, and the publisher hereby disclaims any responsibility for them.

هذا بلاغ للناس و لينذروا به و ليعلموا أنما هو إله واحد و ليذكر أولوا الألباب
سورة إبراهيم – آية 52

This Qur'an is a declaration for all people. And it is thus, to that they may know certainly that He who sent it is, indeed, the One God; and so that those who are endowed with discretion and understanding may heed its admonition and be ever mindful implementing it.

Surat Ibrahim – Ayah 52

....أفحسبتم أنما خلقناكم عبثا و أنكم إلينا لا ترجعون
سورة المؤمنون آية 115

*Did you think, then, that We had created you in vain, and that you would not be returnedto Us for judgement? *

Surat Al-Mu'minun – Ayah 115

يأيها الإنسان ما غرك بربك الكريم*الذى خلقك فسواك فعدلك*فى أى صورة ما شاء ربك*كلا بل تكذبون بالدين*و إن عليكم لحافظين*كراما كاتبين*يعلمون ما تفعلون
سورة الإنفطار آية 6 - 12

O humankind! What has deluded you about your Lord, the All-Gracious the One who alone created you, then fashioned you, then gave you symmetry*and in what awondrous form has He willed to compose you!*No, indeed! You have no excuse for denying faith in One God! Rather, mossurely, you belie the nearing judgment*while indeed, ever vigilant over you are guardian angels*noble ones, writing everything*they know all that you do*

Surat Al-Infitar – Ayah 6-12

Contents

Dedication ix
Prologue xi
Acknowledgment xix

Islam, Science and Atheism 1
Atheism and Science 13
Science 22
Cosmology 66
Islamic Intellectual Heritage 108
The Grand Cosmological Divine Plan 217
Islam and Science 233
Qur'anic Narrative and Modern Scientific Findings 284
The Final Verdict 303

Epilogue 345

Dedication

TO

Dr. Mahathir Mohammad
The unique symbol of Islamic technical and intellectual renaissance that shines from the farthest lands of Islam to illuminate the hearts and souls of aspiring young Muslims everywhere to reach the pinnacles of their fields with competence and integrity

And

To the Muslim Youth that hunger for an Islamic role model

Work on this book goes back many years. Dedication to Dr. Mahathir Mohammad has been a fixture in my mind long before his return to Malaysian Premiership in 2018. It is in appreciation of his gallant efforts in materially modernizing his society while proudly and steadfastly holding unwaveringly to the universal validity of the Islamic outlook on life that prompted me to that dedication. Bearing in mind that I dedicated the previous books of this series to some of the greatest companions of Prophet Mohammad (محمد - عليه الصلاة و السلام) for their impact on the history of Islam and all humanity, the gravity of my consideration of this current dedication should be understood.

Prologue

This is the third book in the series discussing the relationship between "Islam and the West" that spans the entire fifteen centuries of mutual influence. While continuity is the hallmark of Islam and its history, the West vis-a-vis Islam is historically split into two main eras with many subdivisions. There is originally the West as Christendom when the Church overwhelmed all facets of life of the individuals leading to suppression of all aspects of freedom and creating material backwardness. And there is the modern era that began with the renaissance and the enlightenment characterized by rejection of Christianity and the advance of atheism. The second of these books "Islam and the West" subtitled "Bitter Harvest of Ignorance" deals with the contemporary episode of the Western perception of what it calls "Islamic Terrorism". The first book in the series subtitled "Why Do They Hate Us So Much?" tracks the apparent reasons that prompt the West to acquire innate hatred for Islam and uncovers the perpetrators of such fatal approach. This third book covers the atheistic era and responds to Western intellectual fiction that equates Christianity with religion in general and falsely assumes that Islam is fundamentally merely Christianity by a different name. Consequently, just like Christianity, Islam must be irreconcilable with science. After putting this falsehood to rest, the book explores issues of Islamic notions' compatibility with scientific facts as understood in the twenty first century.

Human moral and historical evolution required that the universal message of Islam had to be sponsored by people not tainted by preconceived notions of grandeur while at the same time they had to be in close physical and territorial contact with peoples in major centers of civilizations that are. The Arabs perfectly fulfilled these requirements at the time when Mohammad (محمد - عليه الصلاة و السلام) received his mandate to convey God's (الله - سبحانه و تعالى) message of Islam to all humanity. This thesis is elaborated on, fully explained and can be examined and assessed in the first book of this series. The downside of the pre-Islamic Arabs having been outside history is that they were having very little intellectual foundation, other than their very limited oral tradition, to build on after the passing of Mohammad (محمد - عليه الصلاة و السلام) to continue the process of constructing a unique way of life according to the Creator's (الله - سبحانه و تعالى) guidance. Thus, subtly and rapidly shifting from being an ethnic group (Arabism) into a universal brotherhood encompassing all humanity (Islam), it was inevitable that the Arabs as the premier standard bearers of Islam at the time had to strive to develop under their auspices a naturally unique "Islamic Civilization". Faced with this dilemma however, they were not starting from scratch. They had in their own language the Qur'an which is a unique reference believed to contain absolute facts that can form helpful approaches to all knowledge. Their task (and all Muslims' afterwards) was to diligently uncover and understand its intricacies. Therefore, what is historically known as the "Islamic Civilization" is in essence a social human structure built on and around the Qur'an which is believed to be the literal word of God (الله - سبحانه و تعالى). The prophetic tradition or the Sunnah (السنة) in all its components is fundamentally an elaboration to that reference which expansively explains the relatively very little essential information immediately needed to facilitate organizing the society. On the other hand, it gives brief general hints concerning intellectual ponderings about nature exhibited by the companions of Mohammad (محمد - عليه الصلاة و السلام) leaving the responsibility of finding out answers to these enquiries to the following generations of Muslims till the end of time. Compounding the perceived intellectual problem facing the original Muslims, the Qur'an narrated histories and belief systems of previous ancient nations and peoples that they knew nothing about.

It is fascinating (and an unassailable proof of the validity of Islam) that when the need arose, great personalities in politics, the military, intellect, etc. appeared on the stage to carry out what was necessary. In less than a century Islam dominated the world geographically, politically and militarily but not intellectually. To fulfill God's (الله - سبحانه و تعالى) mandated task, Muslim thinkers had to reach out to other cultures and belief systems to find out explanations to many narratives in the Qur'an for the sake of completion. The other two monotheistic religions of Judaism and Christianity seemed at the time to be natural primary sources for such endeavors. That is when Islamic intellectual tradition was infused with many legitimate accounts as well as forgeries. Since the Jewish Bible was thought of (more or less) as an historical description of human progress, most of the counterfeit tales in the Islamic intellectual tradition can be traced back to that source's various interpretations and is scholarly known as Israeliat (الإسرائيليات). Obviously at the time of their incorporation they were not considered forgeries. This sweeping judgment on the value and viability of tales of the Jewish Bible is a relatively modern phenomenon passed by biblical scholars themselves in accordance with current scientific findings. To the Jewish or Christian believer, these findings are mostly inconvenient and counter intuitive and as such are dismissed as attacks on the faith by fundamentalist Jews and Christians. On the other hand, at the present time reasonable Jews and Christians consider these tales merely parables and allegories rather than facts as was believed for millennia. This is done to keep the faith while enjoying the fruits of science and technology at the same time. Pioneering Muslim thinkers incorporated biblical assertions into their own works as truthful statements due to lack of any other references. However, the Qur'an gave fairly clear outlines to how creation of the universe and therefore humanity took place. Thus, it gives a completely different narration of human history. The Qur'an makes it plain that when the first human "Adam" (آدم - عليه السلام) was created, he was given the highest degree of knowledge as far as God's (الله - سبحانه و تعالى) all other creatures are concerned. Left to his and his progeny's devices after leaving "Al-Jannah - Paradise" (الجنة), somehow humanity lost that knowledge during its pursuit of survival on earth. Gradually it started developing new modes of knowledge while holding firmly to the concept of an all-powerful being overseeing its activities where different local groups gave that entity various locally invented shapes, forms and names. The common thread in all these efforts was the personification of these gods in forms easily detected by human senses. That was offensive to the One God (الله - سبحانه و تعالى) who created humans in the first place. Therefore, *local* messengers and prophets were sent to these groups to correct their ways of thinking spanning the entire human history from "Adam" (آدم - عليه السلام) to Jesus Christ (عيسى – عليه السلام). Finally, the universal message of Islam arrived on the scene and the concept of the One Omnipotent God (الله - سبحانه و تعالى) is restored to human conscious. Deeply ingrained in Muslims' faith regardless of their intellectual capacities is the concept of God's (الله - سبحانه و تعالى) existence beyond human perceptions of space and time as mentioned in numerous places in the Qur'an and the Sunnah (السنة). This is the starkest divergence between Islam and all other religions including monotheistic ones. This explains why for Muslims Christianity's garden variety attempts at personification of God (الله - سبحانه و تعالى) in Jesus Christ (عيسى – عليه السلام) is beyond reason and is untenable. Paradoxically, apart from the clear cut concept of God (الله - سبحانه و تعالى), great numbers of Muslims go along with the conclusions implied in the personification of gods adopted by other systems of faith. For example, they buy into the idea of instantaneous creation (that is the hallmark belief of the major Creationist Christian group) instead of an evolutionary phased process. To them billions of years spent in bringing about creation of the universe is unbecoming of the unlimited power of God (الله - سبحانه و تعالى) unthinkingly disregarding their own unique belief that God (الله - سبحانه و تعالى) is not bound by human perceptions of space and particularly of time. Additionally, and instinctively, unsophisticated Muslims recall the many works and extensive writings of the great ancient scholars which are by default filled with misconceptions of other systems of faith such as the Israeliat (الإسرائيليات) and join in the condemnation of findings of science. Ironically these unsophisticated Muslims staunchly defend arguments that have actually nothing to do with Islam with no roots whatsoever in the faith. The confusion stems from mixing up ideas contained in other belief systems, particularly monotheistic ones, describing details of events and personalities with Qur'anic general approach to these same events and personalities. Nonetheless, this confusion rarely touches on the core tenets and beliefs of Islam even among the least sophisticated which explains the stubborn persistence of these ideas appearing in the intellectual heritage of Islam over the centuries.

Islam presented a formidable intellectual as well as religious dilemma to both Judaism and Christianity because of its basic tenets and universal outlook. By the time Mohammad (محمد - عليه الصلاة و السلام) declared his message, Christianity had already expropriated everything in Judaism, subjugated its adherents and physically and religiously persecuted them. Scholars of both religions invented and found great solace in propagating the imaginary and baseless thesis of a linkage between Islam and their own man-made dogmas. Thanks to Islamic tolerance, these arguments found market during the great translation efforts of the third and fourth Hijrah centuries and still echo in Western perceptions of Islam to the present time. The work at hand is an attempt at sifting through these counterfeit arguments and purging them from the Islamic intellectual tradition. Because science has no religion, **Islam after all these centuries of human knowledge's progress does not need to adopt any alien notions from other belief systems any more** but should directly reconcile itself with this knowledge that Muslim scholars contributed so much to. Therefore, purely Islamic explanations derived from the Qur'an and the Sunnah (السنة) in line with acquired human knowledge should be sought after. That is a task the current endeavor humbly takes upon itself. In attempting to fulfill this most important task, this work calls for an urgent **"Paradigm Shift"** in Islamic intellectual process. This advocated paradigm shift is similar to that undertaken in science when current approaches fail to meet observations. In both cases revolutionary new ideas are pursued. Of particular interest, new global meanings for Qur'anic terms and words that reconcile it with the findings of modern science are called for. It is shown that the Arabic Language is divinely structured to promote that effort. That call does not in any conceivable way infringe upon the fundamental tenets of Islam such as creed, rituals, rules and regulations, etc. since they by definition deal with humans whose nature and basic needs do not change with the passage of time. On the other hand, human knowledge is time dependent. It is an article of faith for Muslims that the Glorious Qur'an as the literal word of the All-Knowing God (الله - سبحانه و تعالى) contains all there is to know and its statements are valid till the end of time. Therefore, its interpretation has to accommodate not only what is conventionally agreed to by humans in their daily life but also what science finds out about the universe especially when the newly acquired knowledge is counter intuitive. All Qur'anic interpretations, ancient and modern, utilize the conventional usage of the Arabic language. In numerous cases this leads to pathetic presentation of Islam and may actually offend the intellectual integrity of Muslims themselves. The book tackles this problem head-on and gives many examples of how global meaning of Qur'anic terminology derived from the Arabic language accommodates what science may affirm.

Modern atheism is based on the essential requirement of science that any theory must explain observables, subject itself to falsification and most importantly make predictions that can be experimentally verified before it can be accepted. As far as atheists are concerned, Existence of God (الله - سبحانه و تعالى) is nothing more than a theory to be verified. Historical failure of both Judaism and Christianity to adequately meet these criteria gave birth to Western Atheism that spread all over the world with the material rise of the West. Innate Western hatred of Islam allowed its dismissal without actually subjecting it to the rigorous scrutiny of science. This work painstakingly carries out that process to clearly show that Islam in its fundamental aspects (unlike Judaism and Christianity) is reconcilable with science. It is shown that the two fundamental pillars of Islam; the Qur'an and the Sunnah (السنة) contain countless predictions. Some came to pass while others are waiting for human knowledge to reach certain levels before descriptions of events such as creation of the universe for example are adequately understood. Many current writings by competent and not so-competent Muslim scientists attempt to prove that latest findings of science especially cosmology as implied in the Qur'an by giving their own interpretations of some of its Ayahs. This is done in the spirit of defending Islam. However, science as opposed to the Qur'an is by definition an evolving system which should eventually render these efforts as futile. The work at hand essentially reverses this approach as a matter of principle. It examines whether there is an acceptable interpretation of the Qur'an and the Sunnah (السنة) in accordance with the global meaning of words of the Arabic Language as opposed to common everyday usage in light of the called for paradigm shift. The global nature of the meaning of Qur'anic and Sunnah (السنة) words is not affected by any modifications that may take place in human knowledge. This represents a more rigorous methodology that can easily withstand the test of time while showing the unassailable nature of Islam at the same time.

Authors of most books popularizing science try to encourage readers by stating that their books do not require special knowledge of the corresponding science but rather a general understanding of its fundamentals. An average level of general education is normally what these books claim to call for to adequately understand the material involved. This book is all about science and its correlation with the conceptual principles of Islam. In it the latest findings of science especially Physics and Cosmology are the main subjects explored. This is done to determine whether or not the close to fifteen centuries old statements of Islam reconcile with these findings. To accomplish such task, only qualitative explanations of the state-of-the-art conclusions in science are given. This approach is more than adequate due to the fact of Islam's inherent impressive simplicity and logical mandates. Therefore, it is easy to demand no particular education or intellectual capacity of any kind to follow this undertaking. What is necessarily essential is merely common sense which normal individuals are naturally endowed with. Most importantly, the reader should have the ability to follow and accept the many counter intuitive ideas that are the hallmark of modern science. It is believed that persons who elevate dogmas of any belief system above common sense and logic will find it hard to agree with any conjectures of this endeavor. They would not be able though to refute them while dismissing them out of hand. Most regrettably, it is anticipated that some Muslim scholars and thinkers may prefer the clearly inadequate stagnant old interpretations over the new paradigms which are based on the irrefutable scientific facts that may be modified in the future as approximations but never rejected as wholly wrong. This position of extreme intellectual conservatism on part of these individuals might be the result of deference to the opinions of the great ancient Muslim scholars or to the desire to preserve a presumed privileged social status. Either way, this is squarely in contravention with the well-established spirit and practices of Islam. To make it easy for the reader to check any of the facts incorporated in the text and to avoid the usual confusion plaguing most books dealing with Islamic and Arabic names of persons, places, events, etc., all such names are rendered in their original Arabic script after given in English. Additionally, when God (الله - سبحانه و تعالى) or His attributes are mentioned, the standard Arabic praise follows. The same is done according to Islamic tradition when Prophets and messengers of God (الله - سبحانه و تعالى) are mentioned.

Acknowledgment

Discussions with Professor Dr. Abd Elgelil Mostafa stimulated many ideas in this book and I am thankful for that. Most of my references and many valuable literature were brought from Scotland by Dr. Ahmad Nassar and his gracious wife Patricia. I would like to express my gratitude for their efforts. As always, contributions and encouragement of my wife Lin Vandenberg are limitless. Her curious and very thoughtful questions as well as her boundless appreciation for Islam opened immense vistas for me and made me think many issues over. That is clearly reflected everywhere in the book. My thanks go also to the team at Authorhouse that helped publish the book.

I would have liked to expand the text much more and include references and recommendations for further reading. However, Authorhouse independently price books based on the number of pages which limited what to be included to make the book available at a reasonable price.

Dr. Wagih Makky
makky.creativity@gmail.com

Friday, First of Rajab 1440 H / March 8, 2019
 Al-Shorouq City, Cairo, Egypt

Islam, Science and Atheism
The Essence of Science

INTRODUCTION

Religious people the world over regardless of their specific belief system dismiss atheistic protestations out of hand without bothering to refute their merits. Muslim intellectuals and Muslim scholars suffer from even more acute failings. They indignantly and uncritically join the swarm for no realistically legitimate reasons. They unthinkingly ignore the maxim that objections of others who do not share their beliefs to any ideas do not necessarily reflect how Islam view most issues. In a wide variety of cases this approach delegitimizes Islam rather than upholds it. Islam differs from all other belief systems in that it strongly encourages reason and abhors dogmatic irrationality. This is true while at the same time giving answers to inquiries about every minute detail concerning either the universe and what it entails or the mundane issues of human beings' daily lives. As such it is convincingly argued by Muslims that Islam is valid in time and space with no exception. Islamic rules and regulations in terms of what is known as the Shari'a (الشريعة) deal with human affairs. Thanks to the negligible history of humans on earth compared to other creatures, human needs to be regulated did not change much if at all over humanity's entire history. That implies the stability of the Shari'a (الشريعة) and its validity under all circumstances or conditions. Contrary to the prevailing idea especially among Westerners, rules and regulations of the Shari'a (الشريعة) constitute only a largely minor part of the Qur'anic narrative.

That is obviously not the case concerning collective human

knowledge which improves with every passing day. On the other hand, the Qur'anic text is singularly dominated by the emphasis on the Oneness (التوحيد) of God (الله - سبحانه و تعالى) that can be deduced through His material creations and their implied signs. This is emphatically stated in Ayah 53 in Surat Fussilat (سورة فصلت). Remarkably, any thoughtful probing of Islam's descriptions of material facts reveals general rather than detailed statements subject to the rules of the Arabic Language. Therefore, these statements are open to interpretations which is exactly what a human being is required to do according to the basic tents of Islam to appreciate God's (الله - سبحانه و تعالى) sovereignty. Thus, Muslims have a duty to ponder their stands for or against any arguments especially those that negate their own beliefs. An excuse that the great ancient scholars settled such issues is patently unacceptable. It goes without saying that intellectual laziness is unambiguously un-Islamic. It may be that dismissing atheistic arguments out of hand as currently practiced by most Muslims is due to ignorance of the underlying intricacies of the subject matter. However, it is clear that in numerous instances atheistic protestations are actually borne by some fundamental Islamic principles. Or at least they do not controvert some basic Islamic notions. It should also be obvious that historically these atheistic challenges to religious arguments were solely and uniquely directed at the Western Church and its doctrines. As such they have no bearing whatsoever on Islam, its tents, its practices or its rational approaches. Muslims, therefore, should not be cheering counter arguments that have nothing to do with them but should discuss what atheists have to say with open mind. They should refute their conclusions while accepting what is valid in their premises.

Critics of "Western Civilization" attribute so many transgressions to it such as the extermination of indigenous peoples during the brutal conquest of the New World, the inhumane treatment of the natives during the age of imperialism, dragging most of humanity into its internal quarrels leading to two savage world wars and numerous other misdeeds. **But in the opinion of Muslims, the worst offense by far has been its legacy of giving birth and respectability to the notion of "Atheism". It did that through its brute force assertion that the word "Religion" is synonymous with "Christianity".** Western

Civilization and Christianity (in its western form) became one and the same in the western mundset. It physically accomplished that using missionary zeal associated with extreme coercion at the expense of defenseless peoples on every continent on the face of the earth. Intellectually Judaism paid the price in the early days and continued to suffer persecution overtly and covertly till the present day. Jews came to the brink of extinction at the hands of Western Christians for their denial of Jesus Christ (عيسى – عليه السلام) even twenty centuries after his proclamations. The religion preaching peace and love has been historically the most brutal inhumane force thanks to its adoption by the Western Civilization. Eastern Christianity showed none of these signs but rather coexisted peaceably with others. After the Age of Enlightenment, atheism's respectability is directly derived from its uncompromising confidence in humanity's ability to find truth through observation. That is the exact opposite of Christianity's dogma. Having imprinted the idea that "Religion" is synonymous with "Christianity" in the minds of people, Western Civilization made science and religion irreconcilable. This is shown in the writings of almost all prominent scientists in every conceivable field. The fact of the matter is simply that religion is not and cannot be synonymous with Christianity. Christianity's dogmas do not contravene the precepts of only science but also those of Islam as well. Therefore, atheism has to be judged not against Christianity but rather against the fundamental conceptions of Islam if religion is to be assaulted. While atheism is essentially built on the supremacy of scientific achievements and the progress of human knowledge, Islam consists of a vast array of ideas and perceptions including great appreciation of human intellect. In essence one of the most basic foundations of faith in Islam is the realization of the power of God (الله - سبحانه و تعالى) the Creator through thought and observation. Obviously, that concept cannot be irreconcilable with science by any stretch of the imagination. However, this is not obvious when one contemplates the current status of Muslims. The seeming contradiction stems from the long stagnation of Islamic thought and the widely perceived self-serving inaptitude of some modern day Muslim scholars and not from the underlying notions of Islam itself.

It is assumed that the fundamental difference between humans and other mammals is language. When over eons Homo-Sapiens communities developed languages, they were able to communicate among themselves and exchange ideas. Presumably, the overriding concern was to understand their universe starting with their own existence. Mythology was apparently the method of choice to explain natural phenomena that could not be interpreted in a straightforward approach. Contemplating nature humanity realized that there are causes to everything taking place around it. Step by step that realization gave impetus to methodical investigation and eventually the rise of what became known as rigorous science. Human senses were the first and foremost judge of any observable phenomenon. However, the thought process itself took precedence and science became more and more abstract. Almost all scientific explanations are reduced to abstract ideas that find their origins in Mathematics which is the most abstract of all. For example, biologists reduce life itself to chemistry and chemical processes albeit very complicated ones. Chemistry is nothing more than relationships between materials' elements which are in essence merely formed by elementary particles arranged in certain ways. These in turn are governed by laws of physics or more precisely quantum physics. These laws are described in mathematical equations and formulations which abstractly present the logical relationships among human conceptual thoughts. The same is true of any other branch of science. Even the origin and structure of the universe is reduced to mathematical representations.

Mathematics itself is a product of pure thought that assumes logical relationships among abstract objects describing natural phenomena, but not the phenomena themselves, such as numbers. These thoughts however, are usually based on fundamental irreducible assumptions as starting points. Major deadlocks in science, especially in physics, are overcome when it is discovered that some of these irreducible assumptions or axioms are wrong, or to put it more precisely, cannot be proven. The most famous one is the assumption in Euclidian geometry that two parallel lines never meet. The fallacy of this arbitrary axiom undermined plane geometry and allowed mathematicians to develop other more realistic higher dimensional geometries. Physicists found it useful to take advantage of these new geometries leading for example to the theory of general relativity which unlike previous theories enabled the description of the universe in terms consistent with observations. But physicists and mathematicians work independently. Mathematicians deal with the purely abstract whether or not the problem at hand has any applications in reality. On the other hand, physicists are only concerned with observations that are extracted from real natural phenomena. Scientists are always surprised to find that solutions to problems blocking their understanding of certain physical facts could be found in abstract mathematical approaches that were established many years or even centuries earlier. However, as was just mentioned, mathematical concepts are based on logical assumptions or axioms that cannot be reduced further and are assumed to be true. Entire body of knowledge is built on these assumptions and is accepted by scientists as formed by self-consistent facts. When real physical phenomena cannot be reconciled with these axioms, they are discarded as absolute facts but so painfully slowly and reluctantly. That is not to say that they remain excellent approximations to reality although they are no longer considered the absolute truth.

The most revealing example is the age old universal acceptance of Euclidian or plane (two dimensional) geometry to describe the universe. This geometry stems from certain fundamental axioms that were taken for granted for millennia. The day-to-day life of every human being is based on its assumptions as it describes the relationships between lines, curves, surfaces, etc. However, when dealing with the universe as a whole, it was found wanting and could not account for the known physical facts. Other multi-dimensional non-Euclidian geometries were already developed in mathematics by George Friedrich Bernhard Riemann (1826-1866) and David Hilbert (1862-1943) without any assumptions of their physical reality. Utilizing these approaches, modern cosmology gives an accurate picture of the universe that is consistent with the observed facts. Euclidian geometry may have turned out not to represent reality but it is still the staple of elementary mathematical teaching the world over. The simple reason is that it gives excellent approximations to what humans experience around them even if it cannot accurately describe the universe itself. In other words, multidimensional geometries are reduced to plane two-dimensional geometry when everyday scale of space is considered. Simply put, *a theory that explains numerous facets of reality but fails when the scales are radically changed is not inherently wrong but rather should be dealt with as an approximation for a more general theory under certain simplifying conditions.* That perception entered public lore with the advent of the theory of relativity. Understanding of space and time was radically altered from the universally accepted Newtonian approach. Newton based his theories on the fundamental ideas of absolute space and absolute time. Adding to these was the idea, which Newton himself found irrational but essential to his theory, of instantaneous action at a distance. A fundamental implication of Newtonian kinematics is that an object can gain limitless speed. In a stroke of genius, Newton succeeded in explaining terrestrial as well as celestial phenomena under the same laws. Scientists for over two centuries held his work in awe and discarded anything that contradicted it. When special relativity put an upper limit to the speed any material object can attain assuming at the same time the constancy of the speed of light, it overturned all Newton's assumptions. That does not mean that Newtonian Physics is wrong but rather that it is an excellent approximation pertaining to everyday's experiences. Relativity is reduced to Newton's theory at speeds much lower than that of light. It is of fundamental importance to understand that relativity brought its own assumptions. If at a future time phenomena are found where relativity cannot account for then these assumptions would have to be modified. Again relativity under these hypothetical conditions could not be considered wrong but rather as an excellent approximation to the more general theory whatever it is.

The important conclusion of this discussion is that *a new more general and more successful theory replacing an old one that failed to account for certain physical facts should imply the old theory under simplified conditioned.* The old theory is by no means wrong but just limited in its scope. It is also essential to understand that new theories come up with their own new assumptions that create new problems as well. What that entails, is the fact that no discernable end to acquiring new knowledge can probably be attained. Anecdotes about Max Karl Ernst Ludwig Planck's (1858-1947) years in school can be found to this effect in the 1990 book by his great protégé's Werner Heisenberg

(1901-1976) 1990 book *Across the Frontiers* which tell the story that his teacher advised him against the study of physics since after all it was essentially finished with, so that for anyone who wanted to do active scientific research it would scarcely be worthwhile to go into this field. Obviously Planck in 1900 proposed the idea of "Quanta" of energy which begat the revolutionary new quantum theory to push through the dawn of modern physics with incalculable implications for almost every field of science not only physics. The situation is elegantly described by another of the great builders of quantum physics during the first half of the twentieth century Richard Feynman (1918-1988). In his 1964 *Messenger Lectures* at Cornell University he asserted that when one speaks of a theory one is never definitely right, scientists can only be sure they are wrong. Therefore, physicists are trying to prove themselves wrong as quickly as possible, because only in that way can they find progress. Speaking of the changing thought patterns that result in scientific revolutions, Heisenberg wonders that such revolutions in science have actually been possible at all but he wanted to know how then have they come about? The answer that comes readiest to everyone is because the new ideas are simply right and the old are wrong. This answer presupposes that in science it is always the right answer that prevails. But that is by no means the case. He supposes that another explanation for the success of revolutions is that they come to pass because scientists gladly defer to the authority of a strong revolutionary personality, such as Einstein. However, the internal resistance against a change in the thought pattern are much too strong to be overcome by the authority of any one man. He then gives the correct answer in his opinion as a perception of scientists that with the new pattern of thought they can achieve greater success in their science than with the old; because the new system proves to be fruitful. It is always stated that Newtonian mechanics is implied in the theory of relativity as long as objects' speeds do not approach that of light. What is less known, is that Newtonian mechanics is also contained in the quantum theory. It is as a limiting case in which the events are completely objectified and the interaction between the object under investigation and the observer can be neglected.

Change from the old to the new more general theory has been exhaustively studied by the American physicist, historian and philosopher of science Thomas Samuel Kuhn (1922–1996) of Massachusetts Institute of Technology (MIT) who has coined the expression **"Paradigm Shift"** to describe the structure of scientific revolutions. History is replete with the necessity of such paradigm shifts. It is probably in the fields of science, physical and social alike that this necessity is starkly clearest. Scientists are known to have changed their standards (paradigms) time and again due to the acquisition of new knowledge. New experimentally verified theories are adopted and old ones are pushed aside. New theories that are not (or cannot be) verified experimentally, due technologies beyond the realm of feasibility, but solve existing problems are also taken up. Major paradigm shifts are designated as revolutions in the scientific thought. This is the type of paradigm shift sought in Islamic thought if it is meant to join the mainstream of human progress.

Paradigm Shifts in Science and Their Relevance to Islam's Outlook

To illustrate what is meant by the inevitability of a paradigm shift in the advancement of science, Kuhn in his landmark 1962 book *The Structure of Scientific Revolutions* describes the situation when a long accepted idea or theory is replaced by a new one when it fails to explain certain observations or it is incapable of solving persisting theoretical problems. He expresses this situation by stating that since new paradigms are born from old ones, they ordinarily incorporate much of the vocabulary [and terminology] and apparatus, both conceptual and manipulative, that the traditional paradigm had previously employed. But they seldom employ these borrowed elements in quite the traditional way. **Within the new paradigm, old terms, concepts, and experiments fall into new relationships one with the other**. If Islamic thought in all its themes is to be regarded, as it must, as merely human scientific activity then, one can readily see the inevitability of a paradigm shift in the usage of the vocabulary and terminology in the traditional Islamic thought for it to survive and thrive. Consequently, expressions and words of the Holy Qur'an itself should be understood in fresh new light.

That is not to understate or undermine the conceptual interpretations or the great works of the glorious past. Borrowing from the history of science, Kuhn is quoted to have written that the laymen who scoffed at Einstein's general theory of relativity because space could not be "curved" – it was not that sort of thing – were not simply wrong or mistaken, nor were the mathematicians, physicists, and philosophers who tried to develop a Euclidean version of Einstein's theory. What had previously been meant by space was necessarily flat, homogeneous, isotropic, and unaffected by the presence of matter. If it had not been, Newtonian physics would not have worked. **To make the transition to Einstein's universe, the whole conceptual web whose strands are space, time, matter, force, and so on, had to be shifted and laid down again on nature whole.** Additionally, consider, for another example, the men who called Copernicus mad because he proclaimed that the earth moved. They were not either just wrong or quite wrong. Part of what they meant by "earth" was fixed position. Their earth, at least, could not be moved. Correspondingly, Copernicus' innovation was not simply to move the earth. Rather, it was *a whole new way of regarding the problems of physics and astronomy*, one that necessarily **changed the meaning** of both "earth" and "motion". Without those changes the concept of a moving earth was mad.

Since the thrust of this study is the relationship between Islam and science, Kuhn's insightful ideas are quite valid in describing what is needed to revolutionize Islamic thought adapting it to the spirit of the twenty first century. Thus, in parallel, fixed traditional concepts and ideas stemming from the universally accepted interpretations of certain expressions and words in the Holy Qur'an and the sayings of the Prophet (محمد - عليه الصلاة و السلام) (which are by definition human undertakings) are not simply wrong but rather outdated due to the newly acquired knowledge. That does not belittle the greatness of their constructs or that of their associated personalities. One can recall Newton's quote when he was praised for his great accomplishments that **"we are in that position because we are standing on the shoulders of giants"**. Kuhn explains the emergence of new physical theories and even physical laws as a result of a competition among various groups adopting different methodologies where the proponents of competing paradigms practice their trades in different worlds. The result is that practicing in different worlds, the two groups of scientists see different things when they look from the same point in the same direction. Again, that is not to say that they can see anything they please. Both are looking at the world, and what they look at has not changed but in some areas they see different things, and they see them in relations one to the other. That is why *a law that cannot even be demonstrated to one group of scientists may occasionally seem intuitively obvious to another.* It is quite intriguing to see the resemblance between this illustration of the situation in science and the current situation in Islamic thought. Certain things have to happen before a new universally accepted paradigm is adopted such that equally, it is why, before they can hope to communicate fully, one group or the other must experience *the conversion that we have been calling a paradigm shift*. Just because it is a transition between incommensurables, the transition between competing paradigms cannot be made a step at a time, forced by logic and neutral experience; it must occur all at once (though not necessarily in an instant) or not at all. Since progress implies moving forward, it is therefore essential that traditional Muslim scholars convert to new methodologies leaving behind the paradigms of the glorious past as it might have been. This process, if it ever takes root, cannot be expected to be easy or to take place in short order. There is nothing unusual about that in the history of science itself as Copernicanism made few converts for almost a century after Copernicus' death. Newton's work was not generally accepted, particularly on the continent, for more than half a century after the *Principia* appeared. Max Planck, surveying his own career in his *Scientific Autobiography,* sadly remarked that a new scientific truth does not triumph by convincing its opponents and making them see the light, but rather because its opponents eventually die, and a new generation grows up that is familiar with it. To anticipate the future of Islamic thought in light of what has happened in science, the following paragraph is quoted from *The Structure of Scientific Revolutions* by Thomas S. Kuhn. (The transfer of allegiance from paradigm to paradigm is a conversion experience that cannot be forced. Lifelong resistance, particularly from those whose productive careers have committed them to an older tradition of normal science, is not a violation of scientific standards but an index to the nature of scientific research itself. The source of resistance is the assurance that the older paradigm will ultimately solve all its problems, that nature can be shoved into the box the paradigm provides....... part of what the acceptance of Ohm's Law {the most fundamental low of electricity} demanded was a **redefinition** of both "current" and "resistance"; if those terms continued to mean what they meant before, Ohm's Law could not have been right that is why it was so strenuously opposed".)

Atheism and Science

MAIN ARGUMENTS OF ATHEISM

Atheism is built on certain foundational issues descending from Western Christianity's dominance over Europe till the "Renaissance" followed by the "Age of Reason". This is true because belief in a higher power is one of the most enduring legacies of humanity. Historically traditions fluctuated between worshiping one or several gods mostly representing the forces of nature. However, denying the existence of such supreme being(s) was expressed by some individuals throughout history that represented a distinct vanishingly small minority among ancient communities. That is until the excesses and abuses of the Western Church exceeded all humanly tolerable dimensions and it had to be reformed. Those who thought that the whole issue of faith in salvation that gave the Church its unlimited ability to exploit and manipulate its followers was a forgery and a charade to enrich the priesthood decided to undermine Church's hegemony by rejecting the idea of God (الله - سبحانه و تعالى) altogether and modern atheism took concrete form. Unfortunately for the Church this was also the time of enlightenment and the rise of rational scientific thinking based on observations. While both the original and the reformed Church(s) survived in one shape or the other, atheism irreversibly dwarfed their methodologies. It is fair to assert that **atheism is Western Christianity's bequest to humanity**. Looking at the issue from the other side one can clearly see that **atheism is also the legitimate child of Western enlightenment**.

Enlightenment bred the scientific method and eventually science and atheism became intertwined with them in the Western mind.

Success of science in explaining many physical phenomena in reasonable and appealing terms overwhelmed Western Christianity. From that point on a considerable segment of individuals who thought of themselves as thinkers and particularly as scientists became atheists rejecting religion wholesale. The militant among them took it upon themselves to destroy any vestiges of faith in the public arena and attributed backwardness to religion and faith in God (الله - سبحانه و تعالى) as the Creator and Sustainer of life. Therefore, it is not unreasonable to claim that **Western Christianity and its institutional Church begat modern day atheism** while Eastern Christianity and its associated Church never went through such convulsions. In Muslim lands one hears about atheistic tendencies at the present time especially among the disenfranchised young. However, the hard core of these tendencies owes its ideas to a desperate attempt to emulate the materially prosperous West with no roots whatsoever in any indigenous tradition. In a very straightforward manner the West exchanged science for God (الله -سبحانه و تعالى). Atheism became the newly found Western religion although atheists cringe at that portrayal. Currently it is very common for scientists once reaching a respectfully meaningful level of competence to declare themselves as atheists. Western scientists who find comfort in faith are forced to allege that the two fields of science and religion are mutually exclusive and to create an unbridgeable barrier between them. To preempt any possibility of science undermining their faith, they erect a "firewall" between the two subjects asserting that both subjects cannot be discussed simultaneously as they are mutually exclusive. Like in any religion, latter day scientists developed their iconic symbols and personalities. It is universally agreed that the most famous and mostly unchallengeable scientific icon during the twentieth century and beyond is the father of the Theory of Relativity "Albert Einstein". He permanently entered the public lore the world over. His personality, career, successes, failings and outstanding contributions shed light on the progress of science and its reflections on public life. His ambivalence regarding religion and the concept of God (الله - سبحانه و تعالى) allowed both religious people and atheists alike to claim him to their side. He rejected religion out of hand but strongly believed in some hidden supreme powers in nature and its deterministic laws. He is known to have freely used the word "God" to support his arguments although he usually meant something completely different from what common folks understood. In addition to the obvious scientific value of his theories, it is instructive to explore how his peers intellectually and philosophically evaluated his work to understand how modern science is perceived, accepted and/or rejected. Therefore, both the special and the general theories of relativity are qualitatively explained in due course in the following sections. The impact of their contributions to physics transcends science into the realm of the future of humanity and of life itself.

Atheism is structured around the principles of science and its basic requirements. First and foremost, the universality of science is undeniable. Scientific facts are invariable irrespective of the language and symbols they are expressed in. The impetus for the progress of scientific investigation is the underlying assumption that fundamental physical laws are the same everywhere in the universe regardless of the scale. Laws of physics governing the behavior of objects are the same whether one is studying subatomic phenomena or galaxies many light-years across. That is basically the driving force behind the search for a "Theory of Everything" combining Quantum Mechanics and General Relativity or put another way; a "Quantum Gravity" theory that currently occupies the interests of the physics community. On the other hand, religions differ from one place to another and from one culture to another. In most cases religions contradict each other's fundamental perceptions. Most ominously, religious beliefs throughout human history caused wars and much misery. It would be utterly absurd to think of scientific arguments leading to such conflicts. Prominent among these many requirements is the indispensable need for measurable observations before passing any judgments or reaching any conclusions regarding any physical phenomena. Another fundamental principle of the scientific approach is for any assumption or proposed theory to be falsifiable by plausible physical means. New theories are normally proposed to solve existing problems however; they should make predictions of phenomena that arise as a consequence of their application. In turn, these phenomena should be verifiable by physical measurements. Adopting this methodology, scientists have achieved marvelous results during the past couple of centuries. As a consequence, humanity acquired more knowledge about itself and its surrounding universe during this brief period than it did during its previous entire existence. Although science in general is built on certain basic assumptions, atheists dismiss out of hand any and all assumptions of faith such as revelation, prophethood, resurrection, afterlife, accounting for one's deeds, etc. Acceptable scientific rules at the present time while tolerating foundational assumptions, demand submission to the process of falsification. It is universally agreed that a theory is no more than a tentative framework, most probably mathematically valid, until it is experimentally verified. The most famous (or rather infamous) example for the application of this rule was the fate of the belief in the existence of the medium "aether" to account for the propagation of force between two objects following Newtonian theories. The failure of the historically significant experiment of Michelson and Morley late in the nineteenth century testifies to that approach and the resultant impasse in physics which was not overcome till the advance of the special theory of relativity.

A very common theme in the history of great breakthroughs in science shows itself in creating new mysteries after helping to explain those that stumped generations of scientists. At the beginning of the European modern era for example it is known that thanks to the Western Church's theological adoption of the Ptolemaic principle that the circle represents perfection, earth became the center of the universe with all other heavenly bodies revolving around it in circles. Kepler followed by Galileo moved the center to the sun. They also noticed but could not explain the fact that planets travel around the sun in an elliptical trajectory rather than circles. According to Church's teachings, everything was providentially determined and had to obviously be perfect regardless of the facts. It took Newton to determine that mystery plays no part in keeping planetary trajectories that way. His inverse square law of gravity was the ultimate reason. For the enlightened mind of that era, that necessarily meant physical law not mystery or providence rules nature. Nonetheless, at the same time Newton considered the fact that planets orbit the sun in practically the same plane a mystery his theory could not explain. Present day astronomers know that this phenomenon is a direct inevitable outcome of the solar system's origin from a spinning dusty disk that condensed into planets. Examples of great theories solving problems to turn around and create others that look as mysterious as their predecessors are abound. Present day cosmologists face a tantalizing verity in the occurrence of so many cosmological coincidences without which the familiar universe could not have been created and life would not have evolved. It is universally accepted that the universe is "fine-tuned" in such a way to make life and consequently humans an end result. These are numerous coincidences to be overlooked or dismissed. One group credits these phenomena to a pre-existing plan to eventually produce humans to observe this universe. This is the "Anthropic Principle" that to some extent intellectually succeeds in explaining the natural flow of such coincidences. Religious overtones are unmistakable in this approach since the existence of a plan inevitably implies the existence of a planner. Most cosmologists feel uneasy about this approach. Other groups get around it by suggesting the plausibility of multiverse where the familiar universe is only unique in the sense that it just happened to have the exact requirements for survival, sustainability and the evolution of life. The infinite numbers of other existing universes are simply governed by different physical laws that would not allow the eventual appearance of life and humans. Clearly that approach precludes any creator. It must be noticed here that both attitudes representing basically Western religiosity and atheism are speculative in nature without concrete evidence one way or the other. They both try to deny or affirm the existence of a creator without basing their arguments on irrefutable observations. Clearly religious individuals whether lay or specializing cannot and most probably will never be able to prove their arguments. On the other hand, atheists' most potent argument in this regard is that according to well established historical evidence, what seems mysterious at one point in time would ultimately be scientifically explained with time and the relentless increase in human knowledge. The universe as well as nature according to this logic are random in essence albeit controlled by specific explicable laws that came about only because they accommodate life without the need for external creative power. Therefore, atheism is steadily permeating Western culture with the spread of education. It is argued by Western thinkers that highly intelligent individuals inevitably and naturally become atheists.

Scientists consider themselves a unique group of individuals and enjoy acknowledging their accomplishments through citations and awards. The most acclaimed such awards are those annually given out under the moniker of "Nobel Prize of …" despite much less associated monetary value than other similar awards. Application of the mentioned rules in awarding these Nobel Prizes elucidates how science is practiced at the present time. The case of the two iconic figures of science in the twentieth and the early twenty first centuries is instructive. Albert Einstein (1879 – 1955) and Stephen Hawking (1942 – 2018) are unique in being both towering scientific personalities and household names at the same time. Despite given practically all major awards related to "Mathematics" where abstract ideas are appreciated for their own sake, Nobel Prize in Physics eluded Stephen Hawking. His enormous contributions to cosmology over five decades that now constitute the basis for most work in the field were not deemed worth the award. That is because his conclusions cannot be experimentally verified with currently available techniques. On the other hand, Albert Einstein earned the prestigious recognition retroactively and with great reluctance but his case is very instructive and deserves some discussion to understand how fundamental science is practiced at the present time which very clearly shows the extreme reluctance of the Nobel Prize Committee to award him the prestigious prize despite his unparalleled contributions in several fields and the scientific weight of the very many scientists who recommended him for that award over more than a decade. The Nobel Prize Committee did not believe any of his contributions were experimentally verified and thus are not worthy of recognition regardless of his dominant reputation within the physics community of scientists and the fact that far less scientifically influential contributions that were experimentally verified received such recognition. Similar to Hawking's case is the fascinating story of the Scottish physicist Peter Higgs who in 1964 developed a theory to describe the fundamental structure of the early universe and its inflationary expansion through the existence of the "Higgs Field". But every field according to modern physics is represented by a particle that ascertains it. Higgs' theory solved many of the problems facing cosmology and as such the discovery of *Higgs Particle* became essential to verify that theory and to support the success of the inflationary model of early universe expansion which was crucial for cosmology. However, due to the anticipated relatively very heavy weight / enormous energy of such particle its discovery was beyond the capability of any elementary particle accelerator (atom smasher) in the world at that time. Thus, the model remained nothing more than an elegant mathematical example that may or may not be true. That situation of the inability to detect the particle frustrated Peter Higgs to the point that he is known to have called it that "God Damn Particle" but in the proper world of the 1960s the word "damn" could not be printed in a decent scientific or any other publication and had to be replaced by dots. Instantly the particle became unfathomably popularly known as the "God Particle" and took on some absurd connotations. Higgs was certain of the validity of his approach but thought the associated particle was never going to be discovered during his lifetime. That is until CERN (the high energy research accelerator in Geneva, Switzerland) refurbished its instrument upgrading it to a level capable of generating enough energy to verify or disprove the existence of the "Higgs Particle". In July 2012, the world of high energy physics celebrated the discovery of the Higgs particle at CERN. Promptly, Peter Higgs was awarded Nobel Prize in Physics in 2013 after waiting for almost half a century. There was no doubt about the validity of his theory anymore.

Enumerating and explaining the extraordinarily stringent rules of modern science and the scientific method is exceedingly important considering the theses promulgated by the work at hand. It should be useful for the purpose of subjecting the Qur'an (which is the singularly unique building block of Islam) later in the book to the scrutiny of science using the tools adopted by its practitioners as is elucidated in the preceding paragraphs. Careful investigation of the development of Islamic scholarship after the seventh century, leads one to conclude that it is rather impressive how Muslims through their nascent scholarship after the death of Mohammad (محمد – عليه الصلاة و السلام) managed to create the tightest possible methodology to collect, classify and interpret the sayings and traditions of the Prophet (محمد – عليه الصلاة و السلام). This scholarly approach that built the foundation for what developed in the following centuries as the Islamic scholarship has been discussed before in other works by the author and many others. The most intriguing aspect of this endeavor is the fact that Arabs collectively (as the vanguard of such effort) maybe apart from poetry were never identified with any cultural or intellectual pursuits till that point in time. It goes without saying that the West in its most admired intellectual contributions has known nothing remotely resembling the acceptable rules of Islamic scholarship for close to a millennium. However, stringent rules of scientific research are presently based on the way Western contributions to human knowledge were carried out since the age of enlightenment. Nonetheless, all profound modern scientific theories are essentially outcomes of abstract unproven assumptions that help scientists and scholars surmount persistent obstacles. This is the unyielding core of all physical, natural and social sciences. On the other hand, no such license was practiced or even tenuously allowed within the scope of Islamic scholarship. Within this context, one has to remember that as strenuously argued in this undertaking that atheism is the inexorable consequence of Western Christianity with all its historical prejudices. Eastern Christianity for a variety of reasons is innocent of such charge. Having affirmed that, it is also argued in this endeavor that Islam has nothing whatsoever to do with the validity of objections raised by atheists against Christianity and religion in general. The Islamic outlook is unique and can only be scrutinized on its own merits. Thus, a legitimate question would be how does Islam deal with these newly discovered and universally accepted cosmological facts and striking coincidences? And what does it say about them in light of its unyielding assertion about God (الله - سبحانه و تعالى) "The Creator" who meticulously laid down the plans for the creation process?

Uniqueness of Islam is manifested in its wholeness. That is to say that its interpretation of any fact in a specific context has to be consistent with its general narrative as a whole. For example, the predictive power of the Qur'an should accommodate common understanding of these facts at any time which is naturally subject to evolution of human knowledge. When new scientific theories create new mysteries while solving existing ones, Qur'anic interpretations to be valid should be easily reconciled with findings of the new theories without undue flights from reason or logic. New unexplained mysteries that science cannot interpret should be accepted within the context of Qur'anic interpretation as acts of God (الله - سبحانه و تعالى) on account of previously proven truths in addition to the countless truths across the entire historical and cultural length and breadth of Islam that stood the test of time and evolution of human intellect. In other words, the authenticity of the whole history and culture of Islam over the millennia should temporarily underpin the acceptance of unexplained phenomena as acts of God (الله - سبحانه و تعالى). Islam persistently never wavered in attributing all natural phenomena to acts of God (الله - سبحانه و تعالى). Clearly alluding in the Qur'an to certain historical incidents that came to pass and the same goes for certain natural phenomena that were later scientifically explained should be enough to uphold that supposition. The Qur'anic narrative is supposed to cover all facts, physical or otherwise, from the beginning of the creation process to its presumed end with reasonable and logically consistent expansion of its interpretation with time. Science by its very limited nature cannot even remotely claim such universality under any circumstances due to its changing validity with time.

Science
The Leap from Classical to Quantum Physics

HISTORICAL CONTEXT

Throughout history thinkers, philosophers and scientists grappled with the idea of whether what they were developing to describe the work of nature were inventions, in the sense of man-made tools, or were discoveries, in the sense of existing realities. Mathematics in particular, in addition to being a reflection of humans' higher intellect, was discovered to accurately represent descriptions of numerous natural phenomena. The ancients (Egyptians, Sumerians, Babylonians, etc.) left clear clues that they have found certain methodologies to deal with the natural world. It is now believed that many ancient civilizations understood and manipulated a host of phenomena in their daily lives. Although their ultimate efforts are shown in the monuments they left, there is no record of how they achieved what they accomplished. The reason lies in the lack of abundant material to commit their science and technology to writing. Current knowledge of ancient civilizations mainly comes from deciphering engravings on stone and other hard surfaces and conjectures as to what these monuments must have meant. The most obvious example is the Ancient Egyptian civilization. It is commonly believed that Ancient Egyptians paid extraordinary attention to religion and the life after death. All these conjectures come from interpreting the monuments they left behind. Almost all of these monuments are *tombs and temples*. The implication is obvious. It is known for example that they measured the distance to the moon and understood earth to be a globe. They also manipulated the seasons and controlled the river. What does this have to do with religion and the afterlife? What modern day scholars and historians advance are their own prejudices. That explains why, in the course of describing the evolution of science, technology and philosophy, scholars (certainly Western ones) currently almost always begin with the Ancient Greeks as if they popped up on the world scene out of nowhere as the most brilliant ancestors of the modern humans. This current attitude is as true in science and technology as it is in the liberal arts and social studies as well as in the arts. As far as the Western civilization is concerned, it was the Greeks that started everything. However, despite these unfounded arguments, it is known that the Greeks learned almost everything from their Ancient Egyptian masters, mentors, patrons and teachers. One may wonder why ancient Egyptian scientific and technological accomplishments were not transmitted, or at least part of them, through oral traditions if written references could not be manufactured. For this, one should realize that a considerable portion of these achievements (e.g. the technique of embalmment) was known, if not practiced, even when this civilization crumbled. Christianity vehemently suppressed the practice of embalmment when it took hold of Egypt and the technology was obviously lost. When Egyptians adopted Arabic as their mother tongue, the entire ancient oral tradition, for all practical purposes, disappeared. Egypt was not unique in this process. All oral traditions of ancient civilizations suffered the same fate. As far as the evolution of science is concerned, it is claimed that the Greek great philosophers bestowed the world its starting points. Philosophy is assumed to be the mother of all thought dealing with nature. These philosophical inspirations were given a concrete form with the help of mathematics. Plato (c. 360 B.C.) is supposed to have advanced the notion that mathematical concepts exist in a timeless ethereal sense. Mathematical theories were proposed as mental concepts in themselves. It was rather amazing to realize that pure mathematics was, in countless cases, actually useful in describing physical phenomena. Natural numbers are a case in point. From them complex numbers evolved. Although the original purpose of introducing complex numbers was to enable square roots to be taken with impunity, by introducing such numbers it is found that the potentiality for taking any other kind of root or for solving any algebraic equation whatever came along as a bonus. More studies of complex numbers resulted in marvelous applications that were never contemplated at the beginning. Vector analysis emanated from the study of complex numbers as an abstract field of study. Pursuing these abstract concepts, Hilbert and Riemann spaces were developed. By the beginning of the twentieth century, Quantum Mechanics found them readily available for explaining its ideas. It is clear that without complex numbers, quantum theory would not have been possible. One has to bear in mind that quantum theory provides a very accurate description for innumerable physical phenomena that could not be explained otherwise. That gives impetus to the argument that complex numbers were actually discovered not invented. They must form an integral part of nature rather than being mere human made tools. The great Mathematician Roger Penrose believed that Mathematicians really uncover truths which are in fact already there and which are truths whose existence is quite independent of the mathematicians' activities. He also indicates that it is remarkable that at the same time, all fundamental theories of nature have proved to be extraordinarily fertile as sources of mathematical ideas.

Humans only have access to the internal experiences of perceptions and thought, so how can one be sure they truly reflect an external world? This is the problem that they have struggled for eons to resolve. Reality is what humans think it is; it is revealed to them by their own experiences. However, since the advent of modern Physics at the beginning of the twentieth century, science has unequivocally shown that human experience is often a misleading guide to the true nature of reality. Breakthroughs in physics have forced, and continue to force, dramatic revisions to humans' conception of the cosmos. Classically, Isaac Newton (1643-1727) using a handful of mathematical equations synthesized everything known about motion on earth and in the heavens. He gave rise to the elaborate structure of classical physics where every object presumably became moving through space according to Newtonian laws of motion. However, even Newton himself could not account for what space is and whether it is a real thing or simply a mathematical expression. He declared space and time to be absolute and immutable entities that provide the universe with a rigid unchangeable arena. For the next two centuries, these conceptions of space and time became dogma. Consequently, classical physics provided a rigorous grounding for human intuition. Gravity was included as a fundamental force in the cosmos in Newton's formulation. In the 1860s, James Clark Maxwell (1831-1879) extended the boundaries of classical physics to include electricity and magnetism. William Thomson (Lord Kelvin 1824-1907) somewhat disingenuously suggested that physics was approaching the point where it could be considered a complete science and all that remained were details of determining some numbers to a greater number of decimal places. In 1900 he noted that only **"two clouds"** *were left hovering on the horizon, the first hd to do with properties of light's motion and the second with aspects of the radiation objects emit when heated.* As it turned out exactly these two issues exploded the revolution that became known as modern physics. These new approaches required rewriting of the laws of nature. The first was addressed by the theories of relativity in 1905 and 1915. The classical conceptions of space, time and reality were overthrown. Far from being the rigid unchanging structures envisioned by Newton, space and time in Einstein's theories of relativity are flexible and dynamic. However, Newtonian physics still provides an approximation that proves to be very accurate and useful in the macro world of every day. The second anomaly which Lord Kelvin referred to led to the 1930s quantum mechanics revolution. A core feature of classical physics is that if one knows the positions and velocities of all objects at a particular moment, Newton's equations together with Maxwell's can tell one their positions and velocities at any other moment past or future. ***Classical physics declares that the past and future are etched into the present.*** This feature is also shared by both special and general relativity which are considered the last great achievements of classical physics. Classical physics approaches failed to account for most phenomena taking place in the atomic and subatomic realms. Only quantum laws could explain a host of phenomena arising from the atomic and subatomic realm. ***Quantum laws treat their objects probabilistically.*** If one can make even the most perfect measurements of how things are at a specific moment, the best one can hope to do is to predict the probability that things will be one way or another at some chosen time in the future or that things were one way or another at some chosen time in the past. The universe according to quantum physics is not etched into the present; *the universe participates in a game of chance*. Things become definite only when a suitable observation forces them to relinquish quantum possibilities and settle on a specific outcome. The outcome that is realized cannot be predicted. Only the probability that things will turn out one way or the other can be predicted. In other words, *reality remains ambiguous until perceived*. That can genuinely be taken as allowing for "Free Will". This fact should have profound implications for humanity and its thoughts and beliefs. Moreover, according to quantum physics, actions can be instantaneously linked regardless of distance which is expressed in terms of "**quantum entanglement**". It is ironic that it was Einstein who pounced and elaborated on this point proposing numerous "thought experiments" to attack quantum physics as nonsensical but by the 1980s it was confirmed that there can be instantaneous bond between what happens to elementary particles at widely separated locations ushering the dawn of the exotic field of "**Teleportation**" in science and science-fiction. Classically speaking, intervening space ensures the absence of a physical connection. Quantum mechanically speaking, at least in certain circumstances, there is a capacity to transcend space; long range quantum connections can bypass spatial separation. Two objects can be far apart in space by distances measured even in light years (or parsecs=3.6 light years) but quantum mechanics may treat them as a single entity. When it comes to the concept of time, the distinction between forward and backward in time is a prevailing element of experiential reality. While everything else in the physical universe clearly exhibit symmetry, time does not despite the fact that the laws of physics do not show that distinction. Nothing in the equations of fundamental physics show any sign of treating one direction in time differently from the other. However, mathematicians suggest that initial conditions at creation might have caused time directionality. This is a major issue in physics known under the designation "**Arrow of Time**". However, Thermodynamics attributes this property to what is known as "**Entropy**"; a quantity that is a measure of chaos or randomness of an enclosed system which is supposed to increase continuously. In simple terms, an object cannot go back to its ordered state (cup made of glass) after it moved into a disordered state (broken cup of glass) because its entropy cannot be reduced. Enclosed systems can assume larger and larger sizes until the entire universe is considered the ultimate enclosed system. That raises the fundamental question as to why the universe started in an ordered state in the first place.

CLASSICAL PHYSICS

By classical one means the theories that held sway in describing the natural world in a rigorous manner after Newtonian physics and before the arrival, in about 1925 - 1928 {through the efforts of Max Karl Ernst Ludwig Planck (1858 – 1947), Albert Einstein (1879 – 1955), Niels Henrik David Bohr (1885 – 1962), Werner Karl Heisenberg (1901 – 1976), Erwin Rudolf Josef Alexander Schrodinger (1887 – 1961), Enrico Fermi (1901 – 1954), Louis Victor Pierre Raymond de Broglie (1891 – 1987), Max Born (1882 – 1970), Ernst Pascual Jordan (1902 – 1980), Wolfgang Ernst Pauli (1900 – 1958), Paul Adrien Maurice Dirac (1902 – 1984) and many less prominent individuals} of the **Quantum Theory** describing the behavior of molecules, atoms as well as subatomic particles. *The fundamental departure from classical to quantum theories lies in the replacement of determinism by uncertainty.* Classical theory presents the future as always completely fixed by the past. Although human's understanding of the physical world evolved over the millennia, there is one single event that ushered what became known as the classical theory. That is the publication in 1687 of Isaac Newton's *Principia*. This momentous work demonstrated how, from a few basic physical principles, one can understand and predict with high accuracy a great deal of how physical objects actually behave. Realizing that nature strictly obeys certain established laws, scientists and technologists quickly grasped the fact that they could manipulate it for humanity's benefit (or sorrow as it turned out). With this realization, the limited and fixed tangible human and animal power were replaced by vastly available and easy to manipulate intangible forces of nature such as steam power and what is known in human history as the "industrial revolution" became a matter of time. While the ancients had excellent grasp of the concepts of statics, understanding motion started only in the seventeenth century with the publication of Galileo's *Discorsi* in 1638. This eventually led to the launching of the dynamical theory by Newton. Newton developed three laws governing the behavior of material objects. The first law stated that if no force acts on a body, it continues to move uniformly in a straight line. One consequence of this law is that uniform straight line motion is physically completely indistinguishable from the state of rest. There is no local way of telling uniform motion from rest. This is historically known as the *principle of Galilean relativity*. The second law stated that if a force does act on a body, then its mass times its acceleration (the rate of change of its momentum) is equal to that force. The third law stated that to every action there is always an opposed and equal reaction. These laws constituted the dynamical theory that governed the behavior of the real world in the classical view point. The fundamental result of this approach is that in principle if the positions, velocities and masses of the various particles are specified at one time, then their positions and velocities are mathematically determined for all times past or future. Masses are obviously taken to be constant. This represents the **deterministic** view of classical physics. For the seventeenth century person, the universe was simply the solar system and it was static and infinite. This view held till the dawn of the twentieth century when certain fundamental physical phenomena could not be explained and the statistical nature of these phenomena (particularly heat) became obvious.

Newton's Contributions

Newton's most famous work was published in 1687 under the short Latin title *Principia*. In English, it is *Mathematical Principles of Natural Philosophy*. His work for the first time explained the material world as controlled by fundamental physical laws that are built into its very fabric. Expressing many natural phenomena, specifically gravity, in an inverse square law gave astonishing results conforming to measurements and observations. The formulation of Newton's approach led to describing the universe as a gigantic "precision clock" endowing nature with a deterministic mechanical structure. Absolute space and absolute time came out of this formulation in a simple commonsensical way. Thus, any event could be exactly determined in space and time without any external effect. In other words, once set in motion, the universe progresses in a predetermined and easily predictable state. Many philosophical and theological theories sprang up out of this way of thinking. It became fashionable to speak of God (الله - سبحانه و تعالى) giving the original push to start the universe then becoming detached without interfering in its goings-on. Now it was logical to think that the only need for God (الله - سبحانه و تعالى) in this mindset is to start the process of creation and promptly get out of the way. The universe only needed an initiator but did not need a sustainer as it was self-sustained. On the scientific plane, after some hesitation, scientists wholeheartedly adopted Newton's equations as universal laws. While Newtonian formulations gave science a huge boost, it handed theology a mortal blow. Materialism and atheism from that instant on stood on much firmer grounds. Cracks in Newton's impressive scheme started to show with the development of James Clark Maxwell's (1831–1879) electromagnetic theory in the mid nineteenth century. The special theory of relativity developed from the study of Maxwell's equations by Henri Poincare (1854 – 1912), Hendrik Antoon Lorentz (1853 – 1928) and Albert Einstein was later given a geometrical description by Herman Minkowski (1864 – 1909) at the beginning of the twentieth century. Physicists now spoke of a four dimensional space-time instead of a three dimensional space. Einstein's general theory of relativity evolved from the special theory to generalize Newton's dynamical theory of gravity and relate gravity to the curvature of the four dimensional space-time. This general theory dismantled the age old belief that the natural world can be accurately described using Euclidean Geometry. This geometry was developed approximately fifty years after Plato (c. 360 B.C.) advanced the view that the objects of pure geometry (straight lines, circles, triangles, planes, etc.) were only approximately realized in terms of the world of actual physical things. Those mathematically precise objects of pure geometry inhabited, instead, a different world; "*Plato's ideal world*" of mathematical concepts. This world consists not of tangible objects but of mathematical things. This ideal world was regarded as distinct from and more perfect than the material world of humans' experiences but just as real.

Classical physics is understood as a picture of the inanimate world ruled by Newton's mechanical laws and Maxwell's electromagnetic ones. It is adequate for comprehending most physical phenomena encountered in daily life. In the nineteenth century, the fundamental forces of nature (before the discovery of the nucleus) were simply gravity and electromagnetism. The outside universe was conceived by scientists to include no more than the solar system. Newton's theory of gravity described the force which holds planets in their orbits around the sun and why things fall. Maxwell's equations of electromagnetism described the behavior of radiation, including light, and the forces that operate between electrically charged particles, or between magnets. These were adequate to explain most phenomena experienced in the daily life. However, in a very fundamental way these two theories are incompatible. Maxwell's equations set a determined speed for light which is the same for all observers, while Newtonian mechanics said that the speed measured for light would depend on the motion of the observer. This dichotomy was one of the reasons why Einstein developed first the special theory of relativity (dealing only with inertial systems where objects are moving with constant velocity) and then the general theory of relativity (where there are no restrictions on object's motion). The general theory is basically an improved theory of gravity that is compatible with Maxwell's equations. Both the General theory and Maxwell's equations treat the universe as a continuum. As such they are considered by scientists to belong to the classical category of theories. Space in the classical view, can be subdivided and measured in units as small as one wishes without limit and energy can come in a quantity as small as one wishes without limit as well. In classical physics there is an objective world "out there". That world evolves in a clear and deterministic way, being governed by precisely formulated mathematical equations. This is as true of the theories of Maxwell and Einstein as it is true of the original Newtonian scheme. The most important thing to emphasize at this point is that Newton's contributions were not personal opinions of his but rather were the inevitable results of his approach. His approach in turn was not a fluke but a very successful description of the world as known observationally to scientists at the time. Newton's was also an extremely successful system in solving numerous nagging problems in describing natural phenomena that stumped scientist up to that era. Newton's (or someone else's) appearance at this specific moment in history was also inevitable to solve the problems that were ripe for solution to keep progress of human knowledge on track. Mathematical formulation of the forces of nature prompted scientists and engineers to reduce their observations to equations that governed behavior of objects in nature. Humanity moved into a new era of wholesale manipulation of its environment. All great scientific works of the next two centuries that followed Newton's theories strictly adhered to deterministic physics. The foundations of natural science that flourished and came of age in the late nineteenth century in the form of the theory of heat, electromagnetism, thermodynamics, etc. were all based on this deterministic approach. That eventually helped usher in the industrial revolution in Europe allowing it to exploit these forces in the service of humanity but also for the making of untold means of evil that could not be even imagined before. It is ironic to know that a landmark designating the dawn of modern physics; the theory of relativity, is in reality actually a classical deterministic theory.

Main Concepts of Classical Physics

Certain ideas were necessary to maintain the validity of classical physics as they naturally evolved from the classical view of the universe. These ideas begat certain physical objects which had to be subjected to falsification and observation to be accepted. As it turned out, some of them represented the kiss of death to classical physics and prompted scientists to look for alternative revolutionary concepts giving birth to the modern scientific era. There were many examples of these ideas but three stood out as stumbling blocks that scientists could not get around using any tricks deploying the conventional classical view of the universe. The concepts of aether, fields and heat conduction represented such insurmountable obstacles to classical physics.

AETHER

Ocean waves are carried by water. Sound waves are carried by air. The speeds of these waves are specified with respect to the medium they are carried by. Therefore, nineteenth century physicists assumed that light, or electromagnetic waves in general, must also travel through some particular medium that has never been observed or detected. This is what was called the *luminiferous Aether* after the ancient expression used by Aristotle to describe the material of which all heavenly bodies are made. In 1690 Christiaan Huygens (1629 – 1695) while developing his theories about the wave behavior of light (based on the work in optics of Al-Hassan ibn Al-Haytham (الحسن إبن الهيثم) four centuries earlier), introduced the concept that light propagates in the same way as sound does. However, propagation of sound waves requires media through which they can propagate. Since light waves propagate through the entire universe, some kind of a medium has to fill what up to that point was considered vacuum. Therefore, to accommodate the new principle of light waves, Huygens re-introduced the ancient idea of "Aether". This is considered to be a medium filling the entire universe. The concept of Aether at absolute rest was introduced in classical physics in 1818 by Augustin-Jean Fresnel (1788 – 1827). Fresnel had predicted that if a liquid is moving through a tube with a velocity "v" relative to the Aether and if a light beam traverses the tube in the same direction, then the net light velocity "c¹" in the laboratory is dependent on the refractive index of the liquid where Fresnel's drag coefficient expresses the fact that light cannot acquire the full additional velocity "v" since it is partially held back by the Aether in the tube. Nineteenth century physics incorporated that concept into its basic theories explaining the propagation of light in space ever since. It is noticeable that the very successful Huygens' wave theory of light and Newton's treatment of light as matter are patently incompatible. However, physicists of that period overlooked such contradictions mainly due to Newton's overwhelming reputation despite the undeniable success of the wave theory of light in explaining most phenomena in nature. As light's motion led to the concept of the *luminiferous Aether*, accelerated motion (supposedly through the aether) prompted Newton earlier to adopt the concept of absolute space. While the concept of absolute space could not be ascertained experimentally, that of Aether could. Several experiments were carried out to measure, and hence prove the existence of the effect of Aether on the propagation of light due to the motion of earth. The most famous and most elaborate one was that performed by Albert Michelson and Edward Morley in 1887. In 1864 Maxwell surmised that scientists have had some reason to believe, from the phenomena of light, that there was an aetherial medium filling space and permeating bodies, capable of being set in motion and of transmitting that motion from one part to another. Following distinguished predecessors, Maxwell assumed that an Aether was necessary for understanding the propagation of electromagnetic waves through space. He had also believed that there could be no doubt that the interplanetary and interstellar spaces were not empty but were occupied by a material substance or body, which was certainly the largest, and probably the most uniform, body of which humans have had any knowledge and that interstellar regions should no longer be regarded as waste places in the universe, which the Creator had not seen fit to fill with the symbols of the manifold order of His kingdom. According to him, scientists would eventually find them to be already full of this wonderful medium; so full that no human power can remove it from the smallest portion of space, or produce the slightest flaw in its infinite continuity. Thus, it extended unbroken from star to star. Maxwell's Aether (the medium through which light is assumed to propagate) therefore is an all-pervasive medium in a state of absolute rest. Consequently, Maxwell's universal velocity "c" is the speed of light relative to this resting Aether. Stars (but not planets), were believed to be at rest relative to this resting Aether. Thus the velocity "c" (which is a fundamental constant in his equations) is the speed of light as measured by a hypothetical observer standing on a fixed star. Since earth moves relative to the fixed stars, an earth-bound observer should expect to measure a light velocity different from "c" depending on the direction of the light beam relative to the direction of the motion of the earth. In 1887 Albert Michelson and Edward Morley (MM) performed their famous experiment to measure that difference. They found none! Their conclusion was that *the velocity of light is independent of the speed with which the light source moves relative to observer.* On April 27,

1900, Lord Kelvin gave a lecture before the Royal Institution describing the (MM) experiment to have been carried out with most searching care to secure a trustworthy result and that the unexpected associated result as a cloud over the dynamic theory of light. Four years later he still insisted that Michelson and Morley have by their great experimental work on the motion of the Aether relatively to the earth raised the one and only serious objection against the then existing dynamical explanations. This last definite position taken by the most prominent British scientist of the time was uttered interestingly enough only one year before the discovery of the theory of special relativity by Albert Einstein who clearly showed that the introduction of a luminiferous Aether has been proved superfluous. A clear paradox arose due to the failure to measure any effect of the presumed Aether on the speed of light due to earth's motion. Classically speaking, if one is travelling at the speed of light, any associated light beam would be stationary relative to one's motion. Consequently, one can study what light is made of. However, Maxwell's equations (which proved to be universally valid) do not allow light to be stationary under any conditions. Either Maxwell's theory or Newton's summation of speeds had to be wrong as far as light propagation is concerned. There was no question of the validity of Maxwell's theory though. That was the impetus for Einstein's theory of special relativity. To explain the requirements of Maxwell's equations and the failure of the (MM) experiment to measure any change in the speed of light, special relativity *postulated* that the speed of light is a universal invariant constant regardless of the motion of the source or the observer. When Einstein visited the United States for the first time to give lectures on the theory of special relativity at Princeton University in May 1921, Dayton Clarence Miller claimed to have found a nonzero Aether drift value. He had performed experiments at Mount Wilson observatory on April 8-21 for the stated purpose of determining the Aether drift. On hearing this news Einstein said his most famous and most quoted statement that *"Subtle is the Lord but Malicious He is not"*. This statement is chiseled in the stone frame of the fireplace of room 202 in what was the Mathematics Department at Princeton University in 1930. Before his departure from the US, Einstein paid a visit to Miller On May 25, 1921 to discuss the results of his experiments.

On April 28, 1925, Miller read a paper before the National Academy of Sciences in which he reported that an Aether drift had finally been established. Later in the same year, he made the same claim in his retiring address as the president of the *American Physical society*. Einstein stated that he never took Miller's results seriously. Thanks to the theory of special relativity, the concept of the *luminiferous Aether* is no more. Michelson was the first American scientist to receive Nobel Prize in 1907. It is remarkable that there is no mention of the Aether wind experiments in his citation. In hind sight, it is easy to see that the highest authorities of the world's top scientific societies rendered foolish conclusions with absolute certainty. By the end of the nineteenth century, most scientists believed that almost all natural phenomena are adequately described and understood. Lord Kelvin; the president of the British Royal Society (the most prestigious scientific association of the period) infamously asserted that all there is to know or discover has already been discovered and only more accurate measurements are left to do. In the 1920s, Max Born (1882 – 1970) was telling people that there would be nothing significant left for theoretical physicists to do within six months! It is notable that afterwards Max Born turned out to be one of the pioneers of the quantum revolution that upended the world of classical physics. As if scientists never learn their limits, paradoxically, in 1980, in his Lucasian lecture, Stephen Hawking (1942 – 2018) again suggested that it is probable that scientists might see the end of physics **"by the end of the century"**. That is to say that physicists would have a complete, consistent and unified theory of the physical interactions that describe all observable phenomena. When the 21st century arrived, that date was pushed back to 2010. Understandably, there were no predictions afterwards. The concepts of empty space and nothingness take a whole new meaning when quantum uncertainty is considered. However, since the discovery of the theory of special relativity in 1905, when the *luminiferous Aether* was done away with, the idea that space is filled with invisible substances kept on coming back in various forms. Key developments in physics have reinstituted various forms of an Aether-like entity, none of which set an absolute standard for motion like the original *luminiferous Aether,* but all of which challenge the naive conception of what it means for space-time to be empty. Modern cosmology established the reality of the Higgs-ocean after determining the existence of the Higgs particle; discovered at CERN in 2012. The condensed Higgs field is assumed to permeate space in much the same way Aether did as a non-removable feature of empty space. However, Higgs-ocean has nothing to do with the motion of light. Moreover, since the Higgs-ocean has no effect on anything moving with constant velocity, it does not pick out one observational vantage point as somehow being special, as the *luminiferous Aether* did. With Higgs-ocean, all constant velocity observers remain on a completely equal footing. Thus, Higgs-ocean does not conflict with special relativity.

ELECTROMAGNETISM

In the Newtonian picture of the world, one thinks of tiny particles acting upon one another by forces which operate at a distance. However, the forces of electricity and magnetism act in a way similar to Newtonian gravitation forces in that they also fall off as the inverse square of distance, though repulsively rather than attractively. That fitted well within the Newtonian scheme. When Michael Faraday (1791 – 1867) introduced the concept of fields as real actual physical substances to explain electrical and magnetic phenomena resulting from his experimental work, it seemed that varying electrical and magnetic fields resulted in waves. He believed that light itself might consist of such waves. This is clearly at variance with the Newtonian world view that considered fields to be mere mathematical tools with no real substance. This situation led James Clark Maxwell to formulate his elegant equations implying that electric and magnetic fields would indeed push each other along through empty space. Maxwell was able to calculate the speed with which this effect would propagate through space. *He found that it would be the speed of light.* In addition to accounting for the properties of visible light, he predicted that electromagnetic waves of other wavelengths should exist. This prediction was confirmed in 1888 after Maxwell's death by Heinrich Hertz (1857 – 1894). Maxwell's equations are deterministic in the sense that they tell what must be the rate of change with time of the relevant quantities (the electric and the magnetic fields) in terms of their values at any given time. Since these equations are field equations rather than particle equations, one needs infinite number of parameters to describe the state of the corresponding system as the field vector exists at every single point in space. A particle theory would require simply six parameters, three for position and three for momentum, to describe the state of the system. The fundamental impact of Maxwell's equations is that fields must be considered real objects which show that when the fields propagate through space as electromagnetic waves, they carry definite amounts of energy with them. As they stand, Maxwell's equations are not really quite complete as a system of equations. They provide description of the way electric and magnetic fields propagate only when the distribution of electric charges and electric currents are given. It is to be noted that these charges are given as particles and their motion produces the electric currents. Given where these charged particles are and how they are moving, Maxwell's equations can describe the behavior of the associated electromagnetic field. On the other hand, Maxwell's equations have nothing to say about the behavior of these charged particles themselves. The motion of charged particles was dealt with in 1895 by the Dutch physicist Hendrick Antoon Lorentz (1853-1928) resulting in the *Lorentz equations of motion for a charged particle*. These equations describe how the velocity of a charged particle continuously changes owing to the electric and magnetic fields at the point where the particle is located. Combining Lorentz and Maxwell's equations, one obtains rules for the time-evolution of both the charged particles and the electromagnetic field. The problem with Lorentz equations is that there will be an electric field due to the particle itself which becomes infinite if the particle is taken as a point to avoid spreading the field over the finite dimensions of the particle which results in distortion that is not experimentally observed. Classical treatment of this problem failed to give any explanation. It was almost half a century before Paul Maurice Adrien Dirac (1902 – 1984) solved this problem contributing fundamental new concepts to the quantum theory. The fundamentals of classical physics became challenged with the adoption and establishment of both the wave and the particle nature of electromagnetic radiation. While particles are described by a small finite number of parameters (six -three for positions and three for momenta), fields require an infinite number of parameters. For a system with both particles and fields to be in equilibrium, all energy gets taken from the particles and put into the fields due to equipartition of energy. At equilibrium, the energy is spread evenly among all the degrees of freedom of the system. Since the fields have infinitely many degrees of freedom, the particles get left with none at all. In particular, classical atoms (as envisioned by Rutherford's (1871 – 1937) planetary model of 1911) would not be stable. As an orbiting electron circles the nucleus it should, according to Maxwell's equations, emit electromagnetic waves of an intensity increasing rapidly to infinity as it spirals inwards and plunges into the nucleus. Therefore, classical physics was shown to be inadequate to describing many observable phenomena by the end of the nineteenth century and the beginning of the twentieth.

BLACKBODY RADIATION

Since a blackbody perfectly absorbs and emits all frequencies, most of radiation should be at high frequencies i.e. the ultraviolet range and upward since the number of states available at high frequencies is far greater than at low frequencies. The main problem with the picture classical physics paints of the universe is that it is unstable. The classical model is based on the stable coexistence of "particles" and "fields". Particles are generally described by a finite number of parameters designating their position and momenta. On the other hand, fields require infinite number of parameter for their description. According to this approach, when a system is at equilibrium, its energy is transferred from the particles into the fields. Since fields have infinite degrees of freedom (number of parameters), particles are left with zero energy. The implication is that a particle revolving around a nucleus would radiate all its energy and eventually falls into the nucleus. This is physically unsustainable. A manifestation of the instability of the co-existence of fields and particles is the phenomenon known as "blackbody Radiation". For example, an orbiting electron circles the nucleus should emit electromagnetic waves of an intensity rapidly increasing to infinity in accordance with Maxwell's equations. In 1900 Rayleigh (1842 – 1919) and Jeans (1877 – 1946) had calculated that all the energy would be sucked up by the field without limit. This implies that energy keeps going into the field, to higher and higher frequencies without stopping. This is what is historically became known as the "**Ultraviolet Catastrophe**". At law frequencies of field oscillation, the energy is as Rayleigh and Jeans had predicted. However, at the high end experimental observations showed that the distribution of energy does not increase without limit, but instead falls off to zero as the frequencies increase. The greatest value of the energy occurs at a very specific frequency for a given temperature. The failure of the Rayleigh-Jeans formula to account for the behavior of radiation at high frequency prompted Max Planck (1858 – 1947) to propose his assumption of discrete emission of radiation in "**quanta**". His suggested equation describing blackbody radiation perfectly fitted experimental results. With the 1905 Einstein's successful accounting for the photoelectric effect using Planck's ideas (which is what the Nobel Committee retroactively cited as the reason for the 1921 Physics award), the quantum revolution was launched in the 1920s. Ironically Einstein never accepted the quantum approach to the day he died. Additionally, Max Planck very conservatively embraced quantum physics but looked forward to a more deterministic theory.

Modern Physics

The maxim normally attributed to Isaac Newton that great individuals (naturally including himself) look very impressive is simply a reflection of the fact that they stand on the shoulders of giants (past great individuals) has most probably been expressed by countless others before and after him. It is a truism that no great new idea is created in vacuum or springs to life all of a sudden regardless of how greatly perceptive its advocate can be. Newton formalized in mathematical form what Copernicus and Galileo experimentally suggested. Their work in turn was nothing more than rearrangement in a rigorous form what centuries before Muslim astronomers and mathematicians developed building on knowledge previously established. Maxwell's electromagnetic theory was another mathematical formalization of Faraday's experimental work and his field concept was advanced to get around the difficulties Newton's action at a distance principle has created. Plank's assumption of energy quantization came about as a brute force numerical solution to heat theory's ultraviolet catastrophe. Einstein's special relativity is wholly derived from Maxwell's theory and represented limitations on Newtonian Mechanics applicability. General relativity is an effort to expand the scope of the special theory to cover non-inertial motion of objects and to apply Riemannian geometry to describe gravity. The aim of modern physics is to explore where humanity came from, where it will end up and in the meantime how it interacts with its surroundings. The beginning of the story gave rise to cosmology dealing with the origin of the universe which is generally accepted nowadays to have occurred according to the "Big Bang theory". The end spurred speculation on the fate of the universe and the associated future of humanity until then. Exploring humanity's surroundings is the subject of the vast majority of scientific endeavors. Science is currently trying to relate the formation of the fundamental building blocks of the universe (elementary particles), formation of stars and galaxies and the emergence of life. Along these trails many theories are proposed with varying degrees of success. However, the two pillars of modern physics are relativity and quantum theories. Relativity has very little impact on technology but it gave birth to many theoretical concepts such as black holes, gravitational waves, etc. On the other hand, quantum theory has far reaching impact on modern day technology that would never have been developed without it. There is practically no convenient everyday gadget that does not owe its existence to the concepts of quantum physics. As is well known, quantum physics deals with the very small and relativity deals with the very large. Quantum physics controls the behavior of the three fundamental forces; strong, weak and electromagnetic forces while general relativity controls the behavior of the fourth; gravity. Most new theories come through marriage of these two basic theories under certain assumptions. Studying the creation of the universe and its later evolution requires combining both fields in what is affectionately known as a "Theory of Everything". This anticipated (but never realized so far) theory is assumed to combine all forces of nature into one single fundamental original force. Other forces split from it during evolution of the universe from one phase to the next.

Special Relativity

According to Maxwell's equations (after the great Scottish scientist James Clark Maxwell) light consists of a combination of an electric and a magnetic field. Maxwell's equations do not allow a stationary solution and they clearly show the speed of electromagnetic waves to be constant. It was soon realized that this speed is the speed of light which is consequently considered as an electromagnetic wave. The very basic idea that prompted Einstein to develop the theory of special relativity is that if one could catch up to one of these light waves, then the light would be standing still. The light wave would then be a standing wave of electric and magnetic fields which is forbidden by Maxwell's theory. It became obvious to him that either the theory is wrong or one can never catch up with a light wave. Maxwell's theory of electromagnetism was unassailable. Therefore, the inescapable conclusion was that *it is not possible for any material object to reach the speed of light*. Thus, in addition to the speed of light (any electromagnetic wave) being a universal constant as is clearly implicit in Maxwell's Equations, it is also a universal upper limit to any attainable speed. The special theory of relativity is wholly derived from Maxwell's theory of electrodynamics. In formulating his special theory, Einstein also *postulated* that it was impossible to determine absolute uniform motion i.e. the non-existence of an absolute reference to measurement of motion in contravention to Newton's assumption of the absolute nature of space and time. One can only refer to motion of an object *relative* to something else i.e. *all motions are relative*. The structure of the "Special Theory of Relativity" comes out of these two postulates. However, the theory as it stands assumes that all motions are relative except that of light which is absolute. As always in science new theories proving the un-tenability of the assumptions of old theories bring about new assumptions of their own that are only based on ***intuition***. After the acceptance of Einstein's theory of relativity, Newton's absolute space and time are no more. Space and time are relative and are interconnected in equations of the special theory of relativity. The term "special" stems from the fact that the theory deals only with "uniform motion". Bringing any type of motion (acceleration, change of direction, etc.) into the analysis resulted in the "General Theory of Relativity". Einstein explains the limited application of the special theory in the sense that in the first place, he started out from the assumption that there exists a reference-body K, whose condition of motion is such that the Galileian law holds with respect to it: A particle left to itself and sufficiently far removed from all other particles moves uniformly in a straight line. With reference to K (Galileian reference-body) the laws of nature were to be as simple as possible. But in addition to K, all bodies of reference K' should be given preference in this sense, and they should be exactly equivalent to K for the formulation of natural laws, provided that they are in a state of *uniform rectilinear and non-rotary motion* with respect to K; all these bodies of reference are to be regarded as Galileian reference-bodies. The validity of the principle of relativity was assumed only for these reference-bodies, but not for others (e.g. those possessing motion of a different kind). Thus, in this sense one speaks of the *special* principle of relativity, or special theory of relativity. Details of Einsteins definitions can be found in the fifth edition of his 1952 book "Relativity by Einstein - The Special and the General Theory". Relativity of space and time resulted in many "paradoxes" since the measurements of one observer are not bound by those of another observer in relative motion with the first except through the transformations given by the mathematics of the theory. While it is shown that causality is nonetheless preserved, *simultaneity lost its meaning*. There is no physical way to assert that two simultaneous events as measured by one observer are also simultaneous as observed by an observer in motion relative to the first one. Relativity theory contributed in no small way to facts of modern physics being in large part **counter intuitive**. Common everyday experiences of individuals are no longer a measure of reality. This is the ultimate legacy of modern physics. The most intriguing part of the history of developing the special theory is that nothing of its conceptual details and even its equations were unknown when Einstein published his thesis. The great French Mathematician "Jules Henri Poincare" (1854-1912) for example has by 1905 independently developed and published exactly the same space-time relationships. They clearly showed that measurements of space and time are not identical in two frames of reference moving in relative uniform motion. Going from one to the other, space and time are transformed according to the already known Lorentz transformations (described by the Dutch Physicist Hendrik Antoon Lorentz (1853 – 1928) in 1892). Associated with speeding objects, time will slow down (*time dilation*) and space will contract (*space contraction*) as measured from another inertial frame of reference in relative uniform motion with respect to the first. However, Poincare thought of his work as a clever mathematical approach devoid of any physical meaning. He thought of these space-time mathematical formulae as *postulates* to deduce the relationship between the two inertial systems in relative motion. That was perfectly consistent with the prevailing belief among scientists at the time in Newtonian concepts of absolute space and absolute time. Even at the present time, scientists do not necessarily see physical meaning in established mathematical theories. Mathematics is generally treated as an abstract discipline allowing mathematicians to establish and study relationships between abstract quantities far removed from nature. On the other hand, Einstein began his analysis giving actual physical meaning to these relationships which turned out to be very insightful and a specific characteristic of nature. Einstein however, originally strongly objected to Minkowski's (Hermann Minkowski (1864 – 1909) who was Einstein's professor who told him that he would not amount to anything because of his lax attitude) concept in 1908 of time being the fourth dimension of a space-time to explain the results of special relativity. This multidimensional approach to geometry is currently an essential tool in physics. Ironically, Einstein himself used that approach to develop the "General Theory of Relativity".

General Relativity

Having restricted the scope of the special theory to inertial systems in rectilinear motion, Einstein spent the next fourteen years trying to expand his analysis to cover all types of motion or simply to generalize the application of the theory. He describes the "General Theory of

Relativity" in simple terms defining motion of an object under no restrictions as due to the fact that all bodies of reference K, K ', etc., are equivalent for the description of natural phenomena (formulation of the general laws of nature), whether may be their state of motion. Therefore, his motivation to work on the general theory was that since the introduction of the special principle has been justified, every intellect which strives after generalization must feel the temptation to venture the step towards the general principle of relativity. Einstein was fond of explaining his analysis using the motion of a train and passengers within its carriages; something that everyone can relate to such that as long as the train is moving uniformly, the occupant of the carriage is not sensible of its motion, and it is for that reason that he can without reluctance interpret the facts of the case as indicating that the carriage is at rest, but the embankment in motion. Moreover, according to the special principle of relativity, this interpretation is quite justified also from a physical point of view. If the motion of the carriage is now changed into a non-uniform motion, as for instance by a powerful application of the brakes then the occupant of the carriage experiences a correspondingly powerful jerk forward. The retarded motion is manifested in the mechanical behavior of bodies relative to the person in the railway carriage. The mechanical behavior is different from that of the case previously considered, and for this reason it would appear to be impossible that the same mechanical laws hold relatively to the non-uniformly moving carriage, as hold with reference to the carriage when at rest or in uniform motion. At all events it is clear that the Galileian law does not hold with respect to the non-uniformly moving carriage. Using the introduction of the concept of "Field" into physics because of electromagnetism to discredit the idea of *action at a distance* which is associated with Newtonian Mechanics, Einstein attributes the fall of a stone to a *gravitational field* rather than the common folksy idea of objects fall because they are attached somehow to earth. Therefore, the basic idea of general relativity is that it is impossible to distinguish the effect of gravity from a non-uniform motion. The example of a person inside an accelerating free falling box in space can be used to illustrate this concept. If the person does not know the fact of the free falling box in space, this person cannot in any physical way determine that the felt effect of *"gravity"* is due to the acceleration of the free falling box. This is what is known as the *principle of equivalence*; the equivalence of non-uniform motion and gravity. If this person releases any objects, they will remain (relative to that person) in a state of rest or in a state of uniform motion independent of any other material characteristics. Then, as in the special case, this person is physically justified in considering being in a state of rest. This has to be a physically natural observation. For if there is even one object which falls differently in a gravitational field than do the others (which is not the case), the observing person would be able to discern by means of it that the observer is falling in it otherwise the observer lacks any objective reason to assume falling in a gravitational field but rather being in a state of rest where the immediate surroundings are field free. The general theory of relativity establishes the laws governing the space and time measurements carried out by observers moving non-uniformly such as one observer is travelling in an accelerated vehicle in space and the other floating in gravity-free space. Although Einstein conceived of the basic principle of general relativity through the equivalence of acceleration and gravity, he could not develop the mathematical framework to describe it and all his attempts were in vain. His problem was that (as he eventually learned) he was utilizing the familiar age old Euclidian two-dimensional geometry (which is not associated with any natural phenomenon) to describe space. His mistake was pointed out by his close friend and old classmate mathematician Marcel Grossman who advised him to use the approach of the curved space (by definition enclosed in a three-dimensional space) as in Riemannian geometry which lends itself to physically applicable descriptions of space. That made the mathematical framework of the general theory of relativity far more complicated but finally furnished the right answers. The implication here is that humans live in a three-dimensional (four-dimensional if time is added) space and two-dimensional geometry representation of space will not be adequate. Consequently, light still as before always travels in a straight line as the shortest distance but as defined in a curved space (as opposed to a two-dimensional Euclidian plane). This has been known in the history of science as a classical example of purely abstract mathematical problems actually having real physical meaning in nature. Following that approach, Einstein *replaced gravity by geometry of curved space*. One has to remember that Newton developed gravity as a "force" that acts at a distance which does not have any physical meaning despite the fact that Newton's ideas were universally adopted for over two centuries thanks to their success in solving so many existing problems at the time. Simply speaking, gravity is geometry and thus it is the curvature of space-time around mass that determines the path of light. These ideas were too much to digest and accept for the world of the First World War. In 1919 the great British astronomer Arthur Eddington (1882-1944) proved the validity of the *predictions* of the general theory in eclipse measurements he carried out in Brazil (and other group at the island of Principe off the west coast of Africa) when he was sent by the Royal Society, where he was a well-known member, to avoid embarrassment and get around refusing obligatory military service as a pacifist. That was great news to the world after the foul mood following the war and it appeared in the headlines of most major newspapers all over the world. This ushered Einstein's fame as an international scientific iconic figure after he was only known within the immediate physics community. Since the advance of general relativity, utilization of multi-dimensional geometries in solving physical problems especially in particle physics and cosmology became the norm. It is a commonly held belief in the popular science lore that Einstein proved Newton wrong. This is grossly misleading and goes against the grain of real science. What Einstein did was to generalize Newton's analysis to include objects travelling with speeds close to that of light which Newton's work naturally never incorporated since such objects were not known at his time. It is of paramount importance at this juncture to emphasize the fact that human thought process and perspectives have to accommodate new knowledge ascertained with observations. This is the theme advanced by this endeavor in dealing with dilemmas encountered by Islamic thought in modern times. As it is emphatically clear in the pages of this book, that approach validates rather than refutes compatibility of Islam and science till the end of time. Additionally, it does not infringe in any way on the fundamental tenets of the faith. That is to say that both atheists and fanatics are at fault when dealing with the conceptual principles of Islam.

It is well known that Newton's equations rather than Einstein's are applied in determining paths for spaceships at the present time because of their relative ease and acceptable accuracy since space exploration deals with objects moving with speeds negligible compared with that of light. Newton based his theory of gravity on the assumption that material objects attract in a specific way and could deduce the trajectories of planets around the sun within the solar system with reasonable accuracy. The fact that all objects fall at the same rate and follow identical orbits and that the force of gravity and the inertia were in exactly the same ratio for all substances represented a mystery to him that his theory could not solve. Every scientist believes that objects follow the *straightest* path in their motion. Einstein assumed that objects do exactly that while moving in a *curved space-time* which identifies gravity with geometry. Therefore, planets trajectories represent their motions in the curved space-time around the sun due to its mass. General relativity *predicts* "time dilation" near large masses. Global Positioning systems (GPS) take such dilation into account and produce extremely accurate measurements which they did not first obtain neglecting relativistic effects. The British Astronomer Royal Martin Rees explains how he came to understand the limits of Einstein's general theory of relativity attending a lecture by the great British mathematician Roger Penrose. He indicates that anything that implodes in an exactly symmetrical way obviously crashes together at a central point: the gravitational force then becomes infinite even in Newton's theory. This infinity is just an artifact of the symmetry. However, if the in fall weren't exactly radial, the pieces would miss each other. But in Einstein's theory a *generic* collapse leads to a singularity. Thus whenever a black hole forms, a singularity must develop inside it. Additionally, for him, Penrose had shown that Einstein's theory *predicts* its own incompleteness. "infinities" in a theory are a signal that some new physics intervene. In this case, what happens is a mystery; space itself may change its nature on very tiny scales; extra dimensions perhaps appear; regions may "pinch off" or even sprout into new universes. Penrose's theorems also imply that there must have been a singularity at the beginning of our universe, even if the big bang were asymmetrical and irregular. He also thinks that General relativity has the virtue of being highly specific. Any single discrepant observation or experiment would be deadly as the theory could not be brought into line by any minor tweaking or small adjustments. This is the motivation for continuing to devise new tests. The discovery of black holes opens the way to testing the most remarkable consequences of Einstein's theory. Martin Rees's discussion of the consequences of the general theory of relativity can be found in his 1997 book *Before the Beginning*. Less than a year after Einstein announced his general theory, the German astronomer Karl Schwarzschild (1873-1916) found a solution to the theory describing how gravity behaved around a spherical mass. This solution is currently the standard approach used by physicists applying general relativity to various physical problems. But in 1963 the New Zealander physicist Roy Kerr found a more general solution to general relativity's equations representing a collapsing rotating object where space behaves like a vortex forcing falling objects to additionally rotate. In a spinning hole stresses become infinite on a ring rather than at a point. Passing this ring both mass and radius become negative. While these results opened exotic vistas for science fiction enthusiasts, they may offer powerful tools to exploring and testing the possibility of the fallacy of the commonly acceptable standard human perceptions of death, resurrection and the afterlife as presented by non-Islamic ideologies. Again these issues are later checked against what the Qur'an and the Sunnah (السنة) say to examine the validity of Islam itself.

BLACKHOLES

It was the great physicist John Wheeler (1911-2008) of Princeton University who coined the term "Black hole" in 1968. An object's gravity is expressed in terms of the speed with which a projectile must be fired to escape its grasp. This is 11.2 Km/s for earth and 600 Km/s for the sun. This value increases with the density of the object. Neutron stars are approximately 1.4 as massive as the sun but only in the range of twenty kilometers across. Therefore, gravity at their surface is trillion times as great as that at earth's surface. Thus, it is a trillion times harder to escape a neutron star than to escape earth's surface. Moving in the straightest path within the curved space-time around a very massive object, light from the surface would be severely bent that a faraway observer would not only see one hemisphere but also part of the backside of the object. With even denser objects, it becomes impossible for any object to escape and even light would wrap around the object and be trapped rather than escaping; giving rise to the idea of a "Black hole". This concept is a direct application of the general theory of relativity. Stephen Hawking (1942-2018) contributed much to the advancement of this field cementing its place in physics. However, he came to realize that according to Thermodynamics, temperature of a black hole could not be zero-degree Kelvin which means there has to be radiation emitted leading to the expectation that a black hole is doomed to eventually disappear. That radiation became known as Hawking's radiation. Clearly a black hole cannot be directly observed since no light escapes its surface. *Indirect measurements* are the only way to confirm whether or not the black hole concept has any validity. Astronomers measure intense gamma ray radiation emitted during the collision of two dying stars to indicate the birth of a black hole. Although cosmologists in general are certain of the existence of black holes everywhere in every galaxy including the familiar Milky way, no definite proof is yet available. Black holes partially solved the problem of lack of mass in the observed universe according to calculations. *Direct detection or observation of black holes is obviously not possible but there is consensus that they do exist.* Astronomers resort, as is very common in experimental physics, to indirect methods. It is believed that black holes form binary systems by capturing stars that are forced to orbit around them. Detection of a star regularly revolving around an unseen object is taken to be an indication of a black hole (the unseen object). Astronomers and cosmologists assume that super-massive black holes exist at the center of galaxies. Such an object is believed to reside at the center of the Milky Way galaxy (which the solar system is but a minuscule part of) and is designated as Sagittarius A* (Sgr A*). Normally measurements of bright bursts of x-rays occasionally emitted by black hole binaries are the definitive clue to the existence of such systems. However, the Milky Way galactic center is too far from earth that strong bursts can only be detected every 100 t0 1000 years. Nonetheless, it is possible to detect the much weaker but consistent emission of x-ray bursts by the binary system. This is how scientists estimated in 2018 that Sgr A* may include hundreds of such binary systems. This is done by extrapolating information gathered by satellites rather than direct measurements though. Black holes give rise to gravitational waves (distortions in the fabric of space-time) due to their mass. Hence, study of black holes advances the study of gravitational waves which in turn is a direct proof of the general theory of relativity. Astronomers believe they directly observed and photographed the black hole at the center of the Milky Way galaxy in early 2019. Neutron stars are known to bend light due to their very large mass squeezed in very small volumes generating enormous gravity. Radiation from a neutron star's surface would reach distant observers with a lower frequency (longer wavelength) due to relativistic effects. This is known as the "gravitational red-shift". With decreasing diameter, light would be unable to escape and a "black hole" is formed. The minimum radius from which light could escape is now known as "Schwarzschild radius" after the scientist who first obtained an exact solution to equations of the general theory of relativity. At that radius the gravitational red-shift becomes infinite creating a **"Horizon"** shrouding the interior from view. A freely falling object could pass the horizon of a black hole without experiencing anything unusual but with no hope of escape. However, moving inwardly time would slow to a standstill and an outside observer would see the object frozen in time at the horizon with total loss of information about what is going on inside the black hole. Information loss within a black hole is currently a major problem in physics that contradicts the principle of conservation of information. Black holes preclude space and time being a seamless continuum. This, prompted scientists to speculate that *black holes may represent gateways to other universes*. Probing the fine structure of matter is possible by increasing the energy of probe's quanta (using shorter wavelength). However, there is a limit to decreasing the distance (wavelength) before black holes are formed because of the large concentration of energy. This is Planck length which light traverses in Planck time that is 10^{-43} seconds indicating that *both space and time are quantized.*

GRAVITATIONAL WAVES

When two massive objects like neutron stars (precursors to black holes) collide, they cause bending in the fabric of the surrounding space-time leading to emission of a gravitational wave. Astronomers believe that for the first time ever they detected such phenomenon in 2018 in combination with electromagnetic radiation. This combination gives cosmologists a very precise tool to measure distances on a cosmological scale. The significance of detecting gravitational waves is that they may provide as yet unknown phenomena that could obviate the need to propose the existence of dark matter and dark energy to account for the mass of the universe and accurately calculate the value of (Ω / Ω_0) (the ratio of the actual density of mass in the universe to the critical density) to determine the fate of the known universe where (Ω_0) is the critical density. Definition of these parameters are given later when discussing cosmology and the origin of the universe according to the big bang theory. Unlike the other three fundamental forces of nature, gravity is extremely weak and acts at tremendous distances. Since these three forces are quantized, physicists instinctively believe that gravity should be quantized as well. The quantum of gravity is called "Graviton". However, detecting gravitons is beyond known technology and some prominent physicists think it would be impossible to achieve in addition to doubting the quantization aspect of gravity itself.

QUANTUM THEORY

Relativity is more popularly known because of its association with a single iconic name; that of Albert Einstein. However, the more prevalent pillar of modern physics is the quantum theory which resulted in countless applications routinely used by almost every human being at the present time. Quantum Mechanics was born as a result of collective effort of numerous brilliant physicists at the beginning of the third decade of the twentieth century. In a wide ranging stroke, it ended forever all simple intuitive explanations given to natural phenomena. Its fundamental ideas and concepts are decidedly counter-intuitive. Its most fundamental concept is the **"Uncertainty Principle"** which negates the possibility of a deterministic physical universe as advanced by Newtonian Mechanics in favor of a probabilistic one in the meantime unintentionally restoring in the process the idea of *"Free Will"* in philosophical and theological discussions. Resorting to quantization of energy by Max Planck was the only way available to get around the insurmountable problems facing the Raleigh-Jeans theory describing black body radiation which by the end of the nineteenth century represented the best approach in Thermodynamics. A black body is theoretically defined as any object that is placed in a light-tight (dark) room such as it could not be seen. Upon heating, it should emit radiation giving red then white glow that can be measured and plotted to give what is known as "Black-Body Radiation Curve". Many experiments were carried out late in the nineteenth century to measure characteristics of black-body radiation. According to the prevailing deterministic heat theory of the time, black-body radiation should increase to infinity with increasing frequency. This obviously cannot happen physically. Measured curves showed the energy increasing with frequency and instead of continuing to increase as predicted by the heat theory of the time it decreased with increasing frequency. These results could not be explained using any known theory. This problem was called the **"Ultraviolet Catastrophe"** in reference to increasing wave frequency beyond visible light. Max Planck in 1900 successfully used mathematical curve fitting techniques to come up with a mathematical expression describing vey precisely the obtained experimental data. He is known to have called his approach an act of sheer desperation. Implied in that description was the assumption that the black body consists of "vibrating oscillators" (atoms were not proven to exist at that time) whose energy exchange with the black-body radiation was *quantized* and not continuous. From the given mathematical expression, he could calculate the amount of discreteness which came to be named **"Plank's Constant"** and is now considered one of the most fundamental constants of nature. Its value is extremely small that cannot be observed in common everyday experiences. Since Newton's and Huygens' time, light was considered a wave despite some unexplained phenomena. Clearly Planck's proposition suggested that light is actually granular rather than continuous and propagates in packets called "*quanta*". The minimum amount of energy exchange was described by Plank's constant or a "quantum" which is currently ascribed to the energy of a photon or the quantum of light. Energy can only be exchanged in integral multiples of quanta. That description veered far from the common held view by most scientists at the time and was promptly dismissed despite Planck's outstanding status as a scientist. *The world of science would not give up determinism according to Newtonian Mechanics.* In 1905 the unknown Swiss patent office clerk Albert Einstein published a paper successfully explaining the "Photoelectric Effect" (light shining on metal surface produces electricity that is proportional to the incident wavelength/frequency and not the beam intensity as was believed) using the concept of quantized energy. Planck immediately understood the support given his work by this paper and brought Einstein physically and figuratively to the attention of the world of physics for the very first time giving him his very first truly scientific and academic position. The quantum theory itself was developed in a unique avalanche of contributions made by outstanding physicists in the decade of the 1920s. The greatest of these contributions came from Werner Karl Heisenberg (1901 – 1976) who showed that quantum mechanics unequivocally implies "***Uncertainty***". The uncertainty arises because in the micro-universe of subatomic elementary particles the act of measurement (observation) inevitably changes particle attributes such as position and velocity for example. That is to say one can never measure both an electron's position and its velocity with certainty independently of the measuring technique. Accuracy in measuring one parameter or attribute is at the expense of measuring the other. This essentially makes the observer (measuring agent whether a person or an instrument, etc.) part of the process of measurement itself. This fact is valid as a fundamental characteristic of nature from the world of subatomic particles to the entire physical universe. Even a cursory glance at this statement shows far reaching consequences beyond science into theology and philosophy. Thus, the conclusion that the universe and its observers (humanity) are inextricably linked is inescapable. In other words, to some thinkers the universe only exists to be observed by humans. Or alternately, humans only exist to observe the universe. The quantum revolution changed the way physicists view the universe. The universe in the quantum view is discontinuous with an ultimate limit on how small electromagnetic energy can be (the quantum or Planck's constant $h = 6 \times 10^{-34}$ joule-second), how small a unit of time can be (Planck's time = 10^{-43} second) and how small a measure of distance can be (Plank's length = 10^{-35} meter). Problems with the interaction of light with matter led to the discovery and development of quantum physics and Maxwell's equations were supplanted by quantum electrodynamics (QED). Mathematical tools of quantum mechanics were extensively deployed to determine the structure and physics of the micro world of elementary, atomic and subatomic particles as well as radiation/matter interactions at that scale. As a direct result, more irreducible particles and more fundamental forces of nature were discovered. Elementary particle physicists constructed what is known as the "Standard Model" to describe the most fundamental building blocks of matter. Studies of the origin and evolution of the universe gave birth to the field of "Cosmology". After the discovery of the nucleus, two more fundamental forces were added to gravity and electromagnetism to account for the stability of the nucleus. These were the "Strong Force" (responsible for holding the particles in the nucleus, protons and neutrons, together) and the "Weak Force" (responsible for radioactive decay). Eventually, electromagnetism and the weak force were found to describe the same phenomena and were combined in the "Electroweak Force". Therefore, at the present time there are three fundamental forces that keep the universe working. Physicists are hard at work to combine these forces in one fundamental force describing the universe. While strong and electroweak forces may be amenable to merging, gravity does not seem to be anywhere close to be incorporated in a Grand Unified Theory (GUT). Gravity is the weakest of all fundamental forces of nature and is extremely long range. It is remarkable that although gravity is the oldest of the forces humans have observed and investigated it is currently the least understood. It has proved most intractable so far when it comes to trying to fit it into the quantum mold. The best current description of gravity is the general theory of relativity with its classical outlook

STANDARD MODEL OF ELEMENTARY PARTICLES AND FORCES

What is currently called the standard model in physics represents the most widely accepted theoretical framework for the basic constituents of the structure of the micro subatomic universe or simply put which particles are truly irreducible fundamental elementary ones. It came about through solving field equations which implies that *particles are simply manifestations of their associated fields*. There is one type of field for each species of elementary particle. The photon is the quantum of the electromagnetic field. The electron is the quantum of the electron field. In this model, protons and neutrons are made of the more fundamental quarks. As such, there are no proton or neutron fields but rather quark fields whose properties are determined by "**Quantum Chromodynamics**". This is a misnomer as the theory has nothing to do with color but was an historical place-holder after-thought that got propagated. The standard model is formulated to explain in a straightforward scientific way how the building blocks of the universe behave and interact. All experimental evidence supports this model albeit in an inelegant way. It was excruciatingly developed over many decades in the twentieth century through attempts to combine the basic forces of nature as determined by modern physics. As is currently well known and universally accepted by physicists, these forces are the "Strong Force" that describes interactions that bind the nucleus together i.e. interactions between and among protons and neutrons and how they take place. Next there is the "Weak Force" which describes radioactivity. These two forces describe the nucleus and its stability. Added to these newly discovered forces is the by now generally familiar "Electromagnetic Force" that Maxwell introduced in the nineteenth century with the revolutionary concept of the *field* that solved the insurmountable physical problem of interaction at a distance. These forces give birth to the most elementary particles in nature that the universe is built from. Stable nuclei combined with electrons form atoms and come naturally out of this model leading to the "Periodic Table of Elements" explaining in the process why more elements do not and cannot exist. However, the Standard model is perfectly understood to be incomplete since it does not account or explain a number of generally accepted concepts such as "dark matter", "dark energy", "gravity", etc. While all known subatomic particles were accounted for in this model, only the "Higgs Particle" was missing due to lack of technical capabilities beyond what was technologically available till it was discovered in 2012 at the "Large Hadron Collider" facility at CERN. Without the existence of Higgs field and its associated particle, all other known particles were calculated to be massless. Higgs Field give all known particles their masses. The long sought after Higgs particle is unique in the sense that it is the very first discovered elementary particle with "spin" zero. Its discovery solidifies the standard model as the most fundamental representation of the constituents of the physical universe. The standard model has been the result of some very impressive theoretical work. However, most of its predictions are experimentally verified which lends it general acceptance among physicists. Its relative success prompted more theoretical work employing very elegant mathematical theories albeit so far with no real experimental approval.

GRAND UNIFIED THEORY (GUT)

The essence of a Grand Unified Theory (GUT) is to unify all the known forces of nature. That means that electromagnetic, nuclear and gravitational forces can be interpreted as manifestations of a single primeval force that existed at the instant of creation when all the physical laws were laid down. Other forces split when the universe went through phase transitions that broke certain symmetries in its structure according to the most acceptable physical theories. In other words, all currently observed fundamental natural forces were embedded in this primeval force at the instant of creation. The very large number of discovered subatomic particles casted a gloomy shadow over particle physics. By the early second half of the twentieth century particles were associated with fields and the fundamental forces in nature. As mentioned before, these forces were reduced to four types; the strong force, the weak force, the electromagnetic force and gravity. Scientists instinctively assumed that nature *should* have but one single fundamental force that all others are derived from by breaking symmetry or decoupling. Symmetry in physics is defined in terms of equations remaining the same when fundamental parameters are exchanged. For example, when space is changed into time or electrons into quarks. In mathematics the study of symmetry is called "Group Theory". Symmetries linked groups of subatomic particles together and that gave rise to the concept of GUT. As always in science, a theory has to be *experimentally verified* and advance some *predictions* that can be tested. As happens very frequently with fundamental ideas in physics, unfortunately the energies needed to verify this theory are far beyond available technology. However, GUT makes a testable *prediction* that protons (made up of three quarks) will decay into electrons through emission and absorption of quarks implying that protons have a finite lifetime and are not elementary particles after all. The theoretical framework of the unified theory successfully linked the strong force with the electroweak force (which links the weak force and the electromagnetic force). On the other hand, it incorporates an unacceptable huge number of arbitrary constants. That is to say in pursuit of simplifying the laws of nature the grand unified theory unsatisfactorily complicates its description. The most fatal strike at the heart of this theory is that it is absolutely incapable of including the gravitational force in its formalism regardless of its claims of grandiose characteristics. It is intriguingly noticeable that all bold attempts to unify the fundamental forces of nature resort to highly complicated mathematical approaches to describe nature rather than the historically usual reverse. The failure of GUT to account for gravity in terms of mathematical theories dealing with symmetries gave impetus to the development of the "Superstring Theory" which went through numerous variations and is still without a final acceptable form.

SUPERSTRING THEORY

All theories that aim to combine all fundamental forces of nature are necessary to explain the evolution and stability of the observable three-dimensional universe as well. They eventually have to contend with singularities (physically meaningless infinities in the solutions of the involved equations). The most famous one is the big bang singularity at the instant of creation. Such approaches attempt to combine the general theory of relativity (which is successful in dealing with phenomena describing very large objects such as galaxies and even the universe) with quantum physics (which is very successful in dealing with phenomena describing very small objects such as subatomic particles). Theoretical Physicists realized that these singularities arise due to treating elementary particles as dimensionless *"point particles"* in the relevant equations. Thus, it was simply natural that someone would try to get around this insurmountable problem by assuming that particles actually have dimensions and are *"strings"* instead of being points. The string theory was born in this way and it went through many modifications to give birth to the *"Superstring Theory"* and many other variations. Mathematically speaking, many unsolvable problems were worked out and the problem of singularities virtually disappeared. The most important underlying principle of the superstring theory is that it could accommodate an infinite number of particles that are nothing more than different resonances of the same string where no particle is more fundamental than any other. The significant implication here is that particles are not assumed before-hand as is the case with all existing theories including GUT. In theories that employ strings, they come to existence as resonances to these fundamental strings. Since particles were assumed to have dimensions, multi-dimensional descriptions of the universe became a must to get around more and more mathematical obstacles that arose. Superstring theory claims to have overcome the big bang singularity. It considers it a consequence of the breakdown of a "ten-dimensional" universe into a four-dimensional one. According to the superstring theory, at the big bang all four fundamental forces were actually combined in one single paramount force. At Planck's time of 10^{-43} seconds after the big bang, energy and matter consisted of unbroken strings. The universe was 10^{-33} centimeter across. Quantum Gravity was the dominant single force at that time. With temperature of 10^{32} K, gravity separated from the other forces. As the universe expanded it cooled and the strong force split from the electroweak force. At approximately 10^{-9} second after creation and a temperature of 10^{15} K, the electroweak force broke into the electromagnetic force and the weak force. Once the hurdle of the big bang singularity is surmounted, the superstring theory follows the generally accepted description of universe evolution and structure obtained by other approaches. It is important to notice that this is done within the context of a *multiverse* rather than a universe. Although there are several other approaches exploring ways to combine the fundamental forces in the universe, string theory and its derivatives (superstring theory, M-theory, etc.) are the most popular ones in the pursuit to formulate a quantum gravity theory to unify all forces of nature. This methodology has had some success in solving many problems but remains fundamentally an elegant mathematical structure without any experimental proof. Many critics are puzzled by its popularity nonetheless. Among its harsher prominent critics is the great British Physicist/Mathematician Roger Penrose. He lambasted its reasoning and formulation in lectures, presentations and writings. In his 2017 book *Fashion, Faith and Fantasy-in the new Physics of the Universe* (section 1.12) after taking it apart point by point, he wonders that if string theory (and its later developments) indeed leads to a higher-dimensional space-time picture which appears to be at such odds with the physics that we know, then why does it continue to have such a fashionable status among this extraordinarily large and exceptionally able community of theoretical physicists? and a few pages later he adds that in a general way, ideas will remain fashionable in science only if they are both mathematically cohesive and well supported by observation. Whether this is true for string theory, however, is debatable at best. He vehemently objects to, in his opinion, its unsubstantiated claims of the existence of higher dimensions.

COUNTER-INTUITIVE RESULTS OF QUANTUM THEORY

What is it that gives a particular person his/her individual identity? Is it, to some extent, the very atoms that compose their bodies? Is one's identity dependent upon the particular choice of electrons, protons and other particles that compose those atoms? There are at least two reasons why this cannot be so. In the first place, there is a continuous turnover in the material of any living person's body. This applies in particular to the cells in a person's brain, despite the fact that ***no new actual brain***

cells are produced after birth. During the life of a person the vast majority of atoms in each living cell (including each brain cell) and indeed virtually the entire material of human bodies would have been replaced many times over since birth. The second reason comes from quantum physics and by an exacting irony is, strictly speaking, in contradiction with the first! According to quantum mechanics, any two electrons must necessarily be completely identical and the same holds for any two protons and for any two particles whatever, of any one particular kind. This is not merely to say that there is no way of telling the particles apart; the statement is considerably stronger than that. If an electron in a person's brain were to be exchanged with an electron in a brick, then the state of the system would be exactly **the same state**. This holds true if the particle were a boson (*Bosons are those particles which have an integer spin 0, 1, 2... All the force carrier particles are bosons, as are those composite particles with an even number of fermion particles - like mesons*). However, if it were a fermion (*A fermion is any particle that has an odd half-integer like 1/2, 3/2, and so forth spin. Quarks and leptons, as well as most composite particles, like protons and neutrons, are fermions*) then, its state would be replaced by its negative state which is physically identical to it anyway. This sign change can be remedied by rotating one of the two particles completely through 360° when the interchange is made and it is as it was before, not merely indistinguishable from it! The same holds for protons and for any other kind of particle, and for whole atoms, molecules, etc. If the entire material content of a person were to be exchanged with corresponding particles in the bricks of his/her house then, in a strong sense, nothing would have happened whatsoever. What distinguishes the person from his/her house is the ***pattern*** of how their constituents are arranged, not the individuality of the constituents themselves.

To many, the term quantum theory evokes merely some vague concept of an *"uncertainty principle"*, which at the level of particles, atoms or molecules forbids precision in descriptions and yields merely probabilistic behavior. Actually, quantum descriptions are very precise. One tends to think of discrepancies between quantum and classical theories as being very tiny, but in fact they also underlie many common physical phenomena. The very existence of solid bodies, the strengths and physical properties of materials, the nature of chemistry, the colors of substances and many other properties are due to quantum effects. In spite of the enormous success of quantum theory, it is still considered a tentative theory providing a partial picture of the world. As a matter of fact, Niels Bohr (1885-1962) and his disciples say that there is no objective picture at all. Nothing is actually "out there" at the quantum level. Somehow *reality* emerges only in relation to the results of *"measurements"*. As such, quantum theory provides merely a calculational procedure and does not attempt to describe the world as it actually is. Humans naturally without a second thought deal with reality in their daily lives as deterministic. In their world causality reigns supreme where cause and effect are tightly bound. This in the final analysis nonetheless is a major delusion since the macro-world of humans is intricately built from micro-world constituents that are strictly governed by the laws of quantum physics. The importance of the micro-world laws cannot be overlooked for the simplest of reasons as it is how the familiar macro-universe began some 13-15 billion years ago. To learn how the universe came to be the way it is at the present time, high energy physics experiments are carried out in major research centers such as Fermi Laboratory in the US and CERN in Europe for example to try to simulate conditions of the early universe when quantum effects controlled everything. In the quantum realm scientists have been grappling with an understanding of what they call *"quantum reality"* since the early days of the quantum theory in the 1920s without the faintest sign of success. As far as human knowledge is concerned, quantum reality is unfathomable. Ironically though, quantum theory is the only theory in the history of science that never failed to explain every associated problem and survived every challenge thrown at it so far. Paradoxically, in the quantum realm everything happens in random where determinism is meaningless. **In the quantum realm there is no direct link between cause and effect. Cause gives only a probability for an effect.** In the language of this undertaking that means that within the quantum realm (which is the definite precursor of the familiar universe) *free will* dominates and its effects are integrally embedded in the development of the macro-universe. The cornerstone of quantum mechanics (equally valid terminology is quantum physics) is that no object (that curiously should include humans and even the entire universe) exists in a definite state until it is observed. That should make it abundantly clear that *reality is a matter of perception*. Put another way a human being for example is not in a definite state (dead or alive!!) until he or she is observed. Observation in this case is external. That is to say from outside their familiar three-dimensional world!!!!! The same goes for the entire universe. Does it truly exist? Or is it just a human illusion since no one is known to have observed it from the outside? That is why quantum reality is counter intuitive and unfathomable but quantum mechanics is nonetheless absolutely correct according to countless proofs. That was not a proposition Einstein could accept. He fiercely dismissed quantum mechanics till the day he died. His requirement for a valid physical theory is that every element of the physical reality must have a counter part in the physical theory. Quantum mechanics fails in this regard; dealing only with group behavior (i.e. statistics), it is a theoretical system that cannot account in detail for individual happenings. Einstein paid much attention to intellectual implications of physics due to his nature and upbringing. However, most scientists ignore these philosophical arguments as irrelevant to their pursuits and limit their work and conclusions to the micro-world of elementary particles and the evolution of the universe since the big bang event. Curiously, atheists on the life sciences side of the debate who prefer to be called strict evolutionists or Darwinists base their arguments about the origin of life on the random collisions of atoms and molecules in the early universe resulting in the origination and evolution of life ending with homo-sapiens and the modern human. Such a recurring random process is absurd and more importantly would take much longer than the known age of the universe. To get around this difficulty, prominent atheists such as the British zoologist "Richard Dawkins" rather convincingly assert that only the very first process of bringing atoms and molecules together to form the essential precursor of life is subject to chance and admittedly should take relatively long time to happen. After that, natural selection takes over and randomness and chance are no more. Although the processes involved are very complicated, they are nonetheless deterministic. That solves both problems just mentioned and such an explanation cannot be dismissed off hand.

*Within the confines of the thesis promulgated by the author in this book, no fundamental objections are raised against the mechanism of evolution as a way to explain the origins of humans after the creation of stars and galaxies. However, all these processes took place in the macro-world after atoms and molecules were formed and after the universe have cooled to a degree enough to allow such interactions. There is not the slightest doubt among cosmologists that these processes were preceded by activities in the micro-world governed by the laws of the quantum realm. As advanced above, that realm maintains **free will** and transmits its implications into the very fabric of the macro-world despite the fact that details of that transition are not currently understood and may never be understood. In other words, atheists arbitrarily choose a starting point that suits their conclusions rather than the universally agreed upon original point of the big bang and the quantum world.*

Quantum Entanglement

From the earliest days of developing the quantum theory, it was clear to everyone that its conclusions are going to be counter intuitive mostly because of its probabilistic nature as complex numbers replace the usual scalar ones. However, one issue stood above all else in its weird implications. That was what is known as the principle of "quantum entanglement" which was first introduced by Schrodinger in a letter to Einstein. This phenomenon arises when considering the linear superposition of the separate states of two presumably independent particles. Due to the strict rules necessary to validate the theory, it was asserted that in some specific cases in linearly superposing states of two objects (elementary particles for example) they behave as a single entity regardless of the space separating them. In this entangled state neither particle has a separate state on its own. Schrodinger asserted that he would not call entanglement as *one* but rather *the* characteristic trait of quantum mechanics, the one that enforces its entire departure from classical lines of thought. In 1935, Albert Einstein seized on this phenomenon to prove that the framework of quantum mechanics is basically flawed or at best incomplete. He enlisted the help of two of his colleagues; the Russian-American Boris Yakovlevich Podolsky (1896-1966) and the American-Israeli Nathan Rosen (1909-1995) in that effort. They formulated what has been historically known as the (EPR) phenomena consisting of a number of thought experiments in the sense that there is no practical reason why they could not be actually performed. Their goal was to show that quantum mechanics has an implication that is physically unacceptable. This implication was that a pair of particles separated in space regardless of the distance involved still have to be considered as a single entity. Therefore, a measurement performed on one of the particles appears to affect the other one *instantaneously* forcing this second particle into a particular quantum state that depends not only upon the result of the measurement made on the first particle but startlingly upon the specific *choice* of measurement that is made on the first particle. Most of these EPR experiments were expressed utilizing elementary particle spin for simplicity. Many such experiments have been put forward during the following decades where divergences between the intuitive expectations of classical physics and the counter intuitive results of quantum mechanics are clearly shown. One interesting actual experiment was carried out in 1993 by Lucien Hardy that unequivocally proved the entanglement principle. The conclusion of these experiments is that in many situations separated quantum objects, no matter how far apart they may be, are still interconnected with one another and do not behave as independent objects like when treated classically. However, such entanglement cannot provide for new information to be transmitted from one object to the other as this would violate the requirements of relativity. It is this incapability of sending actual information via entanglement that allows one to consider that the entanglement is transferred *instantaneously* without violating the tenets of relativity. It is important to understand that it is more accurate to assume that time is not in any way involved in this process of entanglement transference since it makes no difference whether this transference is from the first particle to the second or the other way around. None of these efforts proved Einstein's objections to the approaches of quantum mechanics. On the contrary, they unequivocally validated them. Apart from science and technology, quantum entanglement has far reaching intellectual implications. It will be shown that some of the unusual facts narrated in the Qur'an can easily be explained (if one wills) employing this phenomenon. This issue will be discussed when dealing with subjecting the Qur'an to the scrutiny of science. The above discussion of the divergence of quantum mechanics from the classical approaches may have left the uninitiated unspecialized person with the impression that while any person can relate to the classical picture of reality through common sense and *intuition*, it is to assume that the quantum realm is dominated by chaos and no rules hence the resulting *counter intuition* nature of its conclusions. That would be an unfortunate patently false impression. Both classical and quantum realms are governed by *determinism*. As mentioned several times before, classical (or simply Newtonian) determinism meant that knowledge of parameters of an object at certain instant leads one to determine with *certainty* its parameters in any other instant past or future. On the other hand, *quantum determinism* means that knowledge of the *quantum state* of an object at certain instant completely determines its q*uantum state* at any other instant past or future with the proviso that knowledge of its *quantum state* determines only the **probability** that one or another future will ensue. In this sense, the present (as future of the past) is determined by a specific choice (or observation) to force the object into a specific *certain* state. This is as true in the macro-world as it is in the micro-world. Obviously this is the essence of *free will*.

Cosmology

INTRODUCTION

The universe and where it came from and where it is going to are issues that fascinated the imagination of humans since their early existence. Measurements were carried out by the ancients to determine certain celestial values related to their lives and beliefs but the fundamental questions were left to religion to answer. Empirical studies of the cosmos are very recent phenomena. In 1514 the Polish mathematician Nicolaus Copernicus (1473 – 1543) worked on reforming the Christian calendar at the request of the pope. The result was his study published in 1543 "On the Revolution of Celestial Spheres" which eliminated the prevailing view that earth was the center of the universe. However, he asserted that the sun is that center and that earth orbited it like any other planet. This is the heliocentric universe of Copernicus which replaced the geocentric universe of the ancients. Since scientific knowledge is always tentative, it is always being refined. The mechanical universe of Newton soon replaced that of Copernicus which in turn is currently replaced by the relativistic universe of Einstein. The modern view of the universe gives absolutely no preference to any object over any other as every object great or small moves *relative* to every other object. Therefore, the universe should look the same regardless of the *arbitrary* choice of the reference. It should also be governed by the same physical laws everywhere. Modern physics and its two major pillars of relativity and quantum mechanics spawned interest in studying the universe as a whole. Relativity hinted that the observed universe can be described by its equations. Quantum mechanics suggested that the state of the universe at the instant of creation and its later evolution are controlled by its rules. These new developments stimulated the innate human interest in knowing where the universe came from, how it evolved and where it is headed. The new scientific discipline of cosmology was born out of these contexts. Equations of the general theory of relativity suggested that, unlike what has always been believed, the universe is actually dynamic in nature. General relativity led to modern cosmology which eventually firmly established the "Big Bang" theory as the most probable explanation for the creation of the universe. While general relativity explains the macro world such as stars and the galactic level close to perfection, quantum mechanics does the same to the micro world such as elementary particles and the sub-atomic level. Combining general relativity and quantum mechanics to treat the origin of the universe leads to a singularity and disastrous results at the event of the big bang itself. To reconcile both theories a unified theory has been sought but to no avail yet. Superstring theory is currently the most successful as all species of particles are unified since each arises from a different vibrational pattern executed by the same underlying entity. Instead of the (three) spatial dimensions and (one) time dimension of common experience, superstring theory requires nine spatial dimensions and (one) time dimension. More elaborate expansion of this theory leads to the M-Theory where ten space dimensions and (one) time dimension are required. The reason these dimensions are not seen in everyday experience is, according to the theory, because they are too small for any existing equipment to observe. Or they might be so large to be probed. The small-dimensions approach can explain why the universe has stars and planets whereas the large-dimensions approach introduces the possibility of other universes existing. The discovery of extra dimensions, if it ever takes place, would show that the entirety of human experience had left humankind completely unaware of a basic and essential aspect of the universe.

Cosmology is defined as the study of the origin, evolution and structure of the universe. With the formulation of the general theory of relativity, scientists realized that its equations can be applied to the universe as a whole. That gave impetus to the new science of cosmology. Few proposed theories such as the "Steady State" theory dealt with origin of the universe but the currently most successful and generally accepted one is the "Big Bang" theory. Evolution of the universe analyzes the causes for the formation of galaxies, clusters and super clusters. Most theories in this field take the big bang as the starting point to explain how the universe eventually assumed the current structure. Structure of the universe deals with the formation and distribution of stars. It also describes the violent end of stars. Formation of stars is simply a balancing act between gravity that pulls material constituents together toward the center and pressure that results from such squeezing causing constituent gas to heat up. According to the currently acceptable explanation of how material elements came about, every atom of carbon, nitrogen and oxygen in the solar system (as well as in the universe as a whole) was synthesized in early stars that died before the sun formed. If the universe started completely smooth and uniform, it will remain featureless and life would not evolve even after many billions of years. However, very slight ripples initially departing from uniformity would be greatly amplified with time and expansion allowing complexity to develop. Gravity would cause regions with very slight density of matter to condense and form structures such as stars and galaxies and eventually life. These ripples should show in the *microwave background radiation* which permeates the universe. This radiation is a relic of the big bang and carries information about the initial conditions of creation of the present universe. In 1992 the results obtained by the "Cosmic Background Explorer (COBE)" satellite confirmed the existence of such ripples. That evidence substantially supported the "Big Bang" theory as the most plausible interpretation of how the universe began. These results were of such importance to cosmology that the renowned British scientist Stephen Hawking called them "the greatest discovery of the century, if not of all time." It is curious that NASA only launched that satellite after many years of rejection to maintain its relevance after the disastrous explosion of the space shuttle "Challenger" on January 28, 1986 a mere 73 seconds into its flight killing all seven crew members. George Smoot; the principal scientist of the project received the Nobel Prize in Physics in 2006.

The Big Bang and Inflationary Universe

While Einstein adamantly believed in the static and perpetual nature of the universe, the clear indications of his own general theory suggested the contrary. This is how he committed his "worst blunder" by modifying the formulation of the theory to preserve the static nature of the universe despite the obvious fact that it would be unstable under these conditions. In the 1930s the American astronomer Edwin Hubble (1889-1953) showed experimentally beyond any doubt that the universe is dynamic and expanding at calculable rate. Einstein relented admitting his blunder and accepting the real interpretation of his general theory of relativity offered by other physicists that led to formulating theories about the origin of the observable universe. As is currently accepted, one of the fundamental conclusions of the theory of General Relativity is that the universe is expanding. If that is the case, then there was a point where this expansion has begun. The implication is that humans live in a dynamic universe which goes against the grain of the firmly held belief in a static universe. Humans held that belief from time immemorial. That is why even Einstein himself refused to accept the implications of his own phenomenally successful theory. To restore the static nature to the universe, he added what was called the "cosmological constant. This mathematical device expunged the implications of an expanding universe from the equations of general relativity. Static view of the universe avoids addressing the obvious question of "what happened in the beginning". Later when physicists obtained overwhelming experimental evidence that the universe is actually expanding, it has been known that Einstein admitted that his addition of the cosmological constant as the the biggest blunder of his life. If the universe is expanding, then looking backwards it must have had a beginning. That is the essence of the "Big Bang" theory advanced by the Belgian physicist (who was also a Catholic priest) Georges Lemaittre (1894-1966) in his thesis to obtain his doctorate (who later became the president of the "Pontifical Academy of Sciences" of the Vatican). That theory was practically ignored and some prominent scientists derided and mocked it as nonsensical. However, the discovery of the universal "cosmic microwave background radiation" provided the unassailable evidence of its logic. The theory in its original form could not account for the observable structure of the galaxies and their stars. It also could not explain the uniformity and isotropy of the universe. That was eventually provided by the inflationary theory that the universe in its extremely early stages had experienced tremendous inflationary expansion that preserved its smoothness and uniformity for the rest of its 10-15 billion years so far. Humans have known for a long time that there are star systems just like their own solar system with planets orbiting such stars. They also learned that these stars are not evenly distributed throughout the universe but rather they are clumped together in galaxies such as the disk like Milky Way. With the advances in astronomical observations and studies, it is determined that matter in the universe is highly structured. It is also determined that the shining stars in the dark night sky represent less than one percent of the matter in the universe. Most of the existing matter in the universe is invisible to humans. The night sky is understood to be dark because the universe is not infinitely old hence, light from the more distant stars has not reached earth yet. It is also dark because there is a finite (as opposed to infinite) number of stars in the universe. Many attempts were made to give an explanation to the astronomical discoveries about the structure of the universe. Eventually two major competing theories tried to explain the structure of the universe based on the known astrophysical data. While the "Big Bang" theory adopted the dynamic view of the universe, the "Steady State" theory was championed by Fred Hoyle (1915-2001) who ironically was the one to coin the term "Big Bang" as a derogatory name. It adopted the static view of the universe. The basic premise of the steady state theory was that matter is routinely created from nothing by a creation field in the same sense of gravitational or electromagnetic fields. Accordingly, creation of new matter caused the universe to expand. In other words, the universe was in a steady state always remaining essentially the same as it is at the present time. It was homogeneous in space and time. In contrast, the "Big Bang" theory implied that as space expanded and the galaxies spread outward, the cosmos would gradually dissipate like a cloud. The fundamental problem with the big bang theory from the point of view of mathematicians and physicists was that it required space-time to begin at a single mathematical singularity at a point with zero dimension and infinite density. Physical laws breakdown at singularities, that makes the universe beyond description. On the other hand, the big bang theory advocated the cataclysmic creation of matter as well as space. It is essential at this point to understand that according to the theory, ***the big bang did not move INTO existing space but rather it created space as it expanded***. Cosmologists routinely describe this phenomenon in terms of what happen to two points on the surface of an expanding balloon. The two points are not actually moving themselves when the balloon expands. However, the *space* between them is continuously increasing. Since in this model objects are not physically moving, they are not subjected to the restrictions of the theory of relativity. Therefore, at some point in the distant future space between certain objects would be larger than that which can be travelled by light. In other words, the two objects (separate parts of the universe) under consideration would be moving apart faster than the speed of light. Hence, no information about one would ever reach the other. That necessarily means that the existence of and what happens within one is forever (in terms of time) beyond the knowledge of the other. That is obviously the premise of human experience. This would not be the case if there are other creatures not bound by human perceptions and limitations. That is what the laws of physics and cosmology dictate.

COMPETITION BETWEEN THE STEADY STATE AND BIG BANG THEORIES FOR VERIFICATION

Steady State theory's one clear advantage over the Big Bang theory is that it did not have to explain what happened before creation of the universe. However, the big bang theory implied that all the features of the cosmos (its types of forces, physical constants, etc.) were *arbitrarily* set from the beginning. That is to say they were special *initial conditions* making science incapable of explaining their origin. Both theories made several distinct *predictions* about the age of the universe, the abundance of various elements, the distribution of matter across space and time and the relic radiation of the primeval fireball. Age of the universe is determined to be between 10 and 20 billion years while that of earth is approximately 4 to 5 billion years. Big bang theory *predicted* that closer to the instant of creation (the big bang) the density of the cosmos should be much larger since all matter was created then. Steady state theory *predicted* homogeneity of galaxies across space and time. The most important *prediction* that ultimately tipped the scale was the existence of the cosmic background radiation as a result of the cooling of the universe as it expanded according to the big bang theory while the steady state theory could not *predict* the existence of such radiation or its rationale.

The Cosmic Background Radiation

In August 1959 the United States launched Echo1 satellite as the first step in establishing satellite communications. In 1964 Arno Penzias and Robert Wilson, as researchers at Bell Labs in Holmdel, New Jersey, were trying to measure noise levels that might contaminate communications with Echo. To their puzzlement, they picked up a microwave signal at 7.35 cm regardless of which direction in the sky their antenna array was aimed at. The signal was isotropic i.e. with equal intensity over the entire visible sky. They could not explain their measurement but seeking an explanation with the physicists at near-by Princeton University, they understood that a background microwave signal at 7.35 cm (what physicist equivalently call at temperature 3.5 degrees Kelvin) is what astronomers have been tirelessly looking for as the cosmic background radiation *predicted* by the big bang theory. They received Nobel Prize for their discovery in 1978. That was the most important confirmation of the big bang theory to date and a serious blow to the steady state theory. However, the question of how did the big bang lead to the formation of stars, galaxies, galactic clusters, etc. by the condensation of the products of this cataclysmic event of creation was still unanswered. The cosmic background radiation seemed to suggest a uniform space and energy contradictory to what is observed to be the case. For structures to be formed after the big bang, cosmic seeds or fluctuations in temperature caused by areas of higher density must have existed. According to the big bang theory, matter could have condensed and subsequently formed galactic structures in such areas through gravity. With the proposition of antimatter by Dirac in 1929 as a result of combining special relativity and quantum mechanics (he won the Nobel Prize in 1933 and Carl Anderson (1905-1991) of Caltech won it in 1936 for discovering the positron which is the anti-electron) it became obvious according to the big bang theory that half the mass of the universe must consist of antimatter. The discovery of anti-particles in cosmic rays proved that antimatter exists in the universe as *predicted* by the big bang theory. The puzzle of the great excess of matter over antimatter in the universe was solved by the Soviet/Russian physicist Andrei Sakharov's (1921-1989) hypothesis that processes in the first instant after the big bang produced a slight excess of matter over antimatter. These particles self-annihilated leaving the slight excess of matter. This went on to form the known universe. Energy released in the annihilation process was manifested as cosmic background radiation photons which are presently observed in numbers at least a billion times more than those of the heavy particles or baryons. Grand Unified Theories (GUT) pioneered by Abdu Salam (1926-1996), Weinberg and Glashow attributed this slight excess in matter over antimatter to the unification of the weak, strong and electromagnetic forces at the shortest imaginable time or Planck time (10^{-43} seconds) after the big bang. The implications here are very profound. What is implied is that, **known physical laws do not apply at the immediate aftermath following the big bang** although they majestically confirm all observations throughout the entire life of the known universe.

The general theory of relativity allowed the possibility of a rotating universe raising the inevitable question of "rotating with respect to what?". With the advent of modern physics, it was determined that everything within the universe is rotating from the protons to the galaxies. However, there is no evidence that the universe itself is rotating according to Doppler effects measurements. It was also determined that galaxies are moving in space with extremely high speeds (more than a million miles an hour) which could not be explained except by assuming motion under an enormous gravitational force. That was attributed to the existence of black holes. The universe is therefore more inhomogeneous than could be explained by the sustained simple expansion from the big bang. Some regions of the universe are virtually devoid of galaxies while in others galactic super structures consist of billions of galaxies. These enormous structures could have only developed from cosmic seeds that have existed early on. According to Newton's law of gravitation, objects at the periphery of a rotating structure should have much lower speeds than those close to the center. However, measurements showed that stars in galaxies like the Milky way (where the majority of stars are concentrated at an extremely bright center and are rarer toward the dim periphery) at the far periphery orbited with very much the same speed as those closer to the center. The only possible explanation is that galaxies consist of more than the visible stars detected by astronomical observations. There must be present an unseen enormous mass forming a major component of the galactic structure. This is what is currently known as the "**dark matter**". Although there are numerous aspects of the beginning and evolution of the observed universe still currently unexplained by any existing theory, the consensus among cosmologists is that the big bang theory comes closest to offering such description bearing in mind the many experimental confirmations of its *predictions*. Thus, it is clear that most of what is accepted within the astronomy and cosmology communities is based on speculation and at best indirect observations and measurements rather than actual ones. However, this is typical of scientific endeavors. As emphasized several times in this book, science is tentative and quasi-successful theories are adopted until they are replaced with new revolutionary ones during periods of crises. As with classical physics, modern physics, especially cosmology, depends on certain parameters to be falsified or verified directly if possible or indirectly in the vast majority of times. One of the most interesting parameters emanating from the application of the general theory of relativity to the whole universe to determine its shape and ultimate fate is the parameter Ω.

Ω

In the 1920's Edwin Powell Hubble (1889 – 1953) formulated his law that the steady expansion of space is responsible for the linear correlation between the distance to a galaxy and the speed of its recession. Others used conclusions of his work to calculate the mass of the universe and its density which can give indications about its shape, evolution with time and ultimate fate. The parameter Ω describes that approach stemming from calculations of general relativity equations. It is defined as the ratio of actual to critical density of matter in the universe (critical density is calculated to be in the range of 4.5×10^{-30} gm/cm^3 or about 2.7×10^{-6} nuclear particle per cm^3). It determines whether the universe will collapse or expands forever according to how much it deviates from unity. Taking only known visible matter into account gives it a value of 0.04 which means the universe by a substantial factor will expand forever. However, latest cosmic research strongly suggests that visible matter (galaxies, stars, gas clouds, etc.) actually contribute a small fraction to the mass of the universe. Dark matter and most recently dark energy (emitting no detectable radiation of any kind) constitute most of the material of the universe. It is currently believed that an abundance of black holes in every galaxy does exist and they account for a substantial part of the dark matter. Their existence is inferred from the observed motion of some stars that are too fast to be balanced only by the gravity of the observable stars and gas. The current value of Ω is calculated to be approximately 0.3. This implies that close to the big bang its value was extremely close to unity because of the competition between the kinetic expansion energy and gravitation energy. If Ω started less than unity, expansion energy will completely dominate and Ω would be very small at present. On the other hand, if it started greater than unity, gravity will completely dominate and the universe will not expand as observed. Therefore, the *initial value of Ω must have been extremely well tuned with remarkable precision to start and sustain life in the universe at its state after approximately 13-15 billion years of expansion*. Expansion of the universe could not have started too fast or too slow otherwise conditions to begin life would not have materialized. Therefore, cosmologists calculate that at one second after the big bang Ω could not have differed from unity by more than 10^{-15} to allow the universe to expand and reach its current status with Ω is still not far from unity regardless of any uncertainty in its current value. This represents strong argument for the inflationary theory of the universe. If Ω is more than unity, then there will be a *big crunch* indicating a limited lifespan for the universe and the universe should be bounded. Even adding dark matter would not give Ω

value greater than unity. Most cosmologists at the present time tend to agree that the value of Ω is less than unity albeit they at the same time believe that a comprehensive knowledge of what the universe actually consists of is far from established. Those who believe that the value of Ω to be around unity adopt the as yet vague concept of dark energy to add to the material of the universe. Assuming the value of Ω to be less than unity, the universe will expand forever with all stars eventually dying. The universe should reach *thermal steady state* where its temperature is absolute zero and nothing is moving. This is the grim fate of the material universe as anticipated by cosmologists based on the very little information currently available. On the other hand, these same cosmologists refuse to speculate about life in the same manner. It is curious that almost all well-established cosmologists once they make their fundamental contributions to the story of the universe, they turn their attention to issues of life which to say the least are beyond their competence. What ultimately happens to the universe is determined by its contents. If there is sufficient mass in the universe, gravitational forces will be strong enough to bring the post big bang expansion to a halt and even reverse it, leading to a cataclysmic **"Big Crunch"**. On the other hand, if there is insufficient mass to cause this to happen, expansion will continue forever with the temperature of the universe steadily falling. This is known as the **"Big Chill"**. If the density of mass in the universe is precisely at the boundary between the diverging paths to ultimate collapse or indefinite expansion, then the Hubble expansion may be slowed, perhaps coasting to a halt, but never reversed. This is known as the **"Critical Density"**. It is calculated to be in the range of $4.5 - 5 \times 10^{-30}$ gram of matter per cubic centimeter of space. That is one hydrogen atom in every cubic meter. Knowing this critical density, humanity can figure out its ultimate fate. The universe is extremely inhomogeneous with regions of enormously high densities and others with virtually nothing. *In summary, the ratio of the actual density of mass in the universe to the critical density is known as Omega Ω. An $\Omega < 1$ leads to an open universe (the big chill), and $\Omega > 1$ leads to a closed universe (the big crunch). An $\Omega = 1$ leads to a flat universe.*

The terms "flat", "open" and "closed" refer to the curvature of space. General theory of relativity construes that gravity is caused by the curvature of the four dimensional space-time. The shape, mass and fate of the universe, therefore are inextricably linked. They constitute a single subject, not three. These three aspects are expressed in Omega Ω. The measured visible mass of the universe evident in galaxies and other visible matter, amount to no more than (1%) one percent of the critical density. Adding the estimated mass of the dark matter evident in the measurement of velocity of stars in galaxies the figure rises to (10%) ten percent or less. This clearly points to the big chill. However, when astronomers measured the relative speeds of pairs of galaxies thought to be orbiting each other, they found that the dark halos of galaxies must extend far out nearly as far as the near-by orbiting companion, which implies more dark matter. This approach was extended to clusters of galaxies and super clusters of galaxies, and in each case these celestial structures move as if embedded in larger and larger swaths of dark matter. Moreover, peculiar measured velocities of galaxies deep in space require even more dark matter. The galaxies are being attracted at huge velocities by unseen masses. Therefore, adding all these dark masses to the insignificant amount of the visible mass, one obtains an average density of the universe of close to the critical density. **Ω is very close to 1**. Obviously there are large uncertainties in all these measured values. At any rate, it is tantalizing to realize that 99% (ninety-nine percent) of the material the universe is made of is dark matter. Humans observe and take measurements of an extremely insignificant part of the universe. At the same time most of them (particularly atheists) have no qualms about generalizing their conclusions obtained by their very limited senses and measuring facilities to the whole existence. This is patently unscientific. Looking at the issue from a slightly different angle, the age of the universe is measured to be in the order of 15 billion years since the big bang and if it was formed by gravitational forces around initial seeds which were very small, there are limits to how much matter must be in the universe. If the universe has an average density close to the critical density, then some regions would have an effective Omega Ω greater than (1) and others less than (1). Those regions with Omega Ω greater than (1) would eventually collapse and form structures, while less dense regions would become voids. This is very much the current structure of the observed universe. However, if Omega Ω were well below (1), then very few regions would collapse. On the other hand, if Omega Ω were well above (1), then everything would collapse. ***The closer Omega Ω is to (1), the easier it is to form the structure of the universe as is currently known.*** Calculations of the density of the universe close to the big bang yield the value of the density of water (1) at approximately 3 minutes after the big bang and the value of (10^{-14}) of (1) for Omega Ω. This is the time when hydrogen and helium were formed. At one second, Omega Ω would be within (10^{-16}) of (1). At Plank time (10^{-43} seconds) closest mathematically or physically admissible time to the big bang, Omega Ω would be within about 10^{-60} of (1). In the cooling universe, symmetry breaking could have caused different regions to align improperly. As a result, flaws in space could have formed. These flaws retained the original symmetric superhot, super massive state of the big bang. The flaws could have manifested themselves in various ways, including magnetic monopoles (zero dimensional points), cosmic strings (one dimensional lines), domain walls (two dimensional planes) and three dimensional textures. While measurements of cosmic background radiation unambiguously indicate that the distribution of matter in the early universe was virtually uniform, the structure of the current universe is observed to be extremely clumpy. Mathematicians such as Roger Penrose and Stephen Hawking developed singularity theorems. According to these theorems, at the big bang, space-time must come to a zero dimensional geometrical point. At and just after the big bang, quantum mechanical effects should be dominant. The uncertainty principle requires that matter-energy and space-time must fluctuate. Matter-energy fluctuations increase the effect of gravity. On the other hand, curvature fluctuations of space-time weaken gravity's effects. Therefore, applying quantum mechanics at the aftermath of the creation of the universe leads to the violation of the conditions of singularity theorems allowing escape from that initial singularity. This is the same situation calculated by Hawking to allow the emission of particles in the vicinity of a small black hole which is known as the Hawking radiation. Just after the big bang all forces were unified according to the Grand Unification Theories (GUT). At 10^{-34} seconds, the symmetry breaks separating strong and electroweak forces. As the universe expanded at this early instant, more space was created with high energy density This allows the universe to continue expanding in an inflationary manner. When the energy density drains from the vacuum, it goes into particles and energy. At that point, gravity starts to exert its own attractive effects gradually slowing down the expansion. The accelerating inflationary phase thus comes to a halt. This is consistent with the current phase of the universe where it is coasting in the aftermath of the accelerating phase of expansion. This inflationary scenario requires that the total energy (i.e. the energy of space itself minus the gravitational attraction of the other parts of space) be zero leading to the conservation of energy. This condition allows the universe to start from practically nothing. All inflation needs, is a small region in the right configuration and it will run away to produce a bubble in space vastly bigger than the whole of what is currently observed. Inflation theory explains why the cosmic background radiation is so extraordinarily smooth but it requires the existence of small perturbations (or seeds) to form the universe as currently observed. These seeds were discovered through the "**Cosmic Background Explorer**" project (or **COBE**) headed by George Smoot and carried out by NASA in 1992. According to Smoot, the project had involved more than a thousand people, consumed many years of effort and cost some $160 million. *These results established the "Big Bang" theory as the acceptable explanation for the creation of the known universe.*

Cosmic Microwave Radiation Background

Scientists suggested that if there had not been an intense background of radiation present during the first few minutes of the universe, nuclear reactions would have proceeded so rapidly that a large fraction of hydrogen would have been permanently fused into heavier elements. However, it is known that at the present time approximately three quarters of the universe is hydrogen. This implies that the universe was filled with intense radiation having enormous equivalent temperature (physicists like to describe the intensity of radiation at a given wavelength in terms of an "equivalent temperature" for convenience) at very short wavelength which could blast nuclei apart as fast as they form during the very early minutes of its creation. This radiation should survive the subsequent expansion of the universe with its equivalent temperature falling inversely to the size of the universe. That is to say that presently the universe should have background radiation with vastly reduced equivalent temperature than it started with. Based on well-established cosmological parameters, present day (at least 10 billion years after the big bang) equivalent temperature of the universe should be in the order of a few degrees Kelvin. Because of the red shift in the spectrum due to expansion, this background radiation should be in the microwave region. It should also fill the entire universe regardless of the direction of observation. This is exactly what Bell Labs' Arno Penzias and Robert Wilson accidently discovered in 1965 and won the 1978 Nobel Prize for it. They found out that the cosmic radiation background is actually a perfectly isotropic black-body radiation with equivalent temperature of 3.5 ^0K which gave tremendous credibility to the big bang theory.

Cosmic Background Explorer (COBE)

The COBE experiments aimed at measuring four observables to confirm or disprove the big bang theory. These are, the known cosmic background radiation itself, the dipole which is a slight distortion of the background radiation caused by the peculiar motion of the Milky Way galaxy, the quadrupole which corresponds to the first cosmic distortion and the primordial seeds themselves. The first three are known but have to be correlated to the existence of the fourth. They did confirm these parameters with various degrees of accuracy. If the massive structures in today's universe formed under gravitational collapse during the 15 billion years since the big bang, then evidence of primordial structure *MUST* be visible in the cosmic background radiation. Results of the COBE project showed a pattern of variation in the background radiation. Simulations of the aftermath of the big bang based on inflation theory *predicted* seeds with size distribution as products of quantum fluctuations at the instant of creation. Both measurements and simulations matched. The big bang was correct with inflation theory explaining the initial processes. The pattern of seeds was about right for structure formation by cold dark matter and the size distribution would yield the major structures of today's universe under gravitational collapse through 15 billion years. That implied that the "Big Bang" theory is real and creation of the universe followed the process it envisages. Follow up experiments during the next several years after COBE achieved much higher accuracy leading to the firm belief in the big bang theory as an excellent, albeit open to modifications, description of the creation and evolution of the observed universe. The evolution of the universe is effectively the change in distribution of matter through time. It is matter moving from virtual homogeneity in the early universe to a very lumpy universe today with matter condensed as galaxies, clusters, super clusters and even larger structures such as the presumed attractors. This evolution can be seen as a series of phase transitions in which matter passes from one state to another under the influence of decreasing temperature (or energy). It is currently assumed that at 10^{-43} second after the big bang (Planck time), all the universe was a small fraction of the size of a proton. Space and time had only just begun. It must be understood that the universe *did not expand into existing space after the big bang but rather its expansion created space-time*. The temperature at this point was 10^{32} degrees and the three forces of nature (electromagnetism and the strong and weak forces) were fused as one. Matter was undifferentiated from energy and particles did not yet exist. By 10^{-34} seconds inflation had expanded the universe unfathomable 10^{30} times and the temperature had fallen to below 10^{27} degrees. The strong nuclear force had separated and matter underwent its first phase transition. Quarks (the building blocks of protons and neutrons), electrons and other fundamental particles now existed. The next phase transition occurred at 10^{-2} second when quarks began to bind together to form protons and neutrons as well as antiprotons and antineutrons. Annihilation of particles of matter and antimatter began, eventually leaving a slight residue of matter. All the forces of nature now separated. After about a minute, the temperature had fallen enough to allow protons and neutrons to stick together when they collide forming the nuclei of hydrogen and helium which are the building blocks of stars. This soup of matter and radiation which initially was *the density of water* continued expanding and cooling for another three hundred thousand years but it was too energetic for electrons to stick to the hydrogen and helium nuclei to form atoms. The energetic photons existed in a frenzy of interactions with the particles in the soup. The photons could travel only a very short distance between interactions. The universe was essentially *opaque*. When the temperature fell to about three thousand degrees, at three hundred thousand years, another crucial phase transition occurred. The photons were no longer energetic enough to dislodge electrons from around hydrogen and helium nuclei and so atoms of hydrogen and helium formed and stayed together. The photons no longer interacted with the electrons and were free to escape and travel great distances. With this decoupling of matter and radiation, the universe became *transparent* and radiation streamed in all directions to course through time as the cosmic background radiation that can still be measured at the present time. The radiation released at that instant gives a snapshot of the distribution of matter within the universe at three hundred thousand years of age. Had all matter been distributed evenly, the fabric of space would have been smooth and the interaction of photons with particles would have been homogeneous resulting in a completely uniform cosmic background radiation. The discovery of the small perturbations in this radiation reveals that matter was not uniformly distributed but it was rather structured forming the seeds of the present day universe. Those regions of the universe with a higher concentration of matter exerted more gravitational attraction and therefore curved space positively and vice versa. When radiation and matter decoupled three hundred thousand years later, the suddenly released flux of cosmic background photons bore the imprint of these distortions (perturbations) of space. Radiation travelling from the denser areas looks cooler than the average background and that from less dense areas looks warmer. Matter in the universe is of two kinds; dark matter and visible matter. Dark matter by its nature is unaffected by radiation but responsive to gravity. Therefore, it would have started forming structures much earlier than visible matter which is buffeted by the energetic flux of photons. Molded by the contours of space that originated as quantum fluctuations in the inflationary universe, dark matter could have begun to aggregate under the influence of gravity as early as ten thousand years after the big bang. At three hundred thousand years, the decoupling of matter and radiation freed ordinary visible matter to be attracted to the structures formed by the dark matter. As the visible matter aggregated, stars and galaxies formed. It can be said that visible matter outlines the shape of structures of invisible dark matter to which it has been drawn by gravitational attraction.

CREATION OF ELEMENTS

During the early years of the twentieth century the persistent basic idea since ancient times of matter consisting of atoms which are the smallest indivisible constituents got its scientific confirmation for the very first time by the then unknown scientist Albert Einstein. Contemporaneously, many models existed to describe the structure of the atom itself. The most famous of these were the models proposed by the New-Zealander Ernest Rutherford (1871-1937) in 1911 and the Danish physicist Niels Bohr (1885-1962) in 1913. Rutherford's simple model based on certain experimental results suggested an electronic cloud revolving around a central nucleus emulating the solar system. The electron was already discovered and the discovery of the nucleus as well as the neutron shortly followed suit. However, this approach was not sustainable according to Maxwell's theory as electrons should eventually emit radiation in their way to combine with the nucleus in an inevitable annihilation process. Bohr's model was a quantum physical elaboration on Rutherford's intended mainly to explain the experimentally discovered spectral emission lines of the hydrogen atom or the "Rydberg Formula". Electrons in this model were confined to certain discrete energy levels consistent with the very early assumptions of quantum mechanics. With the establishment of quantum mechanics in the 1930's, it was possible to explore the inner structure of the atom creating in the process the new scientific branch of nuclear physics. In due course it was easy to explain the structure of the periodic table of the elements according to nuclear stability from the energy point of view and the introduction of the strong and weak forces. In this view, the various constituents of the nucleus (protons and neutrons) can have only certain allowed energy levels subjected to the rules of quantum mechanics. The most stable nucleus in the universe is that of **iron (Fe)** consisting of 26 protons and 26 neutrons. Theoretically speaking (subject to many other conditions), elements with atomic numbers (number of protons) higher than 26 can reach more stable states by *fission* to lower their atomic number approaching 26 and those with atomic numbers less than 26 can achieve the same by *fusion* to increase their atomic number approaching 26. However, nothing could explain how these elements came about in the first place. This was left to cosmology in the course of studying the conditions under which the universe was evolving after its creation in the big bang synthesizing the known elements. Nucleosynthesis of the elements was hinted at by Fred Hoyle in 1946 and 1954 but accomplished by the American nuclear physicist/astrophysicist William Fowler (1911-1995) who won the 1983 Nobel Prize for this work. Big Bang theory calculations showed that approximately 75% of the universe consists of Hydrogen (with nucleus consisting of one proton in its most common form) and approximately 25% Helium (with nucleus consisting of two protons and two neutrons in its most common form). All heavier elements constitute less than 1% of the mass of the universe. According to these calculations, elements should be distributed fairly evenly across the universe. On the other hand, "Steady State" theory predicted that elemental abundance would vary across space. Intuitively, cosmologists assumed that all elements could be formed by fusion starting with the simplest nucleus; that of hydrogen and capturing equal number of electrons during the cooling process of the universe after the big bang. However, Enrico Fermi (1901-1954) and coworkers showed in 1949 that elements heavier than hydrogen and helium could not have been formed by the big bang since isotopes with mass numbers of 5 and 8 were unstable and would not have lasted long enough to absorb more protons and neutrons to form heavier elements. Therefore, big bang could explain the origin of light elements (hydrogen and helium) but not heavy ones. On the other hand, the steady state theory could explain the formation of heavy elements as products of nuclear processes within stars but not the formation of hydrogen and helium. These stars eventually exploded and ejected the debris into the galaxy. After millions of years, the debris condensed into new stars and planets. The origin of elements is therefore a two-stage process. Hydrogen and helium are produced in the early stages of a primordial creation event (the big bang) and the heavier elements are subsequently manufactured by nuclear processes within stars (as calculated by the discredited steady state theory). It is ironic that the champion of the steady state theory, Fred Hoyle had hit on the right idea for nucleosynthesis of the elements for the wrong reason and thus did not share in the award of the Nobel Prize. Hydrogen and then Helium could be formed during the early stages following the big bang. Prevailing conditions in the universe during the cooling process would not allow fusion to take place to form heavier elements in the observed amounts. However, gravity forced hydrogen atoms to form stars creating pressure within their interiors. These conditions, which last for millions of years, helped increase the temperature to allow fusion to ensue. Two hydrogen nuclei form *stable* helium nucleus (mass 4) with the help of two neutrons. Adding another hydrogen nucleus (proton) results in a very *unstable* beryllium nucleus (mass 5) that disintegrates before any other process can take place leading to unviable root for nucleosynthesis. Combining two helium nuclei (mass 8) gives the same *unstable* result. The probability of three helium nuclei (mass 12) coming together at exactly the same time to get past this condition is extremely low within the context of the big bang theory and cannot be considered viable either. On the other hand, conditions within stars lasting for millions of years would dramatically improve such probability. Hoyle *predicted* that the Carbon-12 nucleus must be capable to exist in a specific excited state for the three helium nuclei instantaneous collision to be viable. This reaction is enhanced by resonance between the energy state of the three particles and the energy state of the Carbon-12 nucleus. This way the necessary excited energy state of the Carbon-12 nucleus could be calculated to allow for the already observed concentrations of Carbon-12 in stars. Scientists at California Institute of Technology (Caltech) in 1953 experimentally proved the existence of such Carbon-12 excited energy state. Therefore, there was no doubt that as well as burning hydrogen to make helium, stars could burn helium to make carbon surmounting the nucleosynthesis hurdle at element number 8. The process of how elements are made within stars is now understood to proceed by adding helium nucleus (mass 4) to nuclei of other elements. Decays that eject electrons, positrons or neutrons can then form isotopes of all other elements. When some of the old stars that formed these heavy elements eventually exploded as novae and supernovae, heavy elements were scattered across space. They were then included in the formation of new stars at later time. The sun is one of these relatively new stars containing approximately 2% (two percent) heavy elements. This is how earth got its share of such elements including *"Carbon"* that caused the emergence of *"life"*. With the understanding of how helium was formed with the currently observed amount in the universe because of the big bang and the formation of heavier elements with the observed concentrations within stars, science solved the puzzle of the existence of the various elements and their specific properties started with the work of the great Russian scientist Dmitri Mendeleev (1843-1907). Mendeleev arranged the then known naturally occurring elements according to their physical and chemical properties in a table. Ingeniously, he left gaps for elements that were unknown but would later be discovered having very much the *predicted* properties. It is now understood that nuclear structure (number of protons) determines the position of an element in this "periodic table". It is also understood that the number of neutrons in the nucleus determines "isotopes" of the same element. The full table currently has 92 elements starting with hydrogen (one proton) and ending with uranium (92 protons). Trans-uranium elements were made in the laboratory (particularly at The University of California at **Berkeley**) with up to 114 protons in the nucleus. However, these are unstable and tend to decay by *fission* with varying lifetimes. Stars were originally formed as huge masses of hydrogen atoms colliding violently with each other. Collision energy overcame electric repulsion causing hydrogen nuclei to *fuse* forming in the process helium at the center which was subjected to enormous pressure releasing in the meantime large amounts of energy. When all hydrogen in the core was converted to helium, it is pulled inwardly making the surroundings even hotter. This provides enough energy to helium nuclei to overcome the increased electric repulsion due to the increase of the number of protons in the nuclei. Eventually, helium nuclei should start *fusing* as well collapsing the core further and subsequently increasing the temperature. With increasing temperature, *fusion* would theoretically continue. In reality, cores of stars of mass similar to the sun never reach such temperatures to allow these transmutations. However, cores of more massive stars do due to added gravity. The next *fusion* process within cores of these more massive stars lead to carbon (6 protons). Successively, other elements would be manufactured by transmutation and other processes provided that the resulting nucleus is energetically stable according to the laws of quantum physics. According to these laws, an **iron nucleus with 26 protons is more tightly bound than any other**. The reason is that beyond iron, energy must be added rather than being released to build up heavier nuclei. Consequently, an old star core would continue to transmute into iron until it reaches the critical size of approximately 1.4 solar mass where gravity overcomes all other forces and the star implodes into a neutron star creating a *supernova*. At that point in an old massive star while the *outer layers* are still burning hydrogen and helium, the *inner core* debris is thrown into space containing all the processed elements of the periodic table up to iron. The proportions of elements released in this process, taking into account what is currently known about old stars (the ones that are already exploded in supernovae), agree well with their proportions in earth. Elements from 27 to 92 in the periodic table are formed in trace amounts using the energy released during the collapse of these old stars and the blast wave that blows off the outer layers.

EMERGENCE AND EVOLUTION OF LIFE

Human beings since time immemorial wondered about their existence. They longed for answers to questions like "why we are here?", "how did we get here?" etc. Curious humans desperately sought explanations to the physical universe surrounding them and the way it is sustained. They also pondered their origins and how they evolved. Over the eons humanity developed myths that eventually gave way to rigorous scientific investigations and many branches of science thus came to existence. Physics and its allied applied sciences dealt with the universe while biology and its associated applied sciences studied living beings. In pursuing these issues, modern technologies have provided some clues and have established some rough estimates. Before pondering its surroundings, it was natural for humanity to wonder where it came from, what its origins are and how this happened. If humanity originally had information to get clear answers to these questions as is advocated in this endeavor, it obviously lost it in its struggle for survival on this earth against all odds and fierce competition with other creatures. If it did not as is advocated by almost all scientists particularly atheists, it must have *evolved* from some lower intelligence beginnings. One way or the other, with total lack of any appropriate knowledge, human curiosity and eagerness to establish its perceived dominance among all other creatures created myths to account for its origins on this earth. These myths were transmitted through the ages and became well entrenched. Another plausible path to answer these persistent questions in the back of humanity's collective mind was through receiving divine revelations. Religious people assume this process started with the first human while atheists dismiss the whole concept out of hand. Neither group can prove the absolute validity of its doctrine however. Apart from the three monotheistic religions, all other religions can be easily aggregated under the banner of mythology. With the advance of the age of reason and the enlightenment, both Judaism and Christianity could not possibly reconcile their presumably clearly stated facts about the physical world in their corresponding bibles with the unassailable scientific discoveries. Islam is in a completely different situation and tells an absolutely different story not withstanding bigotry and ignorance of its detractors. *This is the central argument of this study.* The presumption of Islam is that all knowledge and facts of what was taking place during and even before the moment of creation till doomsday and even beyond are already included in the Qur'an. The most fundamental difference between the Qur'an and the Jewish and Christian Bibles is that interpretation of the Qur'anic facts is subject to the meaning of words in the Arabic Language. Arabic words are more accommodating to progress of human knowledge unlike other languages, old and new, of these bibles. This issue is extensively discussed by the author in the first book of this series "Islam and the West" subtitled "Why Do They Hate Us So Much?". It is also essential to understand that according to Islamic creed, what goes for the Arabic Language is valid for the **original** language of revelations of the "Torah" received by Moses

(موسى-عليه السلام) and the "Evangel" received by Jesus (عيسى-عليه السلام) which is presumably Aramaic. Not surprisingly all these languages have the very same root. While the **original** Arabic text of the Qur'an survived intact, the **original** Aramaic texts of the Bibles are no more in existence anywhere. In other words, the three monotheistic **original** holy books state the same facts and their interpretations are supposedly subject to the same criteria except that **only** the Qur'an is preserved in its unique **original** text. With this brief discussion in mind, one should therefore rigorously contemplate the difference between the Qur'an and the Bible(s) as far as the origin and evolution of life is concerned and how this relates to scientific results. As far as the current Jewish and Christian sacred texts are concerned, biblical scholars produced numerous studies making it abundantly clear that the veracity of these texts do not survive under scientific scrutiny. As for Islam, no such studies were ever conducted not to mention that any conclusions were ever reached. This is tentatively done in this book as an opening to such studies in the sections subjecting Islam's sacred texts to the scrutiny of science. Now, how did life originate and evolve according to the most reliable findings of science? Geological observations led researchers during the eighteenth and early nineteenth centuries to assume that life began on earth long after its creation. They divided the age of earth into 14 eras which are divided into periods which in turn are divided into epochs. According to this division, life seemed to have appeared during the Cambrian epoch. It is currently assumed that life on earth in its very primitive form is only about 500 million to one billion years old while earth itself is approximately 4.5 billion years old. Conditions on earth were not suitable for the basic elements of life (RNA and DNA, proteins, cells) to form before that epoch. Some prominent scientists even go farther to claim that these elements must have come to earth from somewhere else. Once formed, these elements evolved into more complex life forms subject to the surrounding environmental conditions. This evolutionary approach explains the diversity of life forms through population genetics, speciation processes and extinctions. Since evolution is a continuous random process, there is no reason not to assume that humanity would evolve into a different species given enough time. Although time requirement for evolution is usually measured in geological timescale, some species evolve much more rapidly. At the present time, germs and bacteria resisting antibiotics and insects resisting insecticides are common albeit not existing a few years ago when these chemicals were invented.

ANTHROPOLOGY

As far as the physical cosmos is concerned, it is estimated that the universe came to existence approximately 15-20 billion years ago and earth was formed about 4.5 billion years ago. Life is supposed to have started in its most primitive form approximately close to one billion years ago. Scientists are quite confident that their estimates (unlike those derived from religious dogmas) while are not exactly factual, they represent reasonable ranges to enable them to work out models that eventually lead to the present. When the need arises, these estimates are occasionally reevaluated and new models are created. It is assumed that the universe is work in progress and so must be its understanding. Anthropologists tend to believe that the oldest remains of what is considered the first clear ancestor to modern day humans according to evolutionists, "Australopithecus Afarensis", go back approximately 3.5 million years. Following the evolutionary chain, the very first tools were used by "Homo Habilis" (handy man) that is considered the earliest species to be placed within the genus "Homo" about 2.5 million years ago. "Homo Erectus" appeared approximately 1.8 million years ago and by around 400, 000 years ago would have mastered fire followed by "Homo Erectus - Pre Sapiens" at approximately 200, 000 years ago where huts started to appear. At that juncture, Human ancestry story gets murkier. Old school paleontology mentioned two successive species of human ancestry to have lived during that period. However, "Neanderthal" (the name derives from a valley in western Germany) and "Homo Sapiens" who most current paleontologists believe were two coexisting distinct species rather than successive ones are associated with first burials approximately 40,000 years ago. It is also believed that the two species intermingled. However, for as yet unknown reasons, "Neanderthal" type creatures became extinct. Fossils indicate that

Neanderthals spread across Europe and Asia before becoming extinct an estimated 40,000 years ago. Today, the Neanderthal DNA in each living non-African human is broken up into short segments sprinkled throughout the genome. It is hypothesized that this arrangement is a result of how cells divide. During the development of eggs and sperm, each pair of chromosomes swaps pieces of their DNA. Over the generations, long stretches of DNA get broken into smaller ones. Over thousands of generations, the Neanderthal DNA became more fragmented. Scientists have reconstructed the genome of a man in Siberia who lived 45,000 years ago, by far the oldest genetic record ever obtained from modern humans outside of Africa and the Near East. They also reconstructed the entirety of a Neanderthal genome extracted from a single toe bone. Comparing Neanderthal to human genomes, it was concluded that they share a common ancestor, which is estimated to have lived about 600,000 years ago. According to the most recent generally accepted theory, Modern humans who arrived in Europe about 45,000 - 43,000 years ago and Neanderthals may have coexisted in Europe from Spain to Russia for more than 5,000 years, providing ample time for the two species to meet and mix. Based on carbon dating techniques and mathematical models it is concluded with a high probability that pockets of Neanderthal culture survived until between 41,030 and 39,260 years ago. There is also strong evidence that Neanderthal genes have survived in the DNA of today's humans, suggesting that at least some interbreeding took place. There is additionally evidence that late-stage Neanderthals were culturally influenced by modern humans. It was found that Neanderthal sites include some objects that look like those introduced to Europe by humans migrating from Africa. It is currently estimated that Humans and Neanderthals interbred 50,000 to 60,000 years ago. Approximately 1-4 percent of Neanderthal DNA is present in modern Europeans. "Homo sapiens", considered modern humans own species, appeared in Africa around 200,000 years ago. Studies, both on genes and on fossils, have suggested that they then expanded through the Near East to the rest of the Old World. Mitochondrial DNA analysis has confirmed that the mother of all modern humans (Mitochondrial EVE) came from Africa. Nuclear DNA shows that modern humans came out of Africa and gradually replaced all other hominids. "Cro-Magnon" or "Homo Sapiens - Sapiens" left the first figurative art objects 35,000 years ago. Not much has changed ever since and the chain ends with present day humans. Within the 3–4 million years of evolution, it is estimated that human brain size increased from about 450 cm^3 of the "Australopithecus Afarensis" to the current 1500 cm^3. The very last prehistoric link is when pottery, agriculture, domestication of farm animals and town dwellings took form between 17,000 and 6,000 years ago. That ushered the emergence of civilization along the great river valleys. Bronze and Iron Ages occurred between only 5,500 and 4,600 years ago. Sumerian and Egyptian civilizations go back only to 7,000 – 6,000 years ago. They were the first in history to develop writing. Although some sort of human existence on earth may reach back to 3 – 4 million years, recorded history of humanity is no more than that of the six or seven thousand previous years to the present if even that. What happened before recorded history is not history but archeology. This is the universal conventional wisdom among scholars. It seems that the clear implication prevalent among scholars since the dawn of civilization is that human knowledge started from scratch and evolved with time. Humans acquired knowledge as they went. Most scholars believe that knowledge may move from one geographical place to another but it keeps a linear path with time. The inference here is that human knowledge was, more or less, never lost during the march of time and that of civilization. What is overlooked in that belief is the rise and fall of completely isolated civilizations such as some of which that presumably existed in the pre-Columbian Americas. Additionally, it is assumed, rather without any basis that, since almost nothing is known about the early humans, as they left no discernible traces in the form of records of any type, they must have been only a notch above animals. On the other hand, one can suggest the plausible assumption that they had very advanced knowledge that was lost either due to environmentally adverse conditions or because their strife to survive under extremely difficult conditions that could have befallen them took precedence over transmitting the acquired knowledge to next generations. If these conditions did take place particularly if knowledge was transmitted orally, lack of any traces is no longer surprising. In this respect, knowledge could have been conceivably lost. Ancient humans used only materials readily available in their environment to record what they considered worthwhile. These included clay tablets, stone, animal hides, bones, etc. which are materials extremely prone to decay under various environmental conditions. Whatever knowledge, if any, was thus recorded by early humans must have been lost. Only stone engravings could have plausibly survived leading modern scholars to assume linkages to human beginnings of acquiring knowledge. There is no absolutely acceptable reason for the validity of these assumptions however. In other words, Islamic claim that the first human Adam (آدم-) (عليه السلام) appeared on earth with knowledge of everything there is to know is not necessarily less valid than claims of anthropologists and/or atheists. Both claims (not only the Islamic one) lack any absolute proof.

THE FASCINATING CARBON ATOM

The unlimited success of nuclear physics in applying the laws of quantum mechanics to unfold the secrets of the "Periodic Table of the Elements" prompted scientists to apply the same laws to explore how and why chemical elements interact the way they do. That approach bred the new scientific discipline of "Quantum Chemistry". It turns out that the electronic structure of each atom determines all its physical and chemical properties. This electronic structure consists of energy "shells" surrounding the nucleus. In this paragon the chemical nature and properties of an atom is wholly dependent on the number of electrons in the highest energy shell (outer shell) that is occupied. While lower energy shells (inner shells) are necessarily filled and stable, the highest occupied energy shell (outer) strives to reach stability by somehow cooperating and sharing its electrons with other atoms. This is precisely what is meant by "chemical reactions". Therefore, in a very strict sense the entire subject of chemistry is nothing more than an application of quantum mechanics rules hence the term "quantum chemistry". It is customary to refer to the outer electronic shell in an atom of an element as the "valance band" where the number of electrons determines its affinity to react and form chemical compounds with other elements.

Thus, electrons' sharing is alternatively known as forming a "chemical bond". It was determined that according to the rules of quantum mechanics all chemically active elements occupying the first few rows (beyond the first occupied by hydrogen and helium) of the periodic table need eight electrons in their outer shells through forming chemical bonds with other elements to be stable. Carbon is in the unique position of being the first element in the periodic table with four electrons in its outer/valance shell which enables its atoms to form four bonds at once. Moreover, these bonds can form chains resulting in what is known as biochemical substances. Thanks to the uniqueness of the reactions of the carbon atom, a separate branch of chemistry has been dedicated to their study; **Biochemistry**. This new science of quantum chemistry has been pioneered by the great American biochemist Linus Carl Pauling (1901-1993) who with his wife Ava Helen Pauling are well known as strong advocates of human rights as well. He received Nobel Prize for Chemistry in 1954 and Nobel Prize for Peace in 1963 (for 1962). Early in his career, he worked closely with some of the great names in physics such as Schrodinger, Bohr, Bragg, Sommerfeld, etc. which allowed him to apply the methods of quantum mechanics and crystallography into his field of interest; biology with pronounced results for science and humanity at large. He is one of the founders of the crucial new science of "Molecular Biology" where his work inspired the work of James Watson, Francis Crick, Rosalind Franklin (1920-1958) and Maurice Wilkins on the structure of DNA, which in turn made it possible for geneticists to crack the DNA code of all organisms. Pauling's contributions to life studies go far beyond his research and published results. Undoubtedly his most important contribution was the introduction of the laws of physics, an alien scientific discipline to biologists of that era, into the study of biology. It is very interesting to notice the persistent trend since the advance of quantum physics of highly regarded physicists showing curiosity in the issues of life sciences. That is most probably due to the fact that after learning the secrets of the building blocks of matter, scientists would presumably pay attention to the building blocks of life which is the most captivating puzzle as far as humans are concerned. The most prominent of these physicists in the post second world war era is none other than Erwin

Schrodinger, one of the eminent founders of quantum mechanics. In 1944 he gave lectures that resulted in the publication of a small book titled *What is Life?* which set the stage for young scientists in various fields of physics, engineering and biology to concentrate on discovering the secrets of life starting with the structure of DNA. In his little book he introduced the concept that the most essential part of a living cell (the chromosome fiber) is in fact an aperiodic crystal. That brought the techniques of crystallography into the search for the structure of the molecules of life. In the aperiodic crystal, the number of atoms need not be very large to generate unlimited number of possible arrangements along molecules of DNA. Introducing this concept into genetics, the four bases of DNA could thus be used to write the genetic code of any living thing. Therefore, messages written in the four letter code strung out along molecules of DNA suffice to carry all of the information required for the normal functioning of a living organism. Influenced by the intellectual concepts of quantum mechanics, some biologists believed that genetic mutations is a quantum process involving molecules being pushed over an energy barrier (as in the photoelectric effect) from one stable configuration to a different stable configuration. This way, evolution involving natural selection after mutations can be developed and explained using physical laws. Prominent among those scientists who were greatly influenced by Pauling's and Schrodinger's work at the same time were two. The first was the British physicist Francis Harry Compton Crick (1916-2004) who worked on x-ray studies of proteins for his doctorate degree. The other was the American biologist James Dewey Watson who in 1951 joined the 20 years older Crick at the Cavendish Laboratory in London and convinced him to collaborate on building a model of the DNA structure. Watson had two degrees in zoology: a bachelor's degree from the University of Chicago and a doctorate from Indiana University, where he became interested in genetics. He had worked under the Italian Salvador E. Luria (1912-1991) at Indiana on bacteriophages, the viruses that invade bacteria in order to reproduce—a topic for which Luria received a Nobel Prize in Physiology or Medicine in 1969. Watson went to Denmark for postdoctoral work, to continue studying viruses and to remedy his relative ignorance of chemistry. At a conference in the spring of 1951 at the Zoological Station at Naples, Watson heard Wilkins talk on the molecular structure of DNA and saw his recent X-ray crystallographic photographs of DNA. He was fascinated by this work and decided to join the effort to find the real structure of the DNA molecule. In 1953, Watson and Crick constructed a molecular model representing the known physical and chemical properties of DNA. It consisted of two intertwined spiral strands, resembling a twisted ladder (referred to as the "**double helix**"). They hypothesized that if the two sides split from one another, each side would become the basis for a pattern for the formation of new strands identical to their former partners. This theory and subsequent research led to an explanation of the process behind the replication of a gene and, eventually, the chromosome. Crystallography and the study of X-ray diffraction patterns provided very important tools to study the molecular structure of the complex protein molecules and Pauling was a trail blazer in this regard. The molecules of life which mainly consisted of "fibrous proteins" and "globular proteins" were clearly of particular interest to him. He announced the discovery of the α - helix in 1951 commenting that the α - helix structure was determined not by direct deduction from experimental observations on proteins but rather by theoretical considerations based on the study of simpler substances. The molecule that is the basis for heredity, DNA, contains the patterns for constructing proteins in the body, including the various enzymes. A new understanding of heredity and hereditary disease was possible once it was determined that DNA consists of two chains twisted around each other, or double helixes, of alternating phosphate and sugar groups, and that the two chains are held together by hydrogen bonds between pairs of organic bases—adenine (A) with thymine (T), and guanine (G) with cytosine (C). Modern biotechnology also has its basis in the structural knowledge of DNA—in this case the scientist's ability to modify the DNA of host cells that will then produce a desired product, for example, insulin. Currently biologists define a gene in molecular terms as the length of DNA that a cell uses to produce a protein. Early in the twentieth century scientists discovered that genes were associated with chromosomes which are the string-shaped objects observed in dividing cells and they are mostly proteins. DNA is understood to be a long inert molecule made up of small subunits that store genetic information. Genetic differences are alterations or mistakes in this information which can alter the protein encoded by the gene.

STRUCTURE OF THE DNA

The background for the work of the four scientists James Watson, Francis Crick, Rosalind Franklin (1920-1958) and Maurice Wilkins in conjunction with the discovery of the structure of DNA was preceded by several scientific breakthroughs: the progress made by X-ray crystallographers in studying organic macromolecules; the growing evidence supplied by geneticists that it was DNA, not protein, in chromosomes that was responsible for heredity; Erwin Chargaff's (1905-2002) experimental finding that there are equal numbers of A and T bases and of G and C bases in DNA; and Linus Pauling's discovery that the molecules of some proteins have helical shapes—arrived at through the use of atomic models and a keen knowledge of the possible disposition of various atoms. Crick and Watson were working at the Cavendish Laboratory, London trying to figure out the structure of the DNA molecule. Another group at King's College, London was working on the same problem led by the British Rosalind Franklin (1920–1958) and the New Zealander Maurice Wilkins (1916–2004). During the war Wilkins was shipped out to the United States to work on the "Manhattan Project" but like many other nuclear physicists, he became disillusioned with his subject when it was applied to the creation of the atomic bomb; he turned instead to biophysics, working with his Cambridge mentor, John T. Randall—who had undergone a similar conversion—first at the University of St. Andrews in Scotland and then at King's College, London. It was Wilkins's idea to study DNA by X-ray crystallographic techniques, which he had already begun to implement when Franklin was appointed by Randall. The relationship between Wilkins and Franklin was unfortunately a poor one and probably slowed their progress. Of the four DNA researchers, only Rosalind Franklin had any degrees in chemistry. After the Second World War, through a French friend, she gained an appointment at the "Laboratoire Centrale des Services Chimiques de l'Etat" in Paris, where she was introduced to the technique of X-ray crystallography and rapidly became a respected authority in this field. In 1951 she returned to England to King's College, London where her charge was to upgrade the X-ray crystallographic laboratory there for work with DNA. Rosalind obtained images of DNA using X-ray crystallography, an idea first broached by Maurice Wilkins. Without Franklin's knowledge or permission, Wilkins shared Photo 51 and her data with Watson. Franklin's images allowed James Watson and Francis Crick to create their famous two-strand, or **double-helix**, model. In their published results, Watson and Crick included a footnote acknowledging that they were stimulated by a general knowledge of Franklin's unpublished contributions. In 1962 Watson, Crick and Wilkins jointly received the Nobel Prize in Physiology or Medicine for their 1953 determination of the structure of deoxyribonucleic acid (DNA). Wilkins's colleague Franklin, who died from cancer at the age of 37, was not so honored. The reasons for her exclusion have been debated and are still unclear. There is a Nobel Prize stipulation that states "in no case may a prize amount be divided between more than three persons." The fact she died before the prize was awarded may also have been a factor, although the stipulation against posthumous awards was not instated until 1974. In 1962 Crick wrote *Of Molecules and Men* detailing the recent biochemistry revolution that he had helped to usher in. In 1981, he wrote *Life Itself: Its Origin and Nature* in which he suggested that **life on Earth may have been seeded on another planet**, and his *What Mad Pursuit: A Personal View of Scientific Discovery* was published in 1988. After the "double helix" model, there were still questions about how DNA directed the synthesis of proteins. Crick and some of his fellow scientists, including James Watson, were members of the informal "RNA tie club," whose purpose was to solve the riddle of RNA structure, and to understand the way it builds proteins. The club focused on the "Central Dogma" where DNA was the storehouse of genetic information and RNA was the bridge that transferred this information from the nucleus to the cytoplasm where proteins were made. The theory of RNA coding was debated and discussed, and in 1961, Francis Crick and the South African Sydney Brenner (2002 Nobel Prize in Physiology or Medicine) provided genetic proof that a triplet code was used in reading genetic material. Regrettably, the American

Watson became convinced and propagated racist ideas about the genetic superiority of some races and the inferiority of others.

The Molecules of Life

The countless number of naturally occurring biochemical substances testifies to the importance of the properties of the carbon atom. However, life itself is closely linked to the properties of the carbon atom. In addition to water which constitutes more than 75% by weight of most living things, the rest is half carbon, quarter oxygen and approximately 10% nitrogen. Biologists believe that life must have started in water because of the high concentration by weight of it in all living things. Once started, life evolved using these essential elements (carbon, oxygen, nitrogen and only traces of other elements) because of the way they form chemical bonds. The backbone of all these bonds are those offered by the carbon atom. There are 92 chemical elements which occur naturally on earth. Only 27 of these are essential components of living things. Nonetheless, not all of these 27 elements are essential to all living things. Evolutionists assert that the formation of amino acids was the very first step in the direction of the emergence of life. That is because these acids are the building blocks of proteins which are in turn among the essential building blocks of living things. Biochemically important compounds consist of four categories. These are fats, sugars and starches, proteins and nucleic acids. The role of the first three was well known while the importance of nucleic acids in the evolution of life was not recognized till the 1950s. Only five bases exist in nucleic acids; they are guanine, adenine, cytosine, thymine and uracil referred to by their initials for simplicity as G, A, C, T and U. Researchers identified two subgroups of nucleic acids; RNA (ribose nucleic acid) and DNA (deoxy ribose nucleic acid) each incorporates only four of the five bases. DNA molecules contain G, A, C and T while RNA molecules contain G, A, C and U. It was known by the 1930s that the nucleic acid present in chromosomes was DNA in addition to protein. For historical reasons, protein was more interesting for scientists to study which retarded the discovery of the structure of DNA.

Wagih H. Makky, Ph.D.

Evolution and Natural Selection

Studies of human body is as old as humanity itself for health and healing reasons as the medical profession went through major changes throughout history. Investigators noticed a correlation between the environment and other living things and human wellbeing. That was the impetus to study life on its own merits as a generic subject. The intimate connection between humans and their surroundings necessitated that such studies should be grounded on *direct observations*. That helped bring about rapid advances in acquiring knowledge in these fields. It also made *life scientists confident of their conclusions far more than scientists in other fields for no apparent compelling reason other than direct personal observation*. That may explain why life scientists and biologists are more outspoken atheists than other scientists in even more global fields like cosmology. On the other face of the scholastic coin, figuratively speaking, one finds creationists that are also absolutely convinced of their dogmas and who believe in the creation process as comprehensively given in the Bible especially in the "Book of Genesis". With the passage of time and the acquisition of more observational knowledge, critiques of the biblical account of the process of creation in general and the emergence of humans in particular went from mild discussions to outright insults. Since the age of the earth as calculated according to the Bible was approximately 6000 years, that prompted the French Naturalist Georges-Louis Leclerc, Comte de Buffon (1707-1788) to publish his multi-volume book *Histoire Naturelle* stating that earth is definitely much older than such calculations although he did not refute the biblical narrative. The British anatomist Richard Owen (1804-1892) developed the concept of the "Archetype" as a blueprint to Allow for species change used by a Creator in an apparent effort to preserve some role for a creator albeit a very limited one. However, later scientists started questioning the veracity of the biblical account though. Step by step what amounts to a conceptual abyss between "Atheists" and "Creationists" quickly developed and have been widening and deepening ever since the advent of the theory of evolution as there is no conceivable way to bridge this chasm. Studies of genetics irrefutably linked humans to other living creatures in one evolutionary path making evolution and creationism, hence science and religion, hopelessly irreconcilable. In 1866 meticulous observations of passing specific characteristics/saltations over several generations of peas by the Austrian Monk Gregor Mendel (1822-1884) were published. Life scientists however did not recognize their significance till early in the twentieth century in the work by the British biologist William Bateson (1861-1926) who also gave science the term "genetics".

After a couple of failed stints to become a medical doctor or a clergy and a chance interest in geology, Charles Darwin (1809-1882) was recommended to accompany the captain of the "HMS Beagle" tasked to map the coastline of South America. The trip lasted five long years with Darwin occupied by geological work while collecting enormous number of specimens of everything he encountered. The stop in 1835 at the "Galapagos archipelago" where the variation in birds and tortoises from island to island piqued his interest and led him to speculate that well established ideas (mostly derived from the Bible) about creation are wrong. He was convinced that one species could actually change into another in a descending branching tree form. The main mechanism for that branching was what is later termed "natural selection". Knowing full well how radical his ideas were and the anticipated ridicule he would be subjected to, he did not publish his work for 22 years after the Beagle's return to England in 1836. In 1848, another British naturalist, Alfred Russel Wallace (1823-1913) made a similar expedition to Brazil and the East Indies collecting tens of thousands of specimens and returning in 1862. During his travels, he wrote scientific papers and sent the results to Darwin in 1858. Receiving this letter from Wallace presenting almost identical theory of natural selection to his own, Darwin completed his book *On the origin of species* giving credit to Wallace who is not as well-known as Darwin despite his major contributions to the theory of evolution and natural selection. Results of both naturalists were jointly announced in 1858 to very little attention by the scientific community. That drastically changed when Darwin published his books "On the origin of species by natural selection" and "Preservation of favored races in the struggle for life" where he detailed his theories. The ultimate coup de grace came when he brought humans as well into the theory in the famous 1871 book *The Descent of Man*, *and Selection in Relation to Sex*. In 1900, the group headed by Thomas Hunt Morgan (1866-1945) (Nobel Prize in Physiology or Medicine in 1933) in the United States working with the fruit fly *Drosophila melanogaster* confirmed that regions of the chromosomes (genes) were responsible for heredity in the meantime explaining Mendel's data. Originally Morgan rejected Darwin's mechanism for the origin of species and its associated natural selection. However, he came around to accepting it due to accumulation of experimental evidence supporting the Darwinian approach. From that point on statistics played a major role in analyzing genetics and genetic data as quantitative approaches to genes in populations linked genetics with evolution. The most famous name in this regard is that of the British geneticist/mathematician John Burdon Sanderson Haldane (1892-1964) who later immigrated to India and became an Indian citizen. He is known for "Haldane's Principle" which strictly relates the bodily structure of a living organism to its size. As such, Darwin's ideas gained wide spread acceptance and support through unification of evolutionary biology and genetics. An unfortunate outcome of the firm scientific establishment of the mechanism of natural selection was the development of the infamous racist concept of "Eugenics". Eugenicists among whose ranks are some of the highly respected biologists and scientists of the nineteenth and twentieth centuries advocate improving humanity by regulating breeding to eliminate *inferior races*. In the United States for example, some states enacted programs for the *sterilization of degenerates*. The most infamous of all these trends is obviously the rise of Nazi Germany. On the other hand, issues of "genetic engineering" are among the most controversial scientific and ethical questions at the present time. Most unspecialized people particularly religious types overlook many aspects of Darwin's conclusions in their rejection of his theory. Like all legitimate scientists, Darwin did not start his work from vacuum. He based his assumptions on the classification of species given by the great Swedish botanist Carl Linnaeus (1707-1778) who is described by his contemporary the iconic figure of the Enlightenment Jean-Jacques Rousseau (1712-1778) in glowing terms as he knew of no greater man on earth. Linnaeus's classification, which is still followed to the present time, clearly showed that organic beings have been found to resemble each other in descending degrees, so that they can be classed in groups under groups. This is the starting point of Darwin's work that led him to draw a tree-like diagram describing the evolutionary process of life. This diagram is considered the very first of what is currently called "phylogeny" which is a branching diagram that shows evolutionary links between species currently based on genetic similarity. Genetic interrelatedness convincingly and uniquely establishes which living being descends from which over millions of years.

BASICS OF THE THEORY OF EVOLUTION AND NATURAL SELECTION

The essence of the theory of evolution and natural selection is that life has a single origin and all organisms are actually related to each other. However, the process of speciation or creating new species separated these organisms due to isolation. When a homogenous population is split into subgroups that are then for whatever reason separated from each other, they evolve differently and form independent species that cannot cross-breed. Although mating between members of these two species may be possible, the offspring is doomed to be sterile with no possibility of reproducing. A familiar example is mating between donkeys and horses resulting in sterile mules. Natural selection is the process responsible for the variations within a specific population. While DNA contains all the information needed to build an organism, it may be changed by mutations that take place during duplication in cell division. These mutations can cause changes to the "Phenotype" which is the visible characteristics of a specific organism and as such they form the basic mechanism of evolution by natural selection. Mutations in reproductive cells (sperms and eggs) are passed to the next generation. Variations within the populations are acted upon by natural selection that enhances the survivability and reproduction of the species due to environmental advantages brought about by some specific variations. If a mutation is beneficial in this regard, it becomes predominant in the population. Obviously build-up of mutations in two isolated populations produces speciation and the creation of completely different species. If a mutation is detrimental to the survival of the organism, the produced species variety will quickly disappear and vanish. Natural selection depends to a very large extent on an organism adapting to its environment. Adaptation implies any characteristic that facilitates the survival or reproduction of an organism that has it. That is not necessarily a physiological attribute. Examples of these characteristics are bird migration, cactus spines, leopard spots, etc. Adaptation can lead to divergence between populations and eventually speciation. If two isolated populations of a species are exposed to different selective pressures, natural selection would result in varying adaptations between the two. Over time, these populations may become so different that if they were to come in contact again, they would not be able to mate or reproduce. Consequently, speciation would have taken place and diversity results. It should be understood that many characteristics of an organism are not necessarily resulting from adaptation but are rather products of other characteristics. As an example, development of a circulatory system guarantees the survival of humans but the red color of human blood is simply due to its chemical composition. Special features or adaptations are needed to survive under the prevailing environmental conditions. Organisms living under extreme conditions develop their own specific characteristics to survive what other common organisms cannot possibly tolerate. For example, heat resistant proteins are found in bacteria living in boiling hot springs and cellular anti-freeze is found in cold climate plants. That is to say that wherever organisms exist, there will be environmental challenges of varying severity that have to be overcome to survive. Accordingly, organisms evolve techniques to adapt and as the environment changes over time, they have to continuously adapt or perish. Complex adaptations may build up over thousands or millions of generations of selections. Thus, natural selection is an inescapable force and greater adaptation results. Not all variations among organisms are due to natural selection as some variations can be caused by environmental conditions such as availability of nutritional material, disease, etc. Natural selection can only lead to evolution if the differences it selects are caused by genes. Individuals breed, groups propagate themselves and species speciate. Most biologists believe that natural selection acts at the gene level since adaptations that benefit an individual also promote the spread of its genes and adaptations that benefit groups are already explained by gene level selection.

How Did Life Come About?

It is interesting but not really surprising that the last phases of evolution (i.e. the closest to modern humans) are the more accurately described ones. Similarity between genetic structure of mammals and humans is not in dispute by any scientist. In contrast, how life began, under what conditions and where as well as the path it took to arrive at its currently observed status is very speculative. Many hypotheses with equal probability were proposed to explain these developments. Most researchers at the present time speculate that life on earth descends from an RNA world. However, they cannot ascertain that this is the only path and other forms of life did not predate this RNA-based current form. Biologists now believe that life on earth arose in a transition from non-living organic compounds into primitive living entities and it is simply an on-going chemical reaction that started by chance eons ago and not necessarily on earth itself. Many scientists speculate that conditions on earth could not have facilitated appearance of life when it did. As such, they believe that life started somewhere else in the universe and through some violent cosmic events traces of it landed on earth where conditions were suitable for its evolution to what is currently experienced by humans. Adopting this approach, it is logical to assume that other traces of life exist in other parts of the universe in varying degrees of development. Geological evidence of primitive lifeforms dating back approximately about 4 billion years is found in 2017 in Quebec, Canada. In the same year scientists discovered more evidence of early life on earth in the Pilba Craton of Western Australia. Bearing in mind that earth's age itself is approximately 4.5 billion years, it is clear that life started its march on earth soon after its formation. Evolution went through a number of life explosions and extinctions to arrive at the current status. It is believed that extinctions owe their existence to excessive generation of CO_2 (carbon dioxide) which is what is being observed at the present time due to human activities causing "Climate Change". The transition from non-living into living entities came about through a process of increasing complexity involving molecular self-replication, self-assembly, autocatalysis and cell membranes. Life functions through the specialized chemistry of carbon and water and builds largely upon four key families of chemicals which are: "lipids" (fatty cell walls), "carbohydrates" (sugars and cellulose), "amino acids" (protein metabolism) and "nucleic acids" (self-replicating DNA and RNA). To account for the molecular self-replication stage, in 1962 the American Biologist Alexander Rich (1924-2015) first proposed self-replication of RNA molecules as the chemical path to life. In 1984 the American molecular biologist Walter Gilbert strongly advanced that hypothesis and called it "RNA World". This hypothesis currently has wide acceptance among life scientists because like DNA, RNA can store and replicate genetic information. RNA enzymes or ribozymes can catalyze/start and accelerate life's vital chemical reactions. Additionally, ribosome which is one of the most important components of cells is primarily composed of RNA. Many other observed constituents of coenzymes are known to be surviving remnants of covalently bound coenzymes in an RNA world.

Evolution

Along the islands connecting Australia and Asia, biologist Alfred Russel Wallace noticed an intriguing gradual progression of evolving creatures from the more primitive forms of Australian animals to the more modern Asian forms. Isolated Australia preserves unusual mammals such as the egg-laying platypus which has ten sex-determining chromosomes (compared to humans' X and Y) and the honey possum which gives birth to the smallest offspring of any mammal. This led biologists to conclude that the reason mammals stopped laying eggs and began giving birth due to infection by a virus which left thousands of old broken copies of it littering mammals' including humans' DNA. Evolutionists based their development of Darwin's ideas on the assumption of randomness of mutations which raised powerful objections to the required timescales. However, in 1967 the American biologist Lynn Margulis (1938-2011) proposed the concept of endosymbiosis where smaller partners invade large host cells creating an environment of interdependence. Speculating that some early cells incorporated bacteria that eventually enabled the cells to gain energy by photosynthesis in the case of plants or by processing oxygen in animals and plants. That theory challenged what she called ultra-Darwinian orthodoxy arguing that symbiosis in which organisms that are mutually beneficial may come together to form a single organism was more important at the level of microorganisms and hence in the first 3 billion years of life before complex organisms evolved. Her theory has later received experimental confirmation and became mainstream among evolutionary biologists. Her theory laid the ground work for explaining the Cambrian explosion of life that followed. She also collaborated on proposing the controversial "Gaia" hypothesis of earth as self-regulating entity. This follows the concept of symbiosis but in this case between the living parts of earth and the environment. Evolutionists coined certain terms to deal with their theories. They describe the march of evolution as through "adaptive/fitness landscape" consisting of hills and valleys. This adaptive landscape maps how natural selection leads to the development of some traits and rejection of others. Hills represent well adapted combinations of traits while valleys represent poorly adapted ones. According to this scheme, natural selection can only move a population uphill but never down. Thus, genetic and developmental pathways can make a particular destination inaccessible either because the path crosses a deep valley and intermediate forms die out or because the needed mutations cannot occur and there is no path at all. Living environment determines to a very large extent how an organism evolves. For example, changes in the traits of a prey is usually followed by counter changes in traits of the predator. The same is true of the relationship between host and parasite as well as any mutually symbiotic organisms. Organisms facing the same problem may independently evolve similar solutions and therefore may look alike despite the fact that they have no immediate genetic ancestors.

Islamic Intellectual Heritage

INTRODUCTION

Development and historical evolution of Christian and Islamic thought processes are starkly different. They both obviously stemmed from the advance of their associated faiths but eventually charted remarkably divergent directions. Unique political and social circumstances surrounding each contributed much to that divergence. Dominant Greek culture profoundly permeated early Christian thinking. Preponderance of imperial Roman power determined how it evolved resulting in the identification of Christianity with the power of the empire. When that power waned and the empire then collapsed in the West, the Church choked the cultural life and caused the descent into the dark ages. Eastern Christianity with its Greek underpinnings formulated Christian doctrine where there was no Western Christian thought contribution to speak of. With the rise of Islam, stagnation became the new normal East and West. On the other hand, there was no pre-Islam Arab thought process for all practical purposes. Additionally, there was no dominant power structure to forcibly steer the nascent Islamic thought process in any particular direction. Islamic thought is thus uniquely free of any external influence during its formative years. The only stimulus to its development was the precepts of the faith and the awesome and inspirational personality of Mohammad (محمد - عليه الصلاة و السلام) that guaranteed the appropriate path to be taken. Internal dynamics wholly determined the beginnings of the Islamic thought process. External influences transpired only after the total primacy of Islamic power over the external influencers and at its invitation. These two paths are described in the following sections. The contrast between the two should ultimately show the futility of identifying Islam with Christianity when talking about religion and faith.

It is universally agreed that Christianity as has always been practiced by its adherents is founded on Saint Paul's interpretation of the mission of Jesus Christ (عيسى - عليه السلام). The Christian dogma initiated by Paul inevitably looked down upon reason and promoted blind faith to survive against attacks of its detractors. The renowned author Michael Grant in his 2000 book *Saint Paul* describes the role of Paul as was the one who crystallized in memorable form the *rejection of rational knowledge* which would in due course gain the passionate agreement of many powerful religious groups, and then, eventually, *break down classical culture* and set up medieval irrationalism in its place. Describing how Paul viewed his religion, Grant also states that Paul was emphatic in describing Christianity as a mystery, in the sense of a secret that had been disclosed to him and had brought him the gift of supernatural knowledge. The tide of Paul's irrationality was strong enough among the believers to stamp out the opinions of the great Christian thinkers of the East such as the second century Clement of Alexandria and his contemporary theologian in the Church of Alexandria Origen. The Early Christian Church wholeheartedly adopted the Pauline ideas wholesale. After Constantine had declared Christianity the official religion of the empire, he recognized the authority of bishops in 318 and in 333 placed them on an equal footing with magistrates appointed by the state. That eventually led to bishops, such as Ambrose of Milan, challenging the emperors themselves and raising the status of religious authorities above secular ones. Believing after Paul in the inferiority of reason and the supremacy of irrationalism, the great classical works were ignored and philosophy, which actually helped formulate the building blocks of Christianity and its doctrines, was frowned upon and deemed heretical. Single handedly and deliberately, the Church led the West into the Dark Ages. In the late fourth and early fifth centuries, Augustine wrote in the *City of God* that "***There is another form of temptation, even more fraught with danger. This is the disease of curiosity ... It is this which drives us to try and discover the secrets of nature, those secrets which are beyond our understanding, which can avail us nothing and which man should not wish to learn.***" The story was starkly different in the East where the great centers of Christian thought came under Islam in short order. The region that produced the greatest and probably the only Christian thinkers changed course to produce Islamic thinkers and scholars of the first rate. Classical science and philosophy were rediscovered and the Islamic civilization picked up the standard of science, reason and rationalism. Even after emerging from the darkness of Pauline irrationality, Christian intellectuals could not defend his attitude except by severing the ties between reason and religion. The seventeenth century eminent French scientist Blaise Pascal is quoted to have stated that "Jesus Christ (عيسى عليه السلام) and Saint Paul have the order of charity, not of the mind, for they desired to warm the will, not to instruct". The twentieth century Danish philosopher Soren Kierkegaard stated that "in order to become a Christian one must commit oneself by leaping in the darkness and acting from faith in the absurd." Micheal Grant writes in 2000 "Christianity in particular cannot aim at demonstrating its truth in terms that would be acceptable in a court of law, or a philosophically composed textbook ...Christianity must inevitably rely upon faith in the irrational—upon faith in what Paul knew and declared to be outrageous by any purely intellectual standard ...he (Paul) urged far more frequently the irrelevance of reason, and the overriding, overwhelming need to possess faith in its place." Thus, Christian doctrine was firmly formulated according to Pauline criteria spilling over into Western culture that turned its back to Classical achievement in all fields. However, due to the advent of Islam and its spectacular victories that irrevocably established its dominant culture from China to the Atlantic Ocean and across the Mediterranean into the Iberian Peninsula and Cicely, centuries later the West went through what is known as the "Middle Ages" when westerners came in contact with the Muslims and their thriving cultural life leading to Western awakening and the beginning of the current Western Civilization. Western historians mark the fall of Constantinople as the end of the Middle Ages and the beginning of the modern Western era. After a very long period of cultural and scientific achievements, the Islamic civilization followed the unambiguous natural law of militarization. However, the staggering and stunning military successes of the Ottoman Empire against Western powers instilled irrational fear of Islam in Western hearts and minds. Even today despite the overwhelming material power of the West vis-à-vis the weak and fractured Islamic world, the mind-set of the whole spectrum of Westerners from the average uninformed person to the misinformed decision maker fear of Islam still persists. Bigotry is currently deeply embedded in Western psyche whether religious or secular. Regular anti-Islam convulsions show their ugly faces everywhere in the West under the guise of identity politics and the preservation of Western Civilization. Nature informs historians that militarization is always the last phase in a rising civilization after which decline is inevitable. This is as true in ancient times as it is at the present. Thus, Islamic and Western Civilizations switched roles giving birth to the modern era. Many current thinkers observe the same phenomenon of militarization followed by decline taking place within the Western Civilization. Militaristic tendencies that gave humanity two world wars in the twentieth century and American propensity towards military interventions everywhere in the world for no discernable reason in the twenty first are unmistakable. Signs of Western decline are there for anyone to see according to these thinkers. Therefore, it is argued by many at the present time that the world is at the cusp of witnessing the emergence of a new Eastern Civilization. It can be concluded that one civilization fulfills certain ends suitable to its character and is then replaced by another to accomplish new aims in the relentless march of humanity. It is asserted in this work that this process is simply a recurrent occurrence in the undeniable unfolding of the "Global Cosmological Divine Plan" as well as a clear and emphatic testimony to the existence of such plan according to which the universe is evolving. This is, as will be elucidated in other parts of the book, a distinct refutation of the bedrock notions in atheism of randomness, chaos and chance governing evolution of life and the universe as a whole. It goes without saying that the plan is indicative of a planer. Atheism in a nutshell is the disbelief in the existence of such concepts and the belief that the universe has neither a creator nor a purpose.

After reaching an agreement with Quraysh (قريش) to suspend hostilities for an extended period (the Treaty of Hudaybiyah – صلح الحديبية), Mohammad (محمد - عليه الصلاة و السلام) turned his attention to the international aspect of his mandate from God (الله - سبحانه و تعالى). He sent emissaries to all the rulers of the surrounding regions inviting them to join Islam. Some accepted and became Muslims, some simply ignored the invitations and others became hostile. The two world great powers at the time; the Persian and the Roman Empires were not pleased with this outlandish bold move of an Arab. The Persians heaped verbal abuse on Islam but the Romans moved militarily to stamp it out particularly among their Arab clients. Eventually, Mohammad (- محمد عليه الصلاة و السلام), to the absolute amazement of the rest of the Arab tribes everywhere, went at the head of an army to fight the Romans who on learning his move withdrew. A couple of years later Mohammad (أبو بكر الصديق) (محمد - عليه الصلاة و السلام) passed away and Abu Bakr was selected to head the fledgling nation. He soon passed away too having spent his less than two years at the helm fighting the apostates to consolidate the power of Islam under one system. Most remarkably, he initiated the attack on the Roman Empire's domain in Al-Sham or Greater Syria (الشام) that was intended by Mohammad (محمد - عليه الصلاة و السلام) on his deathbed. He also sent an expedition to meet the Persian Empire's challenge. Omar Ibn Al-Khattab (عمر إبن الخطاب) succeeded Abu Bakr (أ بو بكر الصديق) as the head of what became under his incredible leadership a legitimate far flung unitary state. His reign saw the total obliteration of the Persian Empire and the permanent demise of Roman hegemony over Al-Sham (الشام) and Egypt. After his assassination, the third Caliph was Othman (عثمان إبن عفان) whose state stretched from the Atlantic to China with a foothold in Spain. All this took place in less than a quarter century. All of a sudden Arabs were the masters of lands where ancient great civilizations existed for millennia. They themselves could not boast of any remarkable cultural achievement other than poetry but they had on their side the unequal feature pertaining to the revelation of the Qur'an in their own tongue. Non-Arab new Muslims quickly mastered the language in order to practice Islam. Once the state was established under Omar Ibn Al-Khattab (عمر إبن الخطاب), two urgent and simultaneous problems were at hand to be dealt with. First the state had to be adequately run to survive. The second was that the practice of Islam had to be regulated according to the tents of the religion already established by Mohammad

(محمد - عليه الصلاة و السلام) where brand new situations arose due to the inclusion of multitudes of non-Arab elements with attitudes and norms very different from the usual Arabic ones. It was sound policy for Omar Ibn Al-Khattab (عمر إبن الخطاب) to adopt from the Persian and Roman rules and regulations what suited the proper administration of the new lands as long as they did not conflict with the rules of Islam. His approach became the standard followed by all that came afterwards. This problem was solved with very little disagreement or trouble albeit with tremendous efforts on part of the administrators. Although it was an alien problem to be faced for the first time in the history of the Arabs, the relatively smooth transition was magnificent. Paradoxically, it was the second task involving the practice of Islam that caused much divergence. Supposedly people were very familiar with and absolutely vigilant to protect their religion but they differed in their approaches to achieving their goals. Politics as always formed the platform and background for this divergence of attitudes that quickly deteriorated into civil war. It was inevitable under the circumstances that several factions should appear with each claiming upholding the banner of Islam against the corruption of the others. Naturaly, each came up with interpretations of the Islamic tradition to promote their claims. However, brute force not scholarship settled the dispute and the Omayyad dynasty was born. Many atrocities were committed at its birth, during its century long reign and at its demise when the Abbasid dynasty replaced it. At every turn, each group had its partisans among scholars justifying their deeds. Whatever arguments were advanced they had to comply with Islam even if only superficially. Arguments inevitably solicited counter arguments. That was the dawn of scholarship in the service of politics in Islam. All through this period of political turmoil foreign culture of the non-Muslim citizens of the state seeped into the mainstream of the developing Islamic culture for good or for bad. Whatever remotely touched on the basic tenets of the religion aroused efforts to cleanse it from unacceptable ideas. The result was great works of scholarship that probably would not have been undertaken were it not for these attacks. Therefore, it was clear that the main thrust of scholarship in protecting the religion should be in the fields of basic beliefs and jurisprudence. Many schools of thought sprang up and eventually the famous four schools of Islamic thought were dominant and survived till the present time. In addition to them, the mainstream shi'a approach to jurisprudence with a clear political strand was also developed and thrived. Primarily for historical and political reasons jurisprudence became the most dominant field of scholarship among early Muslim scholars. It is interesting to notice that the Arabs among them were a vanishing minority as the field was overwhelmingly saturated by contributions from non-Arabs and frequently even non-Muslims even in the development of Arabic poetry and literature not to mention rules and grammar of the Arabic language itself. This is a fascinating testimony to the vitality and universality of Islam. The profound importance and centrality of religion and its requirements in the lives of Muslims can easily be detected even today. This is the hallmark of Islam. Islam is not an important issue for its adherents, it is rather "THE" important issue that everything else pales beside it which is something unfamiliar to say the least among non-Muslims especially Westerners who are time and again shocked at the Muslims' persistence in defending it to the point of sacrificing one's own life for the cause. Understanding this point, one should not be surprised to find standardizing Islamic rituals, rules and regulations as the most pivotal topic pursued by individual scholars.

During the span of the twenty-three years of revelations, Mohammad (محمد - عليه الصلاة و السلام) was the ultimate reference everyone sought judgments from. That was appropriate time period for his closest companions (الصحابة) and wives to absorb all the rules and regulations organizing the daily lives of Muslims and to be able to express their own opinions on whatever issues facing the community even during Mohammad's (محمد - عليه الصلاة و السلام) life. After his death, these pillars of the Muslim community continued their tradition of passing judgments on the various issues that arose but always basing these judgments on what they learned from Mohammad (محمد - عليه الصلاة و السلام) himself. In the meantime, a new class of individuals got their scholarly training at the hands of these companions (الصحابة). These individuals formed the next stratum of scholars in Islam and are collectively known as the "Followers" (التابعين). The two categories of the "companions" and the "followers" formed the backbone of Islamic jurisprudence and their opinions and commentaries are never to be dismissed by later scholars. The obvious reason for their elevated status is their closeness in time to the Prophet (محمد - عليه الصلاة و السلام) which did not allow spurious ideas to enter into their work. Like everything else of value in Islam, their lives and narratives are meticulously recorded. As can be immediately construed, they include men and women and there are very many of them to be mentioned here. However, they formed the foundation upon which Islamic scholarship particularly in the field of jurisprudence is built. When these individuals passed, the expanse of the state was immense. It was also now inhabited by peoples of contrasting backgrounds and customs with the non-Arab elements way outnumbered the Arabs. New opinions accommodating new realities started popping up. They generally reflected local circumstances and never clashed with or remotely contradicted the fundamentals of the religion. This is when the expression "schools of thought" became common place. As expected, there were many of them revolving around the work of certain great scholars at certain locations. The ones whose followers wrote down, commented and extended the scope of these works survived and the others regardless of their unique qualities were pushed aside. That situation left the Muslim nation with the famous "Four Schools of Thought" among the Sunnis and the mainstream Shi'a "Twelvers" (الشيعة الإثناعشرية) in reference to the twelve Imams which is dominant in modern day Iran. It is to be noticed that all these scholars were highly appreciative of each other's works especially when they objected to certain opinions. It is also essential to understand that their differences were fundamentally scholarly and only took the politically charged overtones as Sunni and Shi'a, generations after their passing. Therefore, it is worthwhile to give a very brief biography of each of these great scholars as their works and opinions are the ones that form Islamic jurisprudence, which is the most dominant scholarly outcome of early Islam, to this day. This is done chronologically in accordance with their eras. It is fascinating to note, as mentioned before, that the vast majority of scholars of that period were non-Arab and mostly first generation Muslims generally known as Mawali (موالى) meaning descendants of persons who were captured during battles against the Muslims and later freed and associated with certain Arab clans. The leading personalities dealing with issues of Jurisprudence are by default the most influential scholars and founders of wide spread judicial schools of thought. They are by no means the only or even necessarily the best scholars but they were the ones that had an integrated body of opinions covering every aspect of jurisprudence. More importantly, they were the ones that had their work written down and preserved for next generations. The second century of Hijrah witnessed the completion of these monumental efforts carried out by these great scholars. Thanks to them, the world of Islam acquired a highly advanced judicial knowledge based on revelation and acts and sayings of Prophet Mohammad (- محمد عليه الصلاة و السلام). Every aspect of human daily life within the realm of Islam became regulated by the developed rules constituting the Islamic jurisprudence (الشريعة الإسلامية) which is the basis for laws in all Muslim countries that ever existed for the past fourteen hundred years. It has to be perfectly understood that the developed Islamic jurisprudence or "Shari'ah" (الشريعة) served humanity very well during its entire existence. Additionally, apart from the fundamental principles expressed in the Qur'an and the Sunnah (السنة) which cannot be altered, Islamic jurisprudence is the product of scholarly interpretation of these principles. Obviously, this is the reason for the various schools of thought. Therefore, there is absolutely no reason for it not to evolve with the changing social circumstances while maintaining its wholesomeness and integrity. Having perfected the law to a very large degree, Muslims in tandem with their rulers held these scholarly products in great reverence. Unfortunately, in the meantime, bold new interpretations were discouraged even if they were not persecuted. The false impression that one cannot improve on perfection settled in. Stagnation slowly but surely dominated the scene. When Islam was physically challenged by the Western powers, its way of life resting on its rules and regulations were confronted by the Western made to order laws. Muslims are finding out that catching up with Western technology is rather an easy job but this technology is immersed in alien culture resting on sometimes unacceptable laws and ethos. It is next to impossible to acquire technology without having to deal with its social and behavioral baggage. This is the fundamental dilemma facing Muslim societies at the present time. Although the conflict between being able to acquire universal knowledge and maintaining an authentic way of life is not a matter of fate, it is clear that most Muslim societies hesitate and take timid steps into technological revolutions. The paramount importance is attached to the preservation of Islam under any circumstances. Muslims are historically unique in this respect. That seems puzzling to others. That Islamic predicament is cogently captured by the Egyptian intellectual Tareq Al-Bishri (طارق البشرى) in his 2007 book "ماهية المعاصرة" (The definition of Modernization). Present day turmoil in the lands of Islam can be easily defined as a struggle to bring back these rules and regulations into force after their corruption and sometimes complete displacement by Western imposed laws during the infamous imperialist era of the past couple of centuries. Since Islamic jurisprudence or "Shari'ah" (الشريعة) is wholly based on the Qur'an and the Sunnah (السنة), it is revered by Muslims. Therefore, the struggle by the dedicated Muslims to restore it into their lives *will never end* until its success; Western efforts to suppress it not withstanding. ***Western unceasing efforts to squelch Islamic jurisprudence or "Shari'ah" (الشريعة) is perceived (rightly so) as an attempt to erase the unique Islamic character and eventually Islam itself. This can only lead to confrontation and mayhem.***

It was natural for Islamic scholarship during the early days to concentrate on primarily developing rules and regulations to live by. That explains the dominant position of jurisprudence as the main field of scholarship. That does not mean that other fields of scholarship were not pursued. Islamic scholarship during that era covered a very wide range of topics. Works by Islamic scholars (not necessarily Muslims themselves) are well known particularly in the history of science and philosophy as will be later shown. It is interesting to observe that almost all ancient great Muslim scholars in any field are also well versed scholars of Islamic jurisprudence where they characteristically started their studies. This should not be surprising since Islam represents an integrated approach governing all facets of life. As mentioned in Ayah 56 of Surat Al- Zareyat (الذاريات -the Scattering Wind) (و ما خلقت الجن و الإنس إلا ليعبدون) the sole purpose of creating Jinn (الجن) and humans (الإنس) is to worship God (الله - سبحانه و تعالى). It is universally agreed in Islam that every activity whatsoever of a human being is either an act of worship or an act of defiance towards God (الله - سبحانه و تعالى).

In this Ayah (verse) worship is interpreted as getting to know God (الله - سبحانه و تعالى) and His incalculable blessings; He and He alone bestowed on humans. Knowing God (الله - سبحانه و تعالى) is the ultimate purpose of knowledge in Islam. This is done through contemplating His creation and finding out about its details. This is essentially the supreme purpose of learning in the most general sense. To encourage humans to pursue this undertaking, the Qur'an explicitly and pointedly stated the absolute facts which should constitute the definitive final results of human learning / scientific endeavors. In a real sense, God (الله - سبحانه و تعالى) gave humans the answers to help them in their noble quest seeking knowledge on their own as opposed to all other God's (الله - سبحانه و تعالى) creatures animate or inanimate. That designates what Islam considers the uniqueness of humanity among the rest of creation. Expanding into lands of ancient high civilizations afforded Muslims the opportunity to learn about their accomplishments. Thanks to the great emphasis Islam puts on learning and acquiring knowledge, a massive project to translate into Arabic the newly found manuscripts containing the wisdom of these ancient civilizations got underway. It is a testimony to the power of Islam and its self assurance that almost all of these efforts were carried out by non Muslims. Injecting some spurious corrupt ideas and forged texts even into the main body of Islam was thus inevitable. However, it was impossible to get away with anything that touches on the fundamentals since every single Muslim knows them by heart and any fraudulent addition would be easily caught and the perpetrator would be severely punished. When Europe woke up to the advanced intellectual as well as material achievements of the Islamic Civilization through its various contacts with it, it eagerly acquired, digested and absorbed all its fruitsI. Therefore, there is no disputing that Islamic scholarship formed the foundation upon which "Western Civilization" is erected. This new civilization is a by-product of the "Islamic Civilization"; an undeniable fact one can hardly find mentioned let alone acknowledged in any modern Western thinker's work. In developing their civilization, Westerners absorbed what they could through the geographical points of contact with the Muslims who gladly lent a hand and through trade. Contrast that with modern day barriers erected by the West to deprive others, not just Muslims, of the fruits of science and technology that are by definition the heritage of all humanity and know no nationality.

HISTORICAL EVOLUTION OF ISLAMIC THOUGHT

While the West (or more precisely in terms of the era that coincided with the advance of Islam; *Christendom*) was slumbering in blissful ignorance under the supreme hegemony of the Church, Islam firmly established itself in the East. As mentioned in various parts of this book, pre-Islam Arabs have had no culture to speak of. The innate rationalism of Islam prodded Muslims to create their unique culture. While it was natural that the Arabs would form the core of that culture during the very early stages thanks to their mother tongue, non-Arab Muslims quickly realized that *the Arabic Language was no longer the language of the Arabs but rather the language of the Qur'an*. In no time the Arabs became a small minority among the Muslim world community. As exhaustively explained in this and other books by the author, Arabs represented an essential group to receive the final message from God (الله - سبحانه و تعالى) to humanity. However, Islam promulgated equality of all human beings as a fundamental belief and the pride of being an Arab among Muslims became immaterial. Therefore, the vast majority of Islamic scholars are historically non-Arabs albeit making their great contributions in the Arabic Language. Even the second Moses; Musa Ibn Mimoun (موسى إبن ميمون - Maimonides) resurrected the Jewish tradition writing in Arabic. This is a very powerful testimony to the width, breadth and potency of the Islamic cultural tradition. It is thus, instructive to review the various phases that the Islamic thought process went through from its early development till the present time. It is noticeable that there is a symbiotic connection and dialectic correlation between cultural evolution, positive and negative, within the Islamic and the Western Civilizations as they historically interacted with each other. The following sections illustrate how the Islamic thought process developed, reached its pinnacle and declined contemporaneously with that of the Western thought process. The two processes seem to be mirror images in reverse. Consequences of this evolution are far reaching for all humanity; past present and in the future.

THE REVELATION EPOCH (610-632 / -13 TO 11H)

The revealed source and foundation of the entire body of Islamic thought came in the form of the "Qur'an", the "Glorious Divine Sayings" (الأحاديث القدسية) and the "Tradition" or Sunnah (السنة) of Prophet Mohammad (محمد - عليه الصلاة و السلام). It spans 23 years of the life of Mohammad (محمد - عليه الصلاة و السلام) of which 10 were in Makkah and 13 in Madinah as he received revelations and conveyed them to the growing community of Muslims of his time. It is very well known that he could not read or write. The fallacy that he simply plagiarized his teachings from the Jewish and Christian traditions is meticulously refuted by numerous scholars throughout history. However, the single most compelling evidence that is surprisingly overlooked is that Islam unambiguously dissociated itself from the existing Jewish and Christian traditions in the very first days of revelations of the Qur'an as is explicitly mentioned in Surat Al-Fatiha (Opening – سورة الفاتحة). When Mohammad (محمد - عليه الصلاة و السلام) was still not quite sure of what he was embarking on as the final messenger of God (الله - سبحانه و تعالى) to humanity only days after the Archangel Gabriel appeared to him, he received the very first complete Surat of the Qur'an which is Al-Fatiha (الفاتحة) or the Opening. This is the single Surat that is recited in every prayer the omission of which even by mistake invalidates the prayer of a Muslim. At the end of it there is an implied unequivocal censure of both Jewish and Christian practices and existing beliefs. Islam dissociated itself from both Judaism and Christianity as existed then and ever since from the first instance. This is not a politically tinged modern phenomenon advanced by extremists to serve their goals. It is an interpretation by the most outstanding Islamic scholar and interpreter of the Qur'an and Mohammad's (محمد - عليه الصلاة و السلام) cousin Abd Allah Ibn Al-Abbas (عبد الله إبن العباس). Political correctness of the twenty and the twenty first centuries dismisses such interpretation as unwarranted. Debating the merits of this interpretation here is pointless.

However, the historical fact that it did take place and by such an individual is of paramount importance. It irrefutably shows that very early on, Islam dissociated itself from Judaism and Christianity. The idea that Mohammad (محمد - عليه الصلاة و السلام) would formulate his teachings based on these traditions after the narrative of Surat Al-Fatiha (الفاتحة) is the most ridiculous insinuation. Not to mention that at this time, Mohammad (محمد - عليه الصلاة و السلام) did not know much about these religions or was familiar with any of their adherents. As a matter of fact, Islam's dissociation of itself from Judaism and Christianity came long before dissociating itself from the immediate idolaters of Makkah when Surat Al-Kafiroun (سورة الكافرون) or the "Unbelievers" was revealed which earned the ire of the immediate community and prompted the physical persecution of the very few Muslims of the time. It is disingenuous to claim that Mohammad (محمد - عليه الصلاة و السلام) would be concerned with the non-existing (at the time) adversary rather than the real one within his neighborhood. However, the All-Knowing God (الله - سبحانه و تعالى) through His words in the Qur'an established this fact beforehand in anticipation of the unrestrained animosity that would be shown to Islam by the Jews in Madinah and by the bigots among Western Christians ever since.

The Qur'an

To become a Muslim is to perform a very simple act of declaring one's unreserved belief in the Oneness of God (الله - سبحانه و تعالى) and that Mohammad (محمد - عليه الصلاة و السلام) is His messenger to all of humanity. Acknowledging Mohammad's (محمد - عليه الصلاة و السلام) status is essential without which one's declarations about one's Islam are null and void. This is so since he is the person through whom God's (الله - سبحانه و تعالى) literal words; the Qur'an, were conveyed to humanity. To deny him is to deny the revealed word of God (الله - سبحانه و تعالى). The rest of Islam is actually details for this pronouncement. This stems from the fact that from an Islamic perspective, the sole purpose of creation is to submit to and to get to know the Creator. Submission to the will of the Almighty of all animate and inanimate subjects in existence is compulsory. The All Powerful Creator decided to obtain acknowledgement for His ultimate supremacy through the absolute free will of some of His creation and the opportunity was offered to the most formidable created beings which declined to bear that responsibility. The only exceptions were Jinn (الجن) and humanity. The issue of what Jinn (الجن) represent will be discussed when the latest discoveries in physics are dealt with. The existence of Jinn will be assumed for the time being in consistence with the Islamic context. Humanity indicated interest and obtained free will with all the attendant means. That did not set well with the other contestant. This is how Islam explains the animosity between the devil and humans. Islam does not by any stretch of the imagination ascribe forcing humans to commit evil to the devil; only ill advice and persuasion. Humans are wholly responsible for their acts and thoughts and will be held absolutely accountable for them. The Qur'an of which the very first sentence in the first Surat after the opening called Surat Al-Baqarah (سورة البقرة) or "The Cow" emphatically asserts the veracity of the Qur'an using three simple letters from the Arabic alphabet as the raw material in composing the Qur'an inviting any challenger to use such raw material to compose an equivalent text. This challenge is standing till the end of time. This dare was actually taken up by some during Mohammad's (محمد - عليه الصلاة و السلام) lifetime to their utter disgrace among the Arabs believers and unbelievers alike. The most ardent opponents of Mohammad (محمد - عليه الصلاة و السلام) gave their opinion about the Qur'an as something no human being can compose. Although some individuals were familiar with Jewish and Christian sacred writings, they never attributed the Qur'an to plagiarizing these texts albeit they might have claimed adoption of some ancient stories from them. Neither did his contemporary Jews or Christians. This is all well recorded in the annals of history of that epic and not necessarily by Muslims. These references are extremely popular and available to investigation by proponents or opponents. No trace of citation of them though can be found anywhere in any of the extensive body of Western works on Islam over the past fourteen hundred years. As mentioned before, Islamic Thought stands on three pillars the first of which is the Qur'an. It consists of the literal words spoken by God (الله - سبحانه و تعالى) Almighty Himself transmitted to Mohammed

(محمد - عليه الصلاة و السلام) by the agency of the Archangel Gabriel to be conveyed to the rest of humanity in time and space by the successive generations of Muslims till the end of time. The clear implication is that its contents are valid regardless of place or era. Authenticity of the Qur'an has been established beyond the shadow of a doubt by generations upon generations of scholars. Even the most ardent haters of Islam throughout history never challenged its authenticity while rejecting its divine precepts. The issue of the authenticity of the Qur'an has been discussed in details by the author in a previous published works in this series "Islam and the West". In the Islamic tradition, the Qur'an as the literal word of God (الله - سبحانه و تعالى) is a statement of the absolute truth. It contains every fact and description of existence in the here and the hereafter. It tells the events of the past and predicts those in the future but it is not simply a book of history or anthropology. It legislates and establishes the rules and regulations governing humans' relations with each other as individuals and as communities and nations as well as their relations with other creatures and the environment. But it is not merely a book of law. It describes the process of creation of the universe and its demise at the end of time. It also narrates the story of the creation of humanity and its consequences. However, it is not a book of physical and/or natural science. One can find every aspect of knowledge in the Qur'an without limiting it to any one or more category. As was asserted before the Qur'an is a statement of the absolute truth. The dictum that the Qur'an is and only is the Arabic text is promulgated by God (الله - سبحانه و تعالى) in the Qur'an with no exceptions. Although God (الله - سبحانه و تعالى) spoke to many local communities of humanity in their own languages, He used the Arabic language as the medium to speak to all humanity. The Arabic language was chosen by God (الله - سبحانه و تعالى) to convey His final mandates to humanity because of its special unique characteristics. These characteristics are dealt with in the books by the author mentioned before in this section and there is no need for repetition here. It is universally agreed by Muslims that every single word in the Qur'an is in its specific place for a purpose and cannot be replaced by another similar word as commonly used in the daily life. It should be obvious that uniqueness of the Arabic language is not limited to literature or Jurisprudence aspects of the Qur'an. Rather it very accurately conveys scientific and technical facts as well. This is another fundamental reason for the inevitability of a **"Paradigm Shift"** in the current usage of the Arabic language in Islamic Thought.

Because of historical and existential necessities, early Muslim scholars paid far more attention to the rules and regulations included in the Qur'an resulting in an integral and complete body of Islamic Jurisprudence commonly known as the "Shari'ah" (الشريعة). Other fields of scholarship enjoyed relatively less consideration. The Arabic language served jurists very well in reducing the concepts of the Qur'an into detailed laws for the community. With the passage of time, new legal problems faced Muslims requiring different approaches. Instead of coming up with innovative concepts, most scholars resorted to re-interpreting great ancient works in a deliberate way to accommodate new problems. That worked in a majority of cases but in a real sense paralyzed the thought process. In modern times a noticeable segment of the worldwide educated Muslims rejects these stagnant ready-made answers. A gap is thus created between Muslims who want thought-through solutions to modern issues and scholars who dress old solutions in mostly ridiculous new forms loosely using modern technical terms. With time the gap keeps widening giving anti-Islam groups and individuals an opening to promote their hatred especially in the West. Veneration of the ancient great scholars and their works is basically the reason for this unnecessary dilemma. However, unwillingness to develop new usages for expressions of the Qur'an in light of new societal developments stands in the way. For example, one of the most infamous attacks on Islam by Westerners utilizes the explicit Qur'anic mandate of cutting off the hand in cases of thievery. Sober analysis of the associated Ayah illuminates the situation since the punishment (cutting off the hand) is unambiguously stated in the Qur'an and elaborated on by Mohammad (محمد ـ عليه الصلاة و السلام) while the crime itself (thievery) is neither defined nor elaborated on by Mohammad (محمد ـ عليه الصلاة و السلام) this clearly implies that communities over time have the latitude to define what their knowledgeable jurists see fit as thievery. Legal arguments and topics of Jurisprudence are not the subject of this endeavor nor is the expertise of the author to allow further discussions in this domain. However, apart from jurisprudence especially in the realm of science and technology, the situation is vastly different. It is fundamentally the subject of this book. Gradually the Arabic language received an aura of holiness that transcended its being a product of human ingenuity in the first place. The Qur'an and the Arabic language became synonymous with an eternal link. It goes without saying that one cannot approach interpretation of the Qur'an without mastering the Arabic language to perfection. This is clearest when one deals with beliefs, rituals and Jurisprudence for example. In so doing, the great early scholars established certain specific uses and meanings for almost all words of the Qur'an. These became as close as anything can get to sacrosanct deriving their aura from both the divine nature of the Qur'an and the greatness of the scholars. Interpretation of passages of the Qur'an that have nothing to do with beliefs, rituals or Jurisprudence had to utilize the approaches already exhaustively established in the course of elaborating on these issues. These scholarly standardized meanings for words of the Qur'an seem currently inadequately suitable to interpret many explicitly mentioned scientific passages in the Qur'an. This is again another fundamental reason for the inevitability of a "**Paradigm Shift**" in the current usage of the Arabic language in Islamic Thought.

THE "GLORIOUS DIVINE SAYINGS" (الأحاديث القدسية)

The second pillar of Islamic thought concerns what is known as the "Glorious Divine Sayings" (الأحاديث القدسية). These are sayings attributed to God (الله - سبحانه و تعالى) but narrated by Mohammad (محمد - عليه الصلاة و السلام) outside the scope of the Qur'an. They are mostly exhortations to help create the moral infrastructure of Islam and the Muslims by giving examples, parables and allegories. As such, they are, generally speaking, less precise and less eloquent (if one can use such description with the word of God (الله - سبحانه و تعالى) but for lack of any alternatives) than the Qur'an. One can be curious as to why they were not included in the Qur'an since both narratives are the words of God (الله - سبحانه و تعالى) anyway. It is easy to claim that eloquence could be the main reason. However, it is obvious that Almighty God (الله - سبحانه و تعالى) would not be at a loss in finding the most eloquent words to say whatever He wishes to be included in a single revealed text. The apparent reason for the separation is that words of the Qur'an are precisely included to reach a literally intended meaning and therefore cannot be substituted by any other similar words commonly used by the community. On the other hand, the "Glorious Divine Sayings" (الأحاديث القدسية) are intended for giving an impression using examples, parables or allegories and as such they can be quoted using different commonly used words to give the same meaning. Therefore, in the Islamic tradition, one has to quote the Qur'an extremely accurately while quoting the "Glorious Divine Sayings" (الأحاديث القدسية) can be done in one's own words as long as the integrity of the meaning is preserved. Therefore, extrapolating the logic of this argument and the fact that both texts consist of the words of God (الله - سبحانه و تعالى), one can establish an important criterion. It can be concluded that while the "Glorious Divine Sayings" (الأحاديث القدسية) speak figuratively using examples and parables, Qur'anic text does not. That is to say that every word in the Qur'an is strictly meant. Supporting this conclusion, one finds the Qur'an uses the phrase "giving an example" (مثل) frequently when the used word is not strictly meant except to elaborate on a subject employing commonly used words and expressions by the community. There are approximately 83 places in the Qur'an where this expression or its derivatives are used to indicate figurative speech. Otherwise, deeper meanings should be sought for many of the words of the Qur'an when evolving knowledge seems to be at odds with them as commonly understood. However, these new meanings should simply be extensions of commonly held ones and should not alter the integrity of the Qur'anic text in a different passage. This is a groundbreaking thesis never advanced by any scholars before. It is clearly controversial and many may object and reject it. Islamic history shows that early Muslims used classical Arabic literature, fundamentally poetry, to prove the validity of the Qur'an. Once Islam became dominant, all of a sudden the roles were reversed. With several assertions in the Qur'an itself that it is an Arabic text offering the explicit challenge to the most eloquent unbelievers individually or collectively to compose even a single verse, the Qur'an established itself as the most genuine reference for the Arabic language. Poetry was and to a remarkable degree still is the crown jewel of the Arabic language. Now if the Qur'an does not speak figuratively or in allegory and it nonetheless serves as the ultimate reference to the language, then Arabic poetry loses its value in the opinion of these scholars rejecting the new thesis. This is neither a true nor a valid argument. While words and expressions of the Qur'an are eternally valid till the end of time explaining all issues till then, poetry is the transient expression of human understanding of these same issues in an artistic way. Therefore, Arabic poetry can still have a prominent place in the literature with the Qur'an as its definitive reference. This principle would be discussed and explained in detail when some of the Qur'anic passages are scrutinized against the irrefutable findings (consistently explaining natural phenomena even as an approximation) of modern science. As far as this work is concerned, this is the ultimate reason for the inevitability of a **"Paradigm Shift"** in the current usage of the Arabic language in Islamic Thought.

TRADITION OF PROPHET MOHAMMAD
(محمد - عليه الصلاة و السلام). THE SUNNAH (السنة)

The third historic pillar of Islamic thought is the tradition of the Prophet (محمد - عليه الصلاة و السلام). It includes what he said, he did, something he recommended, a practice he did not object to, etc. It is noteworthy that Mohammad (محمد - عليه الصلاة و السلام) while ordering his companions to write down the Qur'an as it was revealed to him, he discouraged them from recording his sayings in any form. An absolute barrier was erected between the word of God (الله - سبحانه و تعالى) and that of His Messenger (محمد - عليه الصلاة و السلام). Approximately a century after his death when the Qur'an was firmly long authenticated by then, Muslim scholars began collecting his tradition in the most rigorous methodology establishing along the way a highly systematized discipline as a branch of science. Restrictions of proof in that discipline far outdo most requirements in modern day advanced sciences. As to be expected, the body of the tradition consists mostly of answers to questions by his companions the vast majority of which dealt with establishing their rituals, rules and regulations. That is not to say that some did not show curiosity about the universe, the creation process and the nature of God (الله - سبحانه و تعالى). There has never been a question that did not get a satisfactory answer albeit in a simple easy to understand way by someone living in the seventh century. Here also one finds careful use of words and expressions by Mohammad (محمد - عليه الصلاة و السلام) to give a meaning that his immediate companions can relate to as well as Muslims and humans in general till the end of time with their continuously improved knowledge. One has to conclude that Mohammad's (محمد - عليه الصلاة و السلام) answers had to conform to the prevailing knowledge of the seventh century's milieu regardless of the nature of the question. That necessarily means including his approaches to legal matters and their applications. It is not in dispute (within the Islamic tradition) that he observed and narrated facts beyond other humans' normal sensory abilities and whether he actually understood how things happened is really irrelevant in this juncture. But his answers put things in a universal perspective in space and time. This shows another potent reason for the inevitability of a **"Paradigm Shift"** in the current usage of the Arabic language in Islamic Thought. Mohammad (محمد - عليه الصلاة و السلام) being the final connection between Heaven and humanity, acquired greatly enhanced capabilities. He could sense things in space and time beyond the detection of others. It has to be understood that Mohammad (محمد - عليه الصلاة و السلام) in this way was an interested observer rather than a participating initiator. In a sense, he did not necessarily understand how things he was sensing happened submitting instead to the belief in the ultimate omnipotence of God (الله - سبحانه و تعالى). He was not by any means unique in this gift but rather the end of a long series of Prophets and Messengers of God (الله - سبحانه و تعالى) in the Islamic tradition who acquired unusual abilities. Nonetheless, it is believed that his acquisitions surpassed all of them. For example, receiving revelation as an acquired ability is an obvious well known characteristic in the Islamic tradition. Again some of these phenomena are explained as natural albeit extremely less probable when discussing the conceptual principles of Islam as related to modern twenty first century knowledge. One of the most fundamental criteria in Islam is that the Qur'an and the tradition of Mohammad (محمد - عليه الصلاة و السلام) left nothing outside their scope to allow humans to lead an enjoyable life till the end of time. They explain everything in every field. When one faces seemingly unexplained puzzling issue and cannot find a solution in these holy references as interpreted by scholars, one should try to look for deeper meanings for words and phrases therewith. In other words, **the fault is not in the texts (Qur'an and Sunnah) but rather in their interpretations**. Here is a compelling reason for a **"Paradigm Shift"** in the current usage of the Arabic language in Islamic Thought. For an unusually extended period of time, Westerners called Muslims "Mohamedans" in the same manner they called themselves "Christians". They pathetically assumed that Muslims worship Mohammad (محمد - عليه الصلاة و السلام) the same way they worship Jesus Christ (عيسى – عليه السلام). While the Qur'an explicitly mentions the infallibility of Mohammad (محمد - عليه الصلاة و السلام), it unequivocally asserts his humanity. There is nothing divine about Mohammad (- محمد عليه الصلاة و السلام) in Islam. He is a man like any other except that he, on his own, reached the status of perfection from whichever point of view. This is how God (الله - سبحانه و تعالى) sees him according to the Qur'an. Since he is infallible, he utters no nonsense as mentioned in Ayahs 3-5 in Surat Al-Najm (سورة النجم). Therefore, Mohammad's (- محمد عليه الصلاة و السلام) infallibility is not something Muslims accorded him out of reverence. It is rather an attribute explicitly mentioned in the Qur'an. The purpose is to eliminate any and all possibilities for second guessing him on issues of belief whether he is alive or long dead. That makes sifting through his sayings and doings by scholars a very daunting task. Traditionalists do not have this problem. Mohammad (محمد - عليه الصلاة و السلام) represents perfection in the eyes of God (الله - سبحانه و تعالى) and the ultimate goal of a believer is to earn God's (الله - سبحانه و تعالى) contentment. Then, one emulates to the letter whatever Mohammad (محمد - عليه الصلاة و السلام) did to the point of even looking like him externally. There is definitely logic in this approach. However, Mohammad's (محمد - عليه الصلاة و السلام) acts in the most general terms are not equal. What he designated as Qur'an and what was unambiguously designated by later scholars as his sayings are not subject to controversy among Muslims of all veins. It is the things that demonstrate his humanity that fall in this category. It is historically known that when suggesting something, he was asked whether it was revelation or personal. It is also historically well known that in rare occasions (the exception that proves the rule) his decisions were overruled by his companions. The implication here is that his merely personal opinions as well as his personal preferences were not obligatory to Muslims. They do not have to dress like him or walk like him, etc. to earn God's (الله - سبحانه و تعالى) contentment and the elevated status promised in the hereafter. It is puzzling that the fact that his own companions did not look like or behave as mere duplicates of his is completely overlooked by these traditionalists. These differences of opinion should obviously be tolerable except when they spill over into politics and cause dissensions and eventually bloodletting which is paradoxically strictly prohibited in Islam. If scrupulously personal acts of Mohammad's (محمد - عليه الصلاة و السلام) are not obligatory to Muslims then, they are not part of his tradition and as such are not part of this undertaking.

PROPHET MOHAMMAD'S (محمد - عليه الصلاة و السلام) ERA

This deals with the formation of Islamic thought within the duration of the life of Mohammad (محمد - عليه الصلاة و السلام) as Prophet and Messenger of God (الله - سبحانه و تعالى). It occupies the last twenty-three years of his life in both Makkah and Madinah. The first part in Makkah covers the establishment of the character of the Muslim individual and it was up to Mohammad (محمد - عليه الصلاة و السلام) to elaborate on what was revealed to him for the benefit of the few individual Muslims at the time. There is practically no other person concerned with what would be called Islamic thought. This period was dedicated to survival nothing more and nothing less. Muslims lived by the prevailing rules and regulations of Makkah or wherever they were. It is noticeable that the Qur'an revealed during this period was wholly concerned with the Oneness and power of God (الله - سبحانه و تعالى) and the associated rituals. Interestingly most issues of creation and statements of cosmological facts belong to this period. Little curiosity was shown among Muslims during that era about these issues since survival came first. The situation became radically different after the Hijrah or immigration to Madinah. A community and a budding nation came into existence. The Qur'an shifted to explaining the new rules and regulations that govern this community and its relation vis-à-vis others. With rules and regulations being promulgated in rapid sequence, many questions have been raised and Mohammad (محمد - عليه الصلاة و السلام) had to respond with explanations. Therefore, matters of jurisprudence took center stage. Physically speaking, the community expanded in numbers and in space. Prayer times were essentially when the community gathered in one place. When external threats had to be dealt with, parts of the community was separated from Mohammad (محمد - عليه الصلاة و السلام) for long periods of time. Thus Mohammad (محمد - عليه الصلاة و السلام) was not always available to answer urgent questions pertaining to the core of personal behavior in most cases. Other companions (males and females) had to step forward and offer their own interpretations. That was a wide spread practice. However, once in contact with Mohammad (محمد - عليه الصلاة و السلام) validity of these interpretations was cross examined. Not every companion regardless of how early joining Islam he or she was volunteered for that task. Only some few notable individuals took up that responsibility. Gradually a body of personal interpretations of the Qur'an as well as various legal opinions began to slowly form under the personal supervision of Mohammad (محمد - عليه الصلاة و السلام) himself. This is the true beginning of a purely Islamic thought process. As to be expected, it was overwhelmingly concerned with issues of jurisprudence with little contribution to other topics such as interpretation of cosmological passages in the Qur'an. It is of paramount importance to understand that *while the companions sought to interpret the Qur'an, **Mohammad** (محمد - عليه الصلاة و السلام) **never offered a complete interpretation***. He gave only answers to questions when asked about the meaning of some passages. If he did, it would have formed the final word and no one else can contribute more. This would have been a stark contradiction in the Islamic maxim that the Qur'an is valid till the end of time regardless of the evolving human knowledge. Muslims through the ages have to keep interpreting it aided by newly acquired human knowledge. This may cause fundamental shifts and probably dislocations concerning issues of physical and natural sciences for example but would have very little influence on jurisprudence and the conduct of Muslims' daily lives. Among the many great companions of Mohammad, one young cousin of his was very eager to learn and express an opinion. That was Abd Allah Ibn Al-Abbas (عبد الله إبن العباس). He was only a teenager during the great struggle to firmly establish Islam in Madinah and beyond. He eventually became well regarded as a scholar by the major personalities of Islam who sought his opinions in many serious matters especially after the death of Mohammad (محمد - عليه الصلاة و السلام). His closeness to Mohammad (محمد - عليه الصلاة و السلام) and eagerness to learn, earned him the designation as the "preeminent scholar of the nation" (حبر الأمة) in spite of his young age. In later years he extensively interpreted the Qur'an and the tradition. Due to the honor of his title, his works gained very prominent status and traditionalists assign them more reverence than any other opinions. Mohammad (محمد - عليه الصلاة و السلام) never said words of his young cousin Abd Allah Ibn Al-Abbas (عبد الله إبن العباس) were infallible which means that one can naturally find mistakes in his interpretations. However, Mohammad's (محمد - عليه الصلاة و السلام) designation of his status as a scholar puts him beyond competition as no one else received such honor. Nonetheless it does not exclude him from making mistakes. His scholarship formed the foundation for later scholarly work which means earning him blessings for every latter effort according to the well known Islamic concept. Therefore, physically speaking no other person can ever attain more blessings than him. That can easily explain why he is "preeminent scholar of the nation" (حبر الأمة) even if it happens that some of his interpretations are wrong.

ABD ALLAH IBN AL-ABBAS (عبد الله إبن العباس)

His name is Abd Allah ibn Abbas Ibn Abd Al-Motaleb Ibn Hashim Ibn Abd Manaf (عبد الله إبن عباس إبن عبد المطلب إبن هاشم إبن عبد مناف). He is also popularly known in the age old Arabic tradition of respectfully calling prominent people as fathers of their sons as Abu Al-Abbas (أبو العباس). He is paternal cousin to Mohammad (محمد - عليه الصلاة و السلام) and maternal cousin to the legendary Khaled ibn Al-Walid (خالد إبن الوليد). He was born three years before the Hijrah when the Hashim clan was ostracized by Quraysh (قريش). He is thought to have been thirteen or fifteen years old when Mohammad (محمد - عليه الصلاة و السلام) passed away. Abu Al-Abbas (أبو العباس) claimed to have personally seen the archangel Gabriel conversing with Mohammad (محمد - عليه الصلاة و السلام) on two different occasions. On one occasion he was with his father approaching Mohammad (محمد - عليه الصلاة و السلام) who did not pay attention to them. Al-Abbas left angry and complained to his son who answered that Mohammad (محمد - عليه الصلاة و السلام) was very busy listening to the other man in the room which puzzled the father as he did not see anyone else. Returning to ask Mohammad (محمد - عليه الصلاة و السلام) who confirmed that the Archangel Gabriel was there and was pleased that his cousin could actually see him. He was designated as "preeminent scholar of the nation" (حبر الأمة) by Mohammad (محمد - عليه الصلاة و السلام) who prayed to God (الله - سبحانه و تعالى) to instill wisdom and knowledge in him. Abu Al-Abbas (أبو العباس) was known among his contemporaries as the "Ocean of Knowledge". He is known to have narrated over 1600 sayings including some of the most famous sayings of Mohammad's (محمد - عليه الصلاة و السلام) and was later quoted by numerous scholars who are known to have established the tradition of Prophet Mohammad (محمد - عليه الصلاة و السلام) as a branch of science. It was narrated by many that when the greats among the companions faced a legal problem during the reigns of the Caliphs Abu Bakr (أبو بكر الصديق) and Omar (عمر إبن الخطاب) they sought his opinion and accepted his judgment. He earned this status by seeking first hand knowledge from the companions even if it meant spending hours waiting for them to be available to talk to him. Later in life, he encouraged people to come to his questions answering sessions on all Islamic issues. He fought alongside Ali Ibn Abi Taleb (علي إبن أبي طالب) against the remnants of the Kharijites (الخوارج) who refused to accede to his arguments defending the Caliph's acceptance of reverting to the Qur'an's judgment over fighting Mo'aweyah (معاوية إبن أبي سفيان). He was appointed by Ali Ibn Abi Taleb (علي إبن أبي طالب) as the Governor of Basra (البصرة) in Iraq. He also unsuccessfully served as a peace maker between Ali Ibn Abi Taleb (علي إبن أبي طالب) and later his sons and the Omayyad partisans. He died at about the age of seventy. It was narrated that Jaber Ibn Abd Allah (جابر إبن عبد الله) who is one of the greatest companions and narrators of Mohammad's (محمد - عليه الصلاة و السلام) sayings on learning of his death has said that the nation has suffered an irreplaceable loss. Of major interest to this undertaking is the fact that Abd Allah Ibn Al-Abbas (عبد الله إبن العباس) made extensive contributions to Qur'anic and tradition interpretation. He gave many explanations to Qur'anic verses that deal with issues currently assumed to be normally covered by scientific arguments such as creation of the universe, human fetus development, etc. It goes without saying that his arguments employed the state of knowledge of his age. That was very much confined to linguistic arguments and logical opinions. It is now clear that some of his arguments leave much to be desired. Since neither he nor anyone else including Prophet Mohammad (محمد ـ عليه الصلاة و السلام) who designated him as "preeminent scholar of the nation" (حبر الأمة) claimed that he was infallible, some of his arguments can be described as mistaken. That does not reduce his status as the leading scholar of Islam. Faulting him on some issues unrelated to fundamentals of Islam constitutes neither an attack on him nor an attack on Islam. In reality refusing to admit his mistakes simply because of who he is represents an attack on the basic concept of truth in Islam. It is detrimental to the principle of Qur'anic validity for all time which undermines Islam itself. Unassailable facts of nature should be respected. Veneration of great personalities does not shield them from making mistakes. It may lead to stagnation and backwardness. To give substance to these arguments, a couple of examples are discussed here. In the Qur'an at the end of Surat Loqman (سورة لقمان) Ayah 34, it is stated that God (الله ـ سبحانه و تعالى) has the exclusive knowledge of when doomsday takes place, He sends down relief (generally understood in Arabic as rain) and *He knows what wombs contain*. Additionally, no soul knows what it will gain in the morrow and no soul knows in what place (the word Ardh - الأرض which is the subject of detailed discussion in this book is mentioned) it shall die indeed God (الله ـ سبحانه و تعالى) is All Knowing. No one argues in interpretation with God's (الله ـ سبحانه و تعالى) omnipotence however, Abd Allah Ibn Al-Abbas (عبد الله إبن العباس) interpreted the highlighted portion of the translation of the Ayah shown above as "knowledge of the gender of the fetus being a male or a female as reserved to the exclusive knowledge of God (الله ـ سبحانه و تعالى) beyond any human capability. That was obviously consistent with the prevailing knowledge of humanity till quite recently. Nonetheless, it is very common practice nowadays to determine the gender of the unborn baby during pregnancy long before birth. To account for this discrepancy, it would be imprudent to assume anything other than Abd Allah Ibn Al-Abbas's (عبد الله إبن العباس) mistaken interpretation. In the meantime, it is essential to note here that a more universal interpretation of the verse would naturally include his insight. It is also narrated in the tradition that someone approached Abd Allah Ibn Omar (عبد الله إبن عمر) with a request to explain what the Qur'an means in Surat Al-Anbia' (سورة الأنبياء) Ayah 30 describing heaven(s) and Ardh (الأرض) but he promptly referred him to Abd Allah Ibn Al-Abbas (عبد الله إبن العباس). The given explanation had to do with the physical phenomenon of rain causing no-rain heaven and no-plants Ardh (الأرض) to change their character. The answer was met with approval by the well versed scholar Abd Allah Ibn Omar (عبد الله إبن عمر) as well. While that statement gives a seemingly rudimentary clarification based on linguistic arguments, modern science and cosmology give an entirely different physical description based on materially verified facts.

THE RIGHTLY GUIDED CALIPHS PERIOD (11–40 H / 631-661)

Abd Allah ibn Abbas (عبد الله إبن العباس) occupied a very prominent position among the earliest creators of a clearly distinguished Islamic thought thanks to Mohammad's (محمد - عليه الصلاة و السلام) endorsement and prayers and his own unrivaled scholarship. But he was by no means the only great name in the field. There are numerous companions who contributed much to the interpretation of the Qur'an and the narration of sayings and acts attributed to Mohammad (محمد - عليه الصلاة و السلام) that were later compiled into what is known as the tradition of the Prophet (محمد - عليه الصلاة و السلام). An Islamic thought process and scholarship has started to take shape under the watchful eyes, guidance and approval of Mohammad (محمد - عليه الصلاة و السلام). By the time of his death, the nation could depend on these scholars and their opinions to formulate its rules and regulations to govern what very rapidly became an ever expanding state. Jurisprudence was the first order of the day simply because the ever present ultimate reference was gone. With the great expansion of the state and the spread of the great companions into the various parts of it, different schools of thought began to coalesce and take shape around the individual companion who happened to move to that part. Abd Allah Ibn Mas'oud (عبد الله إبن مسعود) moved to Iraq in addition to others where an established Islamic thought process took root in the following generations. Abd Allah Ibn Abbas (عبد الله إبن العباس) went back to Makkah and was the fountain of knowledge that everyone there sought. Syria boasted several of these companions such as Obadah Ibn Al-Samet (عبادة إبن الصامت), Al-Meqdad Ibn Amr (المقداد إبن عمرو) and Abu Al-Darda' (أبو الدرداء). Yemen gained Moa'th Ibn Jabal (معاذ إبن جبل) who was first sent by Mohammad (محمد - عليه الصلاة و السلام) himself. However, Madinah retained its prominence as the seat of the Caliphate until Ali Ibn Abi Taleb (علي إبن أبي طالب) left to Iraq. Therefore, the most eminent group existed there under Zayd Ibn Thabet (زيد إبن ثابت) and Obai Ibn Ka'b (أبي إبن كعب) whose efforts in writing down the Qur'an during the time of Abu Bakr (أبو بكر الصديق) and Othman (عثمان إبن عفان) are well documented.

ABU BAKR'S (أبو بكر الصديق) CALIPHATE (11 – 13 H / 632-634)

After the sudden death of Mohammad (محمد - عليه الصلاة و السلام) early in the eleventh year of Hijrah, the great Muslims at Madinah selected Abu Bakr (أبو بكر الصديق) to lead the nation. This selection was an outstanding, inspired and crucial process in the timing and choice as dissension among the new Muslim Arabs showed its ugly face immediately. Abu Bak's (أبو بكر الصديق) very short time at the helm was practically spent fighting these apostates. However, the wars of apostasy (حروب الردة) as they are known in history took a terrible toll on those who memorized the Qur'an. Up to that point Arab tradition and by default the very recently established Muslim one was orally propagated from one generation to the next. But the new realties of the disappearance of a huge number of memorizers of the Qur'an set in. Omar's (عمر إبن الخطاب) vehement appeals to the Caliph Abu Bakr (أبو بكر الصديق) in regard to writing down the text of the Qur'an and the latter's final agreement marked the transition of the Islamic thought process irreversibly from the ancient Arabic oral tradition into the newly appreciated written one. The procedure followed in recording the Qur'an is breathtaking in its scrupulous adherence to rigor and meticulousness. It is truly miraculous for a mostly illiterate people who only a short couple of decades ago lived outside human history to accomplish such an unassailable feat. That was Mos-haf Abu Bakr (مصحف أبوبكر) which is probably the very first full length written text in the Arabic language other than the standard transactional documents handling the affairs of a trading society. It is extremely fortunate for the Arabic speaking people that their first attempt at written literature concerned the Qur'an. This incontrovertibly established the standard to be conformed to in any future literary pursuits. That may explain the very rapid evolution of the Islamic thought process and the greatness of its products in the next few centuries. The divine promise to preserve the Qur'an intact against any distortion gave comfort and supreme confidence to Muslims especially the scholarly leaning ones. It also gave the Arabic language an invincible reference to fall back on. In perfect hind sight, the most interesting aspect of this endeavor was for Muslims to approach foreign cultures and belief systems with utmost self confidence and eagerness to learn. The constant state of fighting against various Arab tribes during these two years created novel legal problems. The apostates raised many issues to justify their misdeeds. These had to be countered with arguments derived directly from Mohammad's (محمد - عليه الصلاة و السلام) tradition. That was not hard for the closest companions who had intimate knowledge of his sayings and doings over the entire twenty-three years of his Prophethood. Obviously not all the opinions of all the companions were identical but a consensus was always reached. However, the situation created a forum for competing legal ideas emanating from the same unique source. That was immensely enriching to the eventual development of an integrated and wholesome Islamic Jurisprudence. The unrelenting struggle took its toll on Abu Bakr (أبو بكر الصديق) and he died approximately two years and three months after taking charge of the affairs of the nation in the middle of the thirteenth year of Hijrah. However, he left a marvelously unrivaled legacy of great accomplishments both religiously and politically that survive to the present day.

Omar's (عمر إبن الخطاب) Caliphate (13 – 23 H / 634-644)

On his death bed, Abu Bakr (أبو بكر الصديق) left a recommendation for the Muslims to select Omar (عمر إبن الخطاب) as his successor at the head of the nation. Omar's (عمر إبن الخطاب) strict adherence to justice was well known and that was the ultimate reason for his selection by Abu Bakr (أبو بكر الصديق) to succeed him. Omar's (عمر إبن الخطاب) election was promptly confirmed by the whole nation. He continued the fight with the Persians and the Romans and unceremoniously defeated both eliminating the Persian Empire from history altogether along the way and forcing the Roman Emperor to Flee Syria for ever as it turned out. Vast tracts of land with their populations came under Muslim rule. For the first time in history, non-Arabic speaking peoples were incorporated into what unquestionably became a fully fledged Muslim state. In other words, Omar Ibn Al-Khattab (عمر إبن الخطاب) succeeded Abu Bakr (أبو بكر الصديق) as the head of what became under his incredible leadership a legitimate far flung unitary state. His reign saw the total obliteration of the Persian Empire and the permanent demise of Roman hegemony over Al-Sham (الشام) and Egypt. Muslims had to absorb the new attitudes and practices associated with the defeated magnificent imperial cultures and formulate their own. In addition to the cultural challenges, new physical cities were built that in time became major Islamic intellectual centers with their own distinct traditions. Basra was built in 635/34 H and Kufa in 638/38 H. Unique among all Prophets and Messengers of God (الله - سبحانه و تعالى), Mohammad (محمد - عليه الصلاة و السلام) had a splendid assortment of companions without whose contributions he could not have succeeded. With his passing, most of them were still around enjoying well deserved veneration by new non-Arab Muslims as well as new Arab generations born after Mohammad's (محمد - عليه الصلاة و السلام) passing. There were those who preferred the austere lifestyle and those who enjoyed the newly acquired fantastic wealth and luxuries. There were also those who occupied the middle ground. All were great companions that Mohammad (محمد - عليه الصلاة و السلام) recommended following the advice and guidance of each and every one of them equally. That illustrates the richness of the Islamic tradition and its diversity in contrast with advocates of a very uncompromising single outlook and interpretation of Islam. On sensing the impending end of his life, Mohammad (محمد - عليه الصلاة و السلام) decided to leave the nation with few final declarations before departing this world. Most famously, he named ten individuals that he was absolutely contended with and grateful to. Understandably, these were the earliest people who embraced his call to Islam and never wavered in their support afterwards. There was only one remarkable exception and that was Omar Ibn Al-Khattab (عمر إبن الخطاب) who joined the fold three years later after a long history of relentless antagonism to Islam. This is undoubtedly a vivid indication of the greatness of the man and his countless contributions to upholding Islam. This group is historically given the splendid appellation as the "ten who were promised Al-Jannah (الجنة) / paradise". During the Caliphate of Omar (عمر إبن الخطاب) there were armies in far away lands composed of old and new Muslims alike. The greats wanted to continue their service to Islam by joining these armies and some of them wanted to educate the newly incorporated peoples about Islam. However, the farsighted Omar (عمر إبن الخطاب) prohibited them from leaving Madinah. In addition to his need for their advice and participation in the decision making process, he realized that their presumably hallowed personalities as intimate companions of the Prophet (محمد - عليه الصلاة و السلام) would unintentionally instill false un-Islamic attitudes among people who till very recently venerated saints and holy men. That decision consolidated the sources of Islamic thought in one spot where arguments and debates can take place developing the foundation for Islamic Jurisprudence for posterity by the best qualified individuals. Omar (عمر إبن الخطاب) himself was a great jurist and true to his faith he promulgated many rules that are quoted even at the present time contributing invaluable substance to the Islamic thought process. For example, he would not cut off the hands of individuals who snatched foodstuff belonging to others to eat during the year when famine swept Arabia due to lack of rain. On the face of it, this is in contravention of the explicit divine mandate in the Qur'an.

Even at the present time some people invoke this fact to suspend Qur'anic commands they find inconvenient. It goes without saying that the great companions in Madinah at the time did not see Omar's (عمر إبن الخطاب) action as subverting the Qur'an, otherwise they would have unhesitatingly risen against him. Neither Omar (عمر إبن الخطاب) nor even Mohammad (محمد - عليه الصلاة و السلام) has the slightest authority to suspend God's (الله - سبحانه و تعالى) mandate. On the other hand, the great companions had proper understanding of the passage in the Qur'an that orders the punishment of a "thief" male or female as cutting off the hand. The punishment is explicit and Mohammad (محمد - عليه الصلاة و السلام) properly indicated what was meant by the "hand". However, he never explained what was meant by the word "thief" but it was left to generations of Muslims to define what it is according to their social circumstances. Another example that causes unwarranted heated arguments between Sunni and Shi'a Muslims at the present time concerns Omar's (عمر إبن الخطاب) abolition of the practice of time limited marriages. While both practically stay away from that approach because of its socially undesirable consequences, unlike Sunnis Shi'a Muslims do not consider it out of bounds of Islamic law. Omar (عمر إبن الخطاب) was a great interpreter of the Qur'an and an accomplished jurist but he was by no means the only one. Those who joined Mohammad (محمد - عليه الصلاة و السلام) for most of his twenty-three years of struggle to propagate and advance the cause of Islam, had plenty to say as well. Abu Hurairah (أبو هريرة) for example who accompanied Mohammad for only three years attentively memorizing his sayings by heart and later earned the title of "narrator of Islam" (راوية الإسلام), kept on giving his opinion on every raised issue citing what he personally learnt from Mohammad (محمد - عليه الصلاة و السلام). Many companions objected to his audacity given his short association with Mohammad (محمد - عليه الصلاة و السلام) but Omar (عمر إبن الخطاب) shied away from silencing him. Without getting mired in the historical details, one can confidently assert that there were numerous divergent points of view, opinions and interpretations expressed by numerous prominent personalities during the time of Omar (عمر إبن الخطاب). It is equally clear that these efforts were almost exclusively dealing with judicious issues of rituals and law. However, thanks to Omar's (عمر إبن الخطاب) confinement of the great companions in Madinah, opinions might have diverged but the final result was a single decision to be enforced. That approach shaped the laudable tendency to seek consensus for the benefit of the overall system. After a magnificent career in the service of Islam as a companion and as a Caliph, Omar (عمر إبن الخطاب) was assassinated during the prayer by a Persian slave in the twenty third year of Hijrah.

OTHMAN'S (عثمان إبن عفان) CALIPHATE (23 – 35 H / 644-656)

Omar Ibn Al-Khattab (عمر إبن الخطاب) survived for a few more days after being repeatedly stabbed during which time he was contemplating who should succeed him in running the affairs of the state as requested by the bulk of the community. He recommended a detailed plan to be followed in selecting the next Caliph from among the individuals he named. These were the surviving six of the "ten who were promised Al-Jannah (الجنة) / paradise". Several of these individuals took themselves out of consideration for their own reasons. In reality everyone realized that the choice fell to either Ali (علی إبن أبی طالب) or Othman (عثمان إبن عفان). Ali (علی إبن أبی طالب) belonged to the same very strict school as Omar (عمر إبن الخطاب), which some looked for a little bit of relief from, to enjoy the fabulous new circumstances after the almost decade of Omar's (عمر إبن الخطاب) reign. On the other hand, Othman (عثمان إبن عفان) was well known for his generosity, easy going manners as well as his wealth and prominent lineage. He was also much older than Ali (علی إبن أبی طالب) and could obviously claim as much kinship to Mohammad (محمد - عليه الصلاة و السلام) as could Ali (علی إبن أبی طالب) having married two of the Prophet's (محمد عليه الصلاة و السلام) daughters. Therefore, it was Othman (عثمان إبن عفان) who was elected to the Caliphate. Under the Caliphate of Othman (عثمان إبن عفان), the Muslim State as a unitary entity under one ruler reached its apex stretching from Spain and the Atlantic Ocean in the West to China in the East. The desert Bedouins of the very recent past acquired formidable sea power and for the first time in their history actually won naval battles. The vast expanse of the Muslim State made the Arabs a very small ruling minority among its citizens. However, they were the guardians of Islam and its Arabic language nonetheless. As new issues concerning the rights and obligations of the new peoples under the state appeared, Jurisprudence occupied center stage in scholarly efforts. By this time a new generation of Muslims who were mostly born after the passing of Mohammad (محمد - عليه الصلاة و السلام) but descending from the original companions came of age and were eager to absorb all knowledge about Islam. Together with non-Arab new Muslims who came in intimate contact with the companions they formed what is known in the tradition as the "Followers" (التابعين). This gave birth to the second generation of Islamic thinkers who vastly expanded the fields of scholarship and left an indelible mark on the Islamic thought process. As the area of the state stretched far afield and more local populations adopted Islam, a novel problem emerged. For the Muslims, Arabs and locals alike, to perform the required rituals, they had to determine prayer times, direction of Makkah and their relationship to the newly established Islamic (Hijri) calendar. Therefore, besides acquiring knowledge about military arts, it was essential to get acquainted with mathematics and other physical sciences. These disciplines were already advanced in Persia, India and beyond where Muslims became recently in intimate contact with their practitioners. Out of sheer necessity, Islamic thought had to branch into new fields of knowledge. As always the case, technical knowledge comes with baggage of culture as well. In addition to need, curiosity pervaded Muslims' spirit and scholars absorbed much of both and quickly started making their own contributions to these fields. In the political arena the formidable criteria established by the previous two Caliphs continued unabated for the first few years of Othamn's (عثمان إبن عفان) Caliphate. However, many changes slowly but surely crept into the system of governance. Bowing to the wishes of the great companions who wanted to spread the teachings of Islam into the far flung parts of the state, he relaxed Omar's (عمر إبن الخطاب) rules about confining them to Madinah. The unintended consequences of this action were drastic albeit eventually proved manageable. It was natural that each companion had his own way of reciting and interpreting the Qur'an. Free to travel without restrictions, these companions went and settled in certain major centers of the state and started teaching their own way of reciting the Qur'an. Muslims in these centers had their accents contributing to the creation of somewhat different approaches to recitation. Since each group took the revered companion who received the Qur'an from Mohammad (محمد - عليه الصلاة و السلام) himself as a reference, it dismissed the others as deviations from the authentic pronunciation. There were no differences in the text but rather in the pronunciations and in a sense in the rhythm of recitation. At the time different Arab communities wrote the same words in slightly different ways. The squabbles between soldiers coming from Syria who recited the Qur'an according to the way Obbay Ibn Ka'b Abu Al-Monthir Al-Ansary (أبى إبن كعب أبو المنذر الأنصارى) did and those coming from Iraq who followed the way preferred by Abd Allah Ibn Mas'oud Al-Hothly (عبد الله إبن مسعود الهذلى) in the army fighting in Armenia and Azerbaijan outraged the great confidant of Mohammad (محمد - عليه الصلاة و السلام) Huthayfa Ibn Al-Yaman (حذيفة إبن اليمان). He travelled back to Al-Madinah relating his anger to Othman Ibn Affan (عثمان إبن عفان) as the head of state. It in turn caused him great anguish prompting his decision, after consulting the other companions, to make Mos-haf Abu Bakr (مصحف أبو بكر) the only permissible copy of the Qur'an in circulation throughout the vastly and quickly expanding lands of Islam. A commission was formed to make as many copies as there are major Islamic population centers of Mos-haf Abu Baker (مصحف أبو بكر). Several copies were made and all other records were destroyed. These copies were dispatched accompanied with a person who is an authority in reciting the Qur'an to teach Muslims accordingly. This is what is called Mos-haf Othman (مصحف عثمان). It is historically established that four more copies were made to make the total nine copies of Mos-haf Othman (مصحف عثمان). He kept one for himself which is the one bearing his blood when he was assassinated. Others were sent later to Egypt, Yemen and Eastern Arabia (then called Bahrain). Some of these copies still exist as well as copies made from them. Private persons all over the Muslim world made their own copies of these standardized ones under strict supervision of the authorities. To narrow any possibility for mispronunciation, the diacritical marks were invented by Al-Khalil Ibn Ahmad (الخليل إبن أحمد) (100-170 H / 718-779). These are used continuously till the present day. When printing became widely available, Mos-haf Othman (مصحف عثمان) was used as the standard. The greatness of the Islamic tradition is most clearly illustrated in the fact that it was Sybawah (سيبويه) the non-Arab Persian who for the first time ever systematized the Arabic grammar with rules that governed it ever since. It was Omar's (عمر إبن الخطاب) habit to deliberately and frequently change governors of the various parts of the state. The only exception was Mo'aweyah Ibn Abi Sofian (معاوية إبن أبى سفيان) governing a district in Syria (which was divided into three districts at the time) due to his alleged familiarity with Roman rules that kept the population contented. Othman (عثمان إبن عفان) shortly after assuming the Caliphate invoked Omar's (عمر إبن الخطاب) approach and changed most of the governors and continued to do so over the years of his Caliphate. In the meantime, consolidating all of Syria under the governorship of Mo'aweyah Ibn Abi Sofian (معاوية إبن أبى سفيان) giving him formidable physical and material power. Almost all of the new governors, although well regarded, were close relatives or associates of his. With the passage of time, some of them became corrupt fearing no reprisals as people complained. Three generations before Mohammad (محمد - عليه الصلاة و السلام) the descendants of the two prominent personalities Omayyah (أمية) and his uncle Hashim (هاشم) became rivals. The Hashim clan took up the great honor of provisioning the pilgrims while the Omayyad clan busied themselves trading and accumulating wealth. By the time Mohammad (محمد - عليه الصلاة و السلام) received the revelation, his clan had all the prestige but exhausted their wealth while the other clan was collectively fabulously rich. Mohammad's (محمد - عليه الصلاة و السلام) Prophethood exasperated the rivalry. The Omayyad clan included some of the most antagonistic opponents of Mohammad (محمد - عليه الصلاة و السلام) and Islam in general. Othman (عثمان إبن عفان) despite belonging to that clan was one of the very first Muslims and an unwavering supporter of Mohammad (محمد - عليه الصلاة و السلام) and his message. The Omayyad clan considered the triumph of Islam an unmitigated disaster that has befallen them but there was absolutely nothing they could do about it. With the ascendance of their kin Othman (عثمان إبن عفان) to the Caliphate, they determined to regain their old lost prominence over others and took full advantage of Othman's (عثمان إبن عفان) unqualified decency and Muslims' love for him. Their misdeeds resulted in widespread discontent that eventually led to his assassination. The rebellion split the community and the great companions still alive into supporters and critics of Othman (عثمان إبن عفان) with each group producing arguments and counter-arguments to promote their case. It was then clear that Islamic thought and Jurisprudence is no longer unanimous as before. Different approaches tried to create different political facts with members of the Omayyad clan particularly the very powerful Mo'aweyah ibn Abi Sofian (معاوية إبن أبى سفيان) in Syria stoking the flames of dissension as the aggrieved kin of the slain Caliph. They put the blame unjustifiably squarely on the shoulders of the newly selected Caliph Ali ibn Abi Taleb (على إبن أبى طالب) the Hashemite. They could not challenge his authority yet but they could undermine it. The seeds of civil war were sown. The assassination of Caliph Othman (عثمان إبن عفان) marked the beginning of politics encroaching on Islamic thought in general but Jurisprudence in particular with opposing factions basing their arguments, legitimately or illegitimately, on interpreting the Qur'an and the Tradition of the Prophet (محمد - عليه الصلاة و السلام).

ALI'S (على إبن أبى طالب) CALIPHATE (35 – 40 H / 656-661)

It was clear that only Ali ibn Abi Taleb (على إبن أبى طالب) could lead the state after these calamitous events. Even the Omayyad clan members could not challenge that approach while leaving no stone unturned to undermine his authority by demanding arresting and punishing the perpetrators of Othman's (عثمان إبن عفان) murder. That was a very devious argument. Due to the chaos engulfing Al-Madinah at the time, it was not clear who committed that atrocity. When the old antipathy against Ali (على إبن أبى طالب) for suggesting years ago to Mohammad (محمد - عليه الصلاة و السلام) divorcing her overcame her sound judgment, the "Mother of the Believers" A'isha (عائشة) decided to raise the same false argument against him and as to be expected many Muslims answered her call to fight Ali (على إبن أبى طالب) in a battle that ended tragically for all involved and for the unity of Muslims. Obviously Ali's (على إبن أبى طالب) hold on the state became tenuous and he could not continue discharging his duties from Al-Madinah anymore but moved to Kufa in Iraq. The situation encouraged Mo'aweyah ibn Abi Sofian

(معاوية إبن أبى سفيان) to openly question and then contest Ali's (على إبن أبى طالب) fitness for the office of the Caliphate. The ensuing fights not only split the nation but pitted one great companion against another. They killed each other. These are some of the very same individuals whom the Prophet (محمد - عليه الصلاة و السلام) highly praised and they are supposed to act righteously. Logically, they could not all be right. The circumstances were perplexing to those who did not participate in the fights. Paradoxically, Ali's (على إبن أبى طالب) fundamental problem sprang from his belonging to the category of giant companions when he was only a teenager. With the passing of Mohammad (محمد - عليه الصلاة و السلام), this fact gave him an unassailable claim to the Caliphate. However, it could not possibly be attained against individuals like Abu Bakr (أبو بكر الصديق), Omar (عمر إبن الخطاب) and Othman (عثمان إبن عفان) who were many years his seniors. Complicating the state of affairs, most of the older generation of the great companions have, by then passed away. While all partisans claimed to uphold Islam, most of them looked forward to the spoils as they never played part in building Islam but were merely soldiers in a formidable expanding state. Therefore, when most of the giants died and his turn to assume leadership came about, he was left with masses that were more interested in material gains than, in their opinion, the moot question of upholding the obviously victorious state. The two dominant factions within the Muslim community of the time were the Syrian disciplined army under the command of Mo'aweyah Ibn Abi Sofian (معاوية إبن أبى سفيان) and the followers of Ali (على إبن أبى طالب) mostly from Iraq who were not of the same mind arguing over every step taken. New ideas about what should and should not be done started popping up among their ranks resulting in more factionalism. It was rather bizarre that the most potent dissension emanated from the originally most ardent supporters of Ali (على إبن أبى طالب) when he accepted arbitration at their urging. The basic idea of arbitration was to enforce what the Qur'an says about each side's arguments. This group, called the Kharejites (الخوارج), created its own fanatical interpretations of the Qur'an and became a menace to the Muslims. Their notions reverberated through the centuries and echo even in present day fanaticism. They eventually fought Ali (على إبن أبى طالب) to their detriment and subsequently caused his assassination. Although the arbitration was politically a fiasco, it gave birth to a most fundamental principle according to Ali's (على إبن أبى طالب) proclamation that the ***Qur'an is what is written between the first and last Surat and it does not speak; only humans claim what it means*** (هذا القرآن إنما هو خط مسطور بين دفتين لا ينطق إنما يتكلم به الرجال). This principle sets in concrete terms the idea that whenever people use the Qur'an to promote their arguments, it must be understood that what they are advancing is simply their own interpretation of the Qur'an and it may or may not be correct. Modern Islamic fundamentalists who suppose that there is only one interpretation of the Qur'an and the Sunnah, apart from the creed and the rituals, are making a grievous mistake and perpetrating an outrageous affront not just to intellectuality but to Islam itself. Ironically, whatever it is this one interpretation, it is not by Mohammad (محمد - عليه الصلاة و السلام) but rather by the great ancient scholars who never claimed infallibility. Appreciation of the greatness of these ancient scholars should not be blindly extended to their works. This truly to a very large extent is the reason for the currently observed stagnation of Islamic thought. The most drastic effect of the civil war was the firming up of tribal loyalties rather than loyalty to Islam with very tragic consequences that extended far beyond the battle field. Tribal prejudices showed their ugly face in the cultural life of the community. Poetry and literature started appearing permeated with falsifications of genealogies and personal lineages as well as historical facts. From the Omayyad side that was an essential development to account for their tardiness in embracing and eventually promoting Islam. This became a recurring theme when political fortunes favored challengers to the ruling class and other dynasties were established. That was reflected in the period's literature. Ali (على إبن أبى طالب) being a great scholar and an accomplished jurist himself served as the chief judge and prominent legal advisor to the previous three Caliphs. His opinions inaugurated an approach to Jurisprudence that is followed by most schools of thought to the present time. He was also one of the closest associates of the most eloquent Arabic speaker of all time; Mohammad (محمد - عليه الصلاة و السلام) since his childhood. His command of the Arabic language and its intricacies was impeccable. Despite spending most of his Caliphate in fighting, great works in literature, poetry and Jurisprudence are attributed to him and were included in curricula of the most prestigious Islamic learning institutions throughout history. When he noticed frequent incorrect usage of Arabic grammar among his associates including members of his household, he requested his close comrade Abu Al-Aswad Al-Doa'li (أبو الأسود الدؤلى) to compile in writing rules and standards of Arabic grammar. These treatises still form the foundations of Arabic grammar to the present day. His move to Iraq contributed in no small measure to the development of major Islamic cultural centers especially in the newly built cities of Basra (البصرة) and Kufa (الكوفة) that lasted for centuries producing distinct school of thought with its associated many of the greatest scholars of Islam. His tragic end and the political turmoil that followed gave birth to the Shi'a group's actual or perceived resistance to the tyranny of both the Omayyad and the Abbasid dynasties. Due to the changing political environment over time this approach evolved into a full-fledged independent interpretation of Islam that is absolutely consistent with its fundamentals but adopts a more revolutionary spirit reflected in its outlook and Jurisprudence. While Westerners as well as some Muslims speak of a religious chasm within Islam between Sunni and Shi'a, the fact is that their differences have nothing to do with the essence and fundamentals of the creed. For example, it is an historical certainty that no one was ever denied joining the Hajj because of their being a Shi'a. In light of the fact that non-Muslims are banned from entering Makkah and Madinah, the implication should be self evident. Different approaches and different schools of thought should be looked at as enriching Islam. When these differences in Islamic thought spill into politics, they undermine rather than uphold Islam.

THE OMAYYAD REIGN (40 – 132 H / 661-750)

Military prowess and trickery deployed against Ali (على إبن أبى طالب) did not help Mo'aweyah ibn Abi Sofian (معاوية إبن أبى سفيان) obtain his coveted Caliphate but the assassination of Ali (على إبن أبى طالب) swung the door wide open to claim it against the wishes of Muslims to nominate Ali's (على إبن أبى طالب) son Al-Hassan (الحسن) for the position. Al-Hassan (الحسن) accepted Mo'aweyah ibn Abi Sofian's (معاوية إبن أبى

(سفيان) Caliphate on condition that it reverts to him as demanded by the community on the latter's death in an attempt to stop the spillage of Muslim blood. However, it was Al-Hassan who mysteriously died first, in 49 H/669 despite his young age. Most historians credit his death by poison to Mo'aweyah's (معاوية إبن أبى سفيان) agents. Mo'aweyah Ibn Abi Sofian (معاوية إبن أبى سفيان) not only proclaimed himself Caliph but also forced the appointment of his playboy son Yazid (يزيد) who grew up in Al-Sham (الشام) estranged from real Islamic upbringing as the next Caliph. That was the beginning of hereditary monarchy in Islam. Succeeding monarchs kept the appearance of obtaining the consent of the citizenry to preserve their legitimacy according to Islamic principles. But with all the misdeeds Mo'aweyah ibn Abi Sofian (معاوية إبن أبى سفيان) committed to secure the Caliphate in his lineage, the other branch of his clan took over after his grandson's death (also named Mo'aweyah) only three months into his reign. Thus, there were only three Caliphs from his line and then ten from Marwan Ibn Al-Hakam's (مروان إبن الحكم) line till the demise of the Omayyad state at the hands of the Abbasids. Almost the entire ninety-two years of Omayyad rule were spent in strife and fighting. In the course of desperately attempting to secure their dynasty, the Omayyad rulers committed many unspeakable heinous crimes against their fellow Muslims. It was the probably most despised character in Islamic history; Yazid Ibn Mo'aweyah (يزيد إبن معاوية) (ruled 60-64 H/680-83) who perpetrated the slaughter of Al-Hussein Ibn Ali (الحسين) in 61 H/680 and almost all the progeny of Mohammad (- محمد عليه الصلاة و السلام) in the infamous battle at Karbela' (كربلاء). The martyrdom of Al-Hussein Ibn Ali (الحسين) has been the highlight of Shi'a celebrations and grievances ever since. It is said that because of Yazid Ibn Mo'aweyah's (يزيد إبن معاوية) reprehensible acts, no Egyptian male has been named Yazid (يزيد) ever despite the fact that Egyptians have been overwhelmingly Sunni throughout history. All his transgressions were committed in the short close to only four years as Caliph. While the Omayyad Caliphs claimed sovereignty over all Muslim lands, in reality only Al-Sham (الشام) was securely under their control. The other provinces were in revolt especially the heartland of Arabia where Abd Allah Ibn Al-Zbair (عبد الله إبن الزبير) in Makkah declared himself Caliph on the death of the third Omayyad Caliph; "The second Mo'aweyah" (معاوية الثانى) who was Yazid Ibn Mo'aweyah's (يزيد إبن معاوية) son. However, in Al-Sham (الشام) the Omayyad clan and their supporters chose Marwan Ibn Al-Hakam (مروان إبن الحكم) as the fourth Omayyad Caliph (ruled 64-65 H/684-85). He worked previously under Othman (عثمان إبن عفان) as the equivalent of modern day chief of staff. Most historians blame him for manipulating his boss and forging his signature on documents that eventually caused Othman's (عثمان إبن عفان) murder. He ruled for a short one year designating his son Abd Al-Malik (عبد الملك) as heir to the Caliphate followed by the other son Abd Al-Aziz (عبد العزيز) in total disregard of Islamic mandates. Although there were two declared Caliphs one in Syria and the other in Makkah, the state remained unitary. Abd Al-Malik Ibn Marwan (عبد الملك إبن مروان) (ruled 65-86 H/685-705) succeeded in pacifying the realm step by step but at a horrendous price. His lieutenants attacked the holy City of Makkah that was made a peaceful sanctuary by God (الله - سبحانه و تعالى) Himself and destroyed the Ka'ba. Wherever they went, they brought death and misery. When Abd Allah ibn Al-Zbair (عبد الله إبن الزبير) who is the son of Asma' Bent Abu Bakr (أسماء بنت أبى بكر) and the nephew of the "mother of the believers" A'isha (عائشة) was captured, they crucified him and his body was left hanging to scare the population until Abd Al-Malik (عبد الملك) ordered it lowered. Consolidating his power and pacifying the land, he could pay attention to other ventures. He and his son after him made a major move to have all administrative actions and correspondence carried out in Arabic and he issued wholly Islamic currency for the very first time in history. During his reign, the current structure of the Aqsa Mosque (المسجد الأقصى) was built in Jerusalem. As to be expected, he wanted to change his father's decree and replace his brother by his own son as heir apparent but prominent scholars warned him against such move. However, Abd Al-Aziz (عبد العزيز) died suddenly and Abd Al-Malik Ibn Marwan's (عبد الملك إبن مروان) son Al-Walid (الوليد) eventually followed him in Caliphate (ruled 86-96 H/705-15). He is known to have built the magnificent Omayyad mosque in Damascus and redesigned the architecture of the Prophet's (محمد عليه الصلاة و السلام) mosque in Al-Madinah. However, a damning stain on his reign was the murder in cold blood by his agent Al-Hajaj Ibn Yosef Al-Thaqafi (الحجاج إبن يوسف الثقفى) of the many tens of thousands of Muslims including one of the great followers (التابعين) the scholar Sai'd Ibn Jobayr (سعيد إبن جبير) in 95 H/714.

Chaos reigned supreme all over the state at this time with rebels everywhere fighting the Omayyad Caliph as well as each other. It is interesting that with the raging turmoil everywhere during that period, Islamic thought took major leaps forward at the hands of the followers (التابعين) especially in al-Madinah. Among them are the sons of some of the great companions as well as new non-Arab Muslims who were protégés of the great companions. The most famous of them are known as the "Seven Scholars of Al-Madinah" such as Sa'id Ibn Al-Mosayab (سعيدإبن المسيب), Orwah Ibn Al-Zobair (عروة إبن الزبير), Abu Bakr Ibn Abd Al-Rahman (أبو بكر إبن عبد الرحمن) who all died in 94 H, Obid Allah Ibn Abd Allah Ibn Otbah (عبيد الله إبن عبد الله إبن عتبة) who died in 98 H, Kharejah Ibn Zayd Ibn Thabet (خارجة إبن زيد إبن ثابت) who died in 100 H, Suleiman Ibn Yasar (سليمان إبن يسار) who died in 107 H and Al-Qasem Ibn Mohammad (القاسم إبن محمد) who died in 108 H. These are the ones that contributed immensely to narrating the sayings of Prophet Mohammad (محمد - عليه الصلاة و السلام), Jurisprudence, interpretation and poetry. They formed the nucleus of what is historically known as Al-Madinah approach. In addition to them, there are dozens of less known albeit no less important followers. Biographies of these individuals are detailed in the most famous book by Ibn Sa'd (إبن سعد) giving the categories or ranks of the great Muslim personalities "Al-Tabaqat Al-Kobra" (الطبقات الكبرى). While politics tore the community apart, scholarship planted the seeds of Islamic thought that would guarantee the eternal survival of Islam. In spite of the apparent peace prevailing in the realm, it was inevitable that rebellions would erupt sometime somewhere because of the decadence and tyranny of the Omayyad Caliphs and they were well aware of this eventuality. To avoid trouble after seven of their tyrannical Caliphs, the clan resorted to electing as Caliph the only truly pious and very well liked Omayyad individual; Omar Ibn Abd Al-aziz (عمر إبن عبد العزيز) who, figuratively speaking, is known in Islamic history as the fifth righteous Caliph (ruled 99-101 H/717-20). He was also a great scholar among the ranks of the followers (التابعين) who contributed much to the rules of governance. His righteousness and curbing the excesses of his relatives did not set well with the Omayyad clan. After only two and half years at the helm, Omar Ibn Abd Al-aziz (عمر إبن عبد العزيز) died in suspicious circumstances and the clan went back to its old ways except for few futile attempts by some rulers to stem the tide of history until the demise of their dynasty in 132 H/750 with the death of Caliph Marwan II (مروان الثانى) in battle against the Abbasid rebels.

As mentioned before, with very minor exceptions the Omayyad clan was in the opposition to Mohammad (محمد - عليه الصلاة و السلام) and Islam until they had to give in. When Mo'aweyah Ibn Abi Sofian's (معاوية إبن أبى سفيان) succeeded in usurping the Caliphate, he and his relations knew full well that they could not base their rule on assumptions of their service to uphold Islam like most clans within Quraysh (قريش) especially the clan of Hashim would. Therefore, they had to find other means to gather powerful individuals and groups around them. They resorted to emphasizing their hard to challenge prominence as an Arab group with very deep roots. In the process they promoted tribal allegiance as opposed to universality of Islam among their supporters and discriminated against their opposition Arab or else. Additionally, they casted doubt on conversion to Islam of non-Arab citizens of the state, to collect more of the taxes that are not levied on Muslims who are obligated to serve in defense of the state. There were many intellectually profoundly damaging consequences to this approach. Fake scholarship was employed to connect the Omayyad clan to other major Arab tribes which muddied the brilliant oral history tradition of the Arabs. Lineages were also tampered with for the same purpose. Major attempts were carried out to rehabilitate Abu Sofian (أبو سفيان), whose animosity to Islam and Mohammad (محمد - عليه الصلاة و السلام) are detailed in every historical narrative concerning the early struggle of Muslims, and his son Mo'aweyah (معاوية) who is the founder of the dynasty. Both embraced Islam only after the total defeat of the opposition to Mohammad (محمد - عليه الصلاة و السلام). Magnifying Arab heritage led naturally to appearance of prominent poets. This era claims some of the greatest Arabic poetry not only by Muslims but also non-Muslims and even atheists testifying to the underlying principles of tolerance and free speech inherent in the Islamic tradition. Towering figures such as Jarir (جرير), Al-Farazdaq (الفرزدق) and Al-Akhtal (الأخطل) composed poetry that is on par with the best of the Jaheli poetry (الشعر الجاهلى) which is normally considered the height of Arabic poetry. However, great poetry was also invented and dedicated to accomplishing the preeminence of the Arabs over other ethnic groups and the Omayyad clan over other Arabs. On the other hand, celebrated foreign prose were translated into Arabic such as the Persian story "Kalilah wa Dimnah" (كليلة و دمنة) which was translated by Ibn Al-Muqaffa' (إبن المقفع) who died shortly after the demise of the Omayyad dynasty. The trouble with the spread of unauthentic poetry in the opinion of the famous twentieth century Egyptian thinker Taha Hussein (طه حسين) made it impossible to authenticate some of the most famous pre-Islam or Jaheli poetry (الشعر الجاهلى). He contends that some were actually composed during that Omayyad period which casts shadows on the viability of the early scholarship supporting the pure Arabic nature of the Qur'an using comparisons with Jaheli poetry (الشعر الجاهلى). It seems therefore inevitable that very powerful ethnic backlash gathered momentum towards the end of the Omayyad dynasty and resulted in what is historically known as the Sho'beyah (الشعوبية) movement. Its members were an assortment of intellectuals drawn from the non-Arab cultures that were far more refined culturally than its Arabic counterpart. Its sole purpose was to denigrate everything Arabic. One has to bear in mind that both groups belong to the newly formed Islamic thought tradition. Despite the political tumult of this era, contact with the old civilizations of Persia, India and Byzantium and mutual exposure to what the other groups had to offer fascinated both the Arabs and the new non-Arab citizens of the state. While many contributed to Jurisprudence, others paid more attention to ancient achievements in science and liberal arts. With the support and sponsorship of the high officials of the state including some of the Caliphs, concerted efforts were undertaken to translate major works into Arabic from a variety of foreign tongues. However, efforts in contributing to Jurisprudence were understandably predominant. It is curious to observe that after the first century almost all great Islamic scholars were not Arabs even in the vital fields of Jurisprudence and Arabic literature. However, most of their works were in Arabic even discussing non-Islamic religious issues testifying to the powerful Islamic intellectual tradition that lends itself to Arab, non-Arab as well as Muslim or non-Muslim.

The unmitigated violence and persisting brutality of the Omayyad Caliphs presented Islamic scholars of the period who were unable to justify these actions with an unprecedented enormous dilemma. Not much time has elapsed since the Prophet (محمد - عليه الصلاة و السلام) and the rightly guided Caliphs' (الخلفاء الراشدون) set the proper Islamic rules and regulations that are being broken at every turn by those who are supposed to be upholding them. Muslims have theoretically consented to the elevation of these tyrants to the Caliphate and thus bear responsibility for their actions. There was no way to get rid of these persons as they wielded the ultimate power of the state but there was no way either to condone their behavior. Clever scholars found a solution. They advanced the thesis that faith and deeds are separated in Islam. The implication is that wicked leaders are Muslims by confession but their deeds are reprehensible and since there were no means to punish them, their accounting should be put off till the Day of Judgment when God (الله - سبحانه و تعالى) will adjudicate. That way scholars, and by default the rest of the community, are absolved of the transgression of not enforcing the law while tyrants are not given a free pass. This school of thought became known as Al-Morge'ah (المرجئة) as they postpone solving the problem. It was thus easy to project back the verdict to cover all good Muslims who fought and killed each other during the civil war. Every time tyrants appeared in Islamic lands, echoes of these opinions came alive. Tyranny and prejudice of the Omayyad reign consolidated the opposition that rallied around the descendants of Ali (علي إبن أبي طالب) that are by definition the only descendants of Mohammad (- محمد عليه الصلاة و السلام) who were given the title of Imam one after the other. This gave form to the revolutionary Shi'a movement in its various strands that survives to the present time claiming to be the true representative of the oppressed masses. These Imams are universally revered by all Muslims and are among the most knowledgeable Islamic scholars especially those first ones who struggled against the early Omayyad dynasty. It was simply natural that as scholars their interpretations of Islamic rules and regulations, apart from the universally agreed foundations of the creed and rituals, would radically differ from the accommodating or tolerating interpretations of other scholars. As with the various Sunni schools of thought, this approach resulted in distinct Shi'a schools of thought with the passage of time. Ideas that affronted fundamental precepts disappeared in short order. There is absolutely no doubt that the differences among all schools of thought are tinged with charged political as opposed to religious adaptations. While history shows that Christian groups staged religious wars of extermination against each other, Muslims' internal fighting took the shape of political wars for domination. Islam in its basic conception is one and the same. With the vast expansion of the state since the time of the four Rightly Guided Caliphs down to the Omayyad era multitude of non-Arab peoples joined the ranks of Islam adopting its language and contributing to its scholarship. Persians were more akin to Arabs than the Romans and as such energetically devolved into Islamic thought. Generations of them moved to Iraq and Kufa (الكوفة) in particular where Islamic thought flourished. Their descendants were by then committed Muslims vying to join the great followers (التابعين) and their students as protégés. Among the most famous ones are Al-No'man Ibn Thabet (النعمان إبن ثابت) who is better known in history as Abu Hanifah (أبو حنيفة) the founder of the school of thought that carries his name among Sunnis and Wasel Ibn 'ata' (واصل إبن عطاء) who is the leading figure in the Mo'tazalah (المعتزلة) school of thought. Cotemporary to them is the Arab Malik Ibn Anas (مالك إبن أنس) whose work and scholarship led to the Sunni school of thought known by his name and Imam Ja'far Al-Sadeq (الإمام جعفر الصادق) whose work is central to the (Twelver) Shi'a (الشيعة الإثنا عشرية) Jurisprudence. They and many others cemented their reputations as eminent scholars during the Omayyad period and lived through the turmoil leading to the establishment of the Abbasid Dynasty. These individuals occupy a very prominent place in Islamic thought and thus, their brief biographies are important to mention at this juncture.

AL-IMAM ABU HANIFA AL-NO'MAN (الإمام أبو حنيفة النعمان) (80-148 H / 702-765)

Abu Hanifa Al-No'man was born in Kufa (الكوفة) belonging to the third generation of Muslims in his family. His grandfather by the name of Zotti (زوطى) was a Persian merchant of noble birth in what is currently modern Afghanistan. After the demise of the Persian Empire, he embraced Islam, moved to the newly built city of Kufa (الكوفة) and had cordial relationship with Caliph Ali (على إبن أبى طالب) who transferred his capital there. His son Thabet (ثابت) grew up as a Muslim with strong interest in the intricacies of the laws of Islam. Al-No'man Ibn Thabet or Abu Hanifa (النعمان إبن ثابت - أبو حنيفة) was therefore born in a well to do merchant family with exposure to the lively intellectual activities of Kufa (الكوفة). However, the young Al-No'man Ibn Thabet or Abu Hanifa (أبو حنيفة) was only interested in his successful trade until he one day was admonished by the most prominent scholar in the city Al-Sha'bi (الشعبى) to utilize his obvious talents in learning rather than making money. He afterwards dedicated most of his time to learning with exposure to many different cultures that merged in the city. He became a protégé for eighteen years of the well known scholar Hammad Ibn Abi Suleiman (حماد إبن أبى سليمان) who was also Persian with his scholarship stretching all the way back to the teaching of the great companions Ali (على إبن أبى طالب) and Abd Allah Ibn Maso'ud (عبد الله إبن مسعود). When his mentor died, Abu Hanifa (أبو حنيفة) took over his position and enjoyed debating other scholars and even his own students. He created a unique approach to Jurisprudence where he used his personal judgment for decisions on matters that were not clearly defined in the Qur'an or the Sunnah. Because of his piety he adamantly refused any payment for his legal work even as an official of the state which was made easier thanks to his successful business. It is known that Abu Hanifah Al-No'man (أبو حنيفة النعمان) travelled numerous times to Makkah and Madinah where he met, befriended and debated issues with many of the followers (التابعين). It is interesting to know that although he is considered by Sunnis as the "Grandest Imam" (الإمام الأعظم), he was a friend and co-scholar of Ja'far Al-Sadeq (جعفر الصادق), Zayd Ibn Ali Zain Al-A'bideen (زيد إبن على زين العابدين) and Abd Allah Ibn Hassan Ibn Hassan Ibn Ali

Ibn Abi Taleb (عبد الله إبن حسن إبن حسن إبن على إبن أبى طالب) and his son Mohammad Al-Nfs Al-zakeiah (محمد النفس الزكية). These are the pinnacles of Shi'a scholarship. He was particularly on good terms with Ja'far Al-sadeq (جعفر الصادق); the most prominent jurist of the Twelvers Shi'a (الشيعة الإثنا عشرية). He wholeheartedly supported the revolt against the last Omayyad Caliph and was imprisoned and tortured as a consequence. He fled to Makkah in the hundred thirtieth year of Hijrah where he lived for six years during which the Abbasids took over. He spent a few years on good terms with the Abbasid Caliphs but when tyranny became their method he openly endorsed their enemies among the descendents of Ali Ibn Abi Taleb (على إبن أبى طالب) who revolted against their kin. Although both descended from the immediate cousins of Prophet Mohammad (محمد - عليه الصلاة و السلام), the Abbasids mercilessly massacred them. That did not set well with the Abu Ja'far Al-Mansour (أبو جعفر المنصور) the Abbasid Caliph but could do nothing due to the high regard people accord the deep knowledge of Abu Hanifah Al-No'man (أبو حنيفة النعمان). He tried in vain to bribe him with the job of the Chief Jurist of the Empire. Abu Hanifah Al-No'man (أبو حنيفة النعمان) was finally imprisoned again in Baghdad, tortured and died in prison in the hundred fiftieth year of Hijrah at the age of seventy. The great mosque in Baghdad bearing his name was desecrated and partially destroyed by the Americans during their invasion of Iraq in the first decade of the twenty first century. Abu Hanifah (أبو حنيفة النعمان) lived fifty years under the Omayyad Dynasty and eighteen under the Abbasid. He witnessed many revolts but never joined one devoting his time to spreading knowledge. His sympathy was clearly with the descendants of Ali (على إبن أبى طالب) especially when their cousins the Abbasids persecuted them. Because of his strong vocal stands against the tyranny of the Abbasid rulers, he was imprisoned and tortured leading to his traumatic death. Although there is a major school of thought named after him, he did not leave many written works. It was his numerous students that spread his opinions. The primary reason nonetheless for cementing his approach as a distinct school of Islamic thought concerning Jurisprudence is the extensive writings of his long time protégés Ya'qoub Ibn Habib known as Abu Yusuf (يعقوب إبن حبيب – أبو يوسف) and Mohammad Ibn Al-Hassan Al-Shaybani (محمد إبن الحسن الشيبانى). The school of thought following Abu Hanifah (أبو حنيفة النعمان) is considered the mildest most liberal among the host of Islamic Jurisprudence approaches. By the end of the first century of Hijrah, so many schools and intellectual approaches sprang up indicating the richness and fertility of the tradition. There were the Kharejite (الخوارج), the Shi'a (الشيعة), the Morje'h (المرجئة) and the Zydiah (الزيدية) firmly established approaches. Knowing that pre-Islamic Arabs lacked in intellectual activity except in the field of poetry, it is clear that Islam opened the flood gates of curiosity and breathed new life into the Arabic spirit. The core conceptual Islamic belief in the equality of human beings enabled Islamic thought to successfully engulf all of its contributors regardless of race or religion. Islam is the only belief system in history, whether heavenly revealed or human-made, that transcended its religious tents to become a culture, a way of life and a civilization encompassing non-Muslims as well.

AL-IMAM JA'FAR AL-SADEQ (الإمام جعفر الصادق) (80-148 H / 702–765)

He is born in Al-Madinah blessed with parentage that reach all the way back to Ali (على إبن أبى طالب) and Fatimah (فاطمة) the daughter of the Prophet (محمد - عليه الصلاة و السلام) on his father's side and to Abu Bakr (أبو بكر الصديق) on his mother's side. After the vile massacre at Karbala' (كربلاء) at the hands of the agents of the Omayyad ruler Yazid Ibn Moa'weyah (يزيد إبن معاوية), only one sick young son of Al-Hussein (الحسين) was unintentionally spared on account of his sickness and the protection of his aunt Zaynab (زينب). That was Ali Ibn Al-Hussein or Zain Al-A'bedin (على إبن الحسين – زين العابدين) whose piety earned him the honorific description. He rejected engaging in politics unlike his father Al-Hussein (الحسين), his uncle Al-Hassan (الحسن) and later his son Zayd (زيد) after his passing away; all dying unceremoniously fighting the tyranny of the Omayyad dynasty. His other son Mohammad Al-Baqer (محمد الباقر) followed his way becoming an outstanding scholar in Al-Madinah. In time, his son Al-Imam Ja'fara Al-Sadeq (الإمام جعفر الصادق) also strenuously advised against his uncle Zayd's (زيد) rebellion as they were of the same age. Al-Imam Ja'fara Al-Sadeq (الإمام جعفر الصادق) is the sixth Imam of the Shi'a (Twelvers) (الشيعة الإثنا عشرية) and is known for his unbounded knowledge and wisdom. He is Ja'far Al-Sadiq (الإمام جعفر الصادق) the son of Mohammad Al-Baqir (محمد الباقر) (the **fifth** Imam) son of (the **fourth** Imam) Ali Zain Al-A'bideen (على زين العابدين - إبن الحسين) son of (the **third** Imam and the most famous shi'a personality and Martyr) al-Hussein (الحسين) son of (the very **first** Imam of the "Twelvers" shi'a) Ali Ibn Abi Taleb (على إبن أبى طالب) who is the fourth Caliph and Mohammad's (محمد - عليه الصلاة و السلام) first cousin as well as the husband of his daughter Fatimah(فاطمة). He was related to, met and dealt with a host of the major Followers (التابعين) whom he engaged in many discussions. He is a contemporary and friend of both men of letters such as Al-Imam Abu Hanifa Al-No'man (أبو حنيفة النعمان), Al-Imam Malik Ibn Anas (مالك إبن أنس) and Al-Jahez (الجاحظ) who is a foremost writer of Arabic prose (150-255 H/767-868) and those of the physical sciences such as his student Jaber Ibn Hayan (جابر إبن حيان 101-110 H / 721-815) who is considered the founder of the rigorous science of Chemistry according to the great renaissance philosopher Francis Bacon (1561- 1626). Abu Hanifah Al-No'man (أبو حنيفة النعمان) and Malik Ibn Anas (مالك إبن أنس) are the founders of the first two Sunni schools of thought who held him in the highest esteem. Later Sunni scholars had nothing but praise for his knowledge and scholarship. As late as the first half of the twentieth century the Grand Imam of Al-Azhar (الأزهر الشريف), the unrivaled highest institution of Sunni scholarship, and a highly regarded renowned twentieth century Islamic scholar in his own right Sheikh Mahmoud Shaltout (الشيخ محمود شلتوت) acknowledged him as a trusted source for Sunni Islam. In addition to being an esteemed scholar, Ja'far Al-Sadiq (الإمام جعفر الصادق) was a political lightening rod around whom opposition to the newly founded Abbasid dynasty gathered. Unlike the other great scholars who supported the armed opposition, he was a direct descendant of Ali Ibn Abi Taleb (على إبن أبى طالب) and as such people considered him the legitimate head of the nation of Islam. However, he refused to take up arms against the rulers while vehemently condemning their tyranny. He even openly opposed the revolutions led by his uncle Zaid Ibn Ali Zain Al-A'bideen (زيد إبن على زين العابدين) (whose followers can currently be found mainly in Yemen forming the Zaidi (الشيعة الزيدية) Shi'a sect) and that led by his son Ismai'l (إسماعيل) (whose followers can be currently found mainly in Oman, Afghanistan and Pakistan forming the Ismai'li (الشيعة الإسماعيلية) shi'a sect). His refusal to lead or endorse any armed struggle was based on his well founded doubts about the sincerity and loyalty of the followers of those calling for it. It is universally known that he resolutely turned down the offer of the Abbasid messenger to endorse their revolt against the late Omayyad dynasty despite their claim that it is carried out in the name of their cousins the descendants of Ali (على إبن أبى طالب) of whom he at the time was among the most prominent. He additionally refused to go along with the rebellion of his direct cousins the descendants of Al-Hassan Ibn Ali (الحسن إبن على إبن أبى طالب) against the Abbasid when the latter reneged on their promise and began arresting and murdering in cold blood the heads of the two branches of the descendants of Ali (على إبن أبى طالب). His stands did not stem from fear or cowardice but from the fact that he did not think the time was ripe for these moves and from a healthy distrust of the seriousness of support of the same Iraq partisans who in the past betrayed his forefathers when they accepted their urgings. On the other hand, he dedicated his life to scholarship, learning and teaching. When he died, a great body of work was left for posterity in his name. A great deal of mythology was created around his personal traits outside the pail of Islam that offended him and he disowned the perpetrators of such ideas. His approach in Jurisprudence was eventually formalized and became the cornerstone of the Shi'a (Twelvers) (الشيعة الإثنا عشرية) law. It should be understood that the Shi'a schools of thought crystallized around movements protesting tyranny and as such had to be radically different in their outlooks from what is historically known as the Sunni four schools of Jurisprudence. However, there is absolutely no daylight among them when it comes to the fundamental aspects and tenets of Islam. Nonetheless, some fanatics on both sides create dissension by invoking trivial and semi-serious issues that even if true would not cause Sunni and Shi'a beliefs to diverge. The differences are not in the primary or even the secondary fields of belief and rituals but rather in the tertiary zone of details. Ja'far Al-Sadiq's (الإمام جعفر الصادق) opinions were eventually adopted by the Islamic Republic of Iran as the basis for its fundamental law. His followers form the largest body of Shi'a Islam and he was designated as the sixth Imam of the "Twelvers" shi'a (الشيعة الإثنا عشرية). Imam Ja'far Al-Sadiq (الإمام جعفر الصادق) died the one hundred forty eighth year of Hijrah. He is believed to have been poisoned by agents of the then Abbasid Caliph Abu Ja'far Al Mansour (أبو جعفر المنصور).

THE ABBASID REIGN (132 – 656 H / 750-1258)

The continuous strife that plagued almost the entire Omayyad reign sapped its energy and made its demise a matter of time. Tyranny and misconduct of most of the Omayyad Caliphs deprived them of popular support. However, their avowed enemies were principally the bulk of the community who rallied around the descendants of the Prophet (محمد - عليه الصلاة و السلام) son after the murder of the father and generation after generation who relentlessly revolted. That contributed in no small part to the Omayyads having to brutalize both the opposition leaders and the population at large. Having championed Arabism to sustain their dominion, they antagonized other races who actually constituted the vast majority of the population. Because of the unreliability of the Arab elements of the state to stage a rebellion, recruitment moved far to the east. It was not difficult to convince the population of the cause of *restoring the Caliphate to the legitimate heirs of the Prophet* (محمد - عليه الصلاة و السلام) since that is what they were used to under the Persian Empire. The circumstances eventually facilitated a marriage of convenience between the non-Arab rebels and those claiming to represent the progeny of the Prophet (محمد - عليه الصلاة و السلام). The most formidable character was Abu Muslim Al-Khorasani (أبو مسلم الخراساني 100-137 H / 718-755) commanding the anti-Omayyad onslaught. As his name shows, he hailed from Khorasan (خراسان) which is modern day Afghanistan. The movement went from the underground phase into an open revolt rolling from the East. The final stage of the assault was very bloody ending with the murder of the last Omayyad Caliph Marawan II (مروان الثاني) and the overthrow of the regime. Abu Al-Abbas (أبو العباس السفاح) was pronounced the Caliph in Kufa in 132

H/750. He claimed to be working to regain the Caliphate rights of his cousins the descendants of Ali (علی إبن أبى طالب). He is known in history, properly so, as Abu Al Abbas - the Mass Murderer. He went about methodically exterminating every Omayyad individual even reaching to their graves which he systematically desecrated and even burnt their decayed corpses. Only one single young man escaped the carnage fleeing to the farthest western part of the realm and then crossing into Spain. This is Abd Al-Rahman Ibn Mo'aweyah Ibn Hisham Ibn Abd Al-Malik (عبد الرحمن إبن معاوية إبن هشام إبن عبد الملك) better know in history as "Abd Al-Rahman Al-Dakhil" (113-172 H/731-788 عبد الرحمن الداخل) and "The Falcon of Quraysh" (صقر قريش). He spent the next thirty-two years in building the foundation of the magnificent Islamic Spain after seceding from the newly established Abbasid Dynasty. The new Caliphs in Iraq and all their descendants could do nothing to bring Spain back under their control. The new masters dropped all pretenses of working for their cousins and declared their own dynasty

They eventually severely persecuted and when needed assassinated the major figures of the descendants of Ali (علی إبن أبى طالب), who by then took the title of "Imam", to eliminate any possibility of losing their status as monarchs. In order to claim legitimacy, they pretended to rule in the name of the Prophet (محمد - عليه الصلاة و السلام) in contravention of all norms of Islamic tenets. They were not concerned with any protests against openly enforcing hereditary monarchy since Mo'aweyah Ibn Abi Sofian (معاوية إبن أبى سفيان) made it the grudgingly acceptable form of government over the objections of all. In this vein they named their dynasty after their ancestor Al-Abbas Ibn Abd Al-Motaleb (العباس إبن عبد المطلب) who was Mohammad's (محمد - عليه الصلاة و السلام) uncle. It was a cynical but inevitable choice for people claiming to uphold Islam. Al-Abbas although protecting Mohammad (محمد - عليه الصلاة و السلام) as his kin, was a very late comer to Islam; no better than Abu Sofian (أبو سفيان) the father of the founder of the Omayyad dynasty. His own son and illustrious cousin of Mohammad's (محمد - عليه الصلاة و السلام), Abd-Allah Ibn Abbas (عبد الله إبن العباس) was a far more formidable Muslim but he was a very young man when associated with Mohammad (محمد - عليه الصلاة و السلام). On the other hand, the father was only three years older than Mohammad (محمد - عليه الصلاة و السلام) and of great wealth and renown. Obviously these facts were not necessarily commonly known to all Muslims at the time except for the small involved group. That required re-projecting Al-Abbas's personality in a more glowing light and history repeated itself in rehabilitating him with false assertions as the Omayyad clan did with Abu Sofian (أبو سفيان) in the past. Some historians refer to the son of Abd-Allah Ibn Abbas (عبد الله إبن العباس) who is another person by the name of Al-Abbas (العباس) and is more contemporary to the events that led to the establishment of the Abbasid dynasty. Be that as it may, that was the bloody beginning of the Abbasid dynasty. It is ironic that the Abbasid Caliphs did not learn the hard lessons that ruined the Omayyad dynasty such as making sure to designate an heir apparent followed by another one. This tradition was repeatedly broken by the one who wanted his son to succeed him rather than the designated next in line. "Abu Al Abbas - the Mass Murderer (أبو العباس السفاح) designated his brother Abu Ja'far Al-mansour (أبو جعفر المنصور) as his successor followed by his nephew Isa Ibn Musa (عيسى إبن موسى). However, Abu Ja'far Al-mansour (أبو جعفر المنصور) (ruled 136-158 H/754-775) made sure his son Al-Mahdi (المهدى) took over with a promise to the nephew to succeed him. As to be expected, Al-Mahdi (المهدى) (ruled 158-169 H/775-785) after abusing Isa Ibn Musa (عيسى إبن موسى), designated his sons Al-Hadi (الهادى) and Haroun (هارون) to succeed him. When Al-Hadi (الهادى) became Caliph (ruled 169-170 H/785-786), he conspired to get rid of his brother but failed and died shortly. Haroun (هارون) became Caliph in 140 H/786 and ruled till 194 H/808 and is known to history as the legendary Haroun Al-Rashid (هارون الرشيد). He bypassed his eldest son Al-Ma'moun (المأمون) whose mother was Persian in favor of the younger one Al-Amin (الأمين) (ruled 194-198 H/808-813) whose mother was the famous Zobaydah (زبيدة) the daughter of Abu Ja'far Al-mansour (أبو جعفر المنصور) and in time reinstated him to succeed Al-Amin (الأمين) and later added the third son to the succession. It is known that prosperity and magnificence of the state reached its zenith during the reign of Haroun Al-Rashid (هارون الرشيد) but as so many powerful monarchs throughout history did, he divided the realm among his sons. When Al-Amin (الأمين) tried to pull the same trick on his brother, fighting erupted between the two camps and since Al-Ma'moun (المأمون) was a well regarded and well liked military leader, his forces defeated his brother's who was killed in the ensuing battle. This episode marked the first time two Muslim brothers fought each other to dominate the state. Al-Ma'moun (المأمون) ruled the vast prosperous state from 198/813 till his death during a campaign against the Byzantines in 218 H/832. Although the followers of the descendants of Ali (على إبن أبى طالب) or the Shi'a never ceased to revolt against the Abbasid Caliphs, their fortunes declined and they never succeeded in overthrowing the dynasty. The reigns of the first three Abbasid Caliphs are characterized by extreme violence but in the end the dynasty was firmly in charge. The fourth; Al-Hadi (الهادى) was a decadent monarch and did not last long to have any effect. Haroun Al-Rashid (هارون الرشيد) was the fifth and the most famous Abbasid Caliph who still faced Shi'a trouble at home in addition to the Omayyad dynasty in Spain. Contact with the Byzantines continued unabated since the time of the Prophet (محمد - عليه الصلاة و السلام) mostly on the battle field but also occasionally in cordiality. However, when Charlemagne (742-814) made himself emperor in the West an unholy alliance materialized between him and Haroun (هارون). The former wanted to challenge Islamic Spain to extend his Christian realm over all of Western Europe and in so doing would eliminate the potential Omayyad threat to the Abbasid dynasty once and for all. The latter seems actively destroying the Orthodox Christian Byzantine Empire step by step and in so doing would be potentially achieving Charlemagne's fantasy of restoring the glory of the bygone Roman Empire making him the only master of Christendom. Neither of these dreams came to pass but the friendship between the two sovereigns solidified showing in no uncertain terms that politics and empire building considerations overrode any religious pretenses.

Prosperity, intimate contact with foreign cultures and systematic assertions of legitimacy gave birth to astounding cultural developments and widening of knowledge in many fields. It goes without saying that in the early couple of centuries after the Arabs burst into world stage in a very significant and dominating role, their only cultural claim to fame was the Arabic language and its poetic tradition. Thus, it was poetry that responded first and foremost to external influences with the added advantage of reacting from a relative position of strength. The lofty poetry of the Omayyad period was more than matched by very impressive contributions in the field during the early Abbasid era. Abbasid sponsorship of talented poets gave the world personalities rivaling in stature the highly distinguished poets of the bygone dynasty. Just to mention a few, one finds names as Ibn Al-Moqaffa' (إبن المقفع 106-142 H/724-759) who translated great Persian literature in addition to serving as advisor to the Caliph, Ibn Ishaq (إبن إسحاق) (85-151 H / 704-767) who is the original and most reliable biographer of the Prophet (محمد - عليه الصلاة و السلام), Abu Nawas (أبو نواس) (145-199 H / 756-815) who composed memorable poetry that crossed every taboo, Al-Asma'i (الأصمعى) (121-216 H / 740-828) who was a first rate grammarian and lexicographer in the meantime collecting old poetry, Abu Tammam (أبو تمام) (188-231 H / 805-845) who shared in collecting old poetry while composing according to the old style, Abu Al-'atahiah (أبو العتاهية) (130-213 H / 748-828) who was personally close to Haroun (died c.828), Abu Al-A'la' Al-Ma'ari (أبو العلاء المعرى) (363-449 H / 973-1057) and many others. It should be mentioned that Al-Khalil Ibn Ahmad (الخليل إبن أحمد) lived through this time. He is the one who invented the important diacritical marks which standardized Arabic pronunciation especially of the Qur'an. These are crucial for the phonetically oriented language and are still being used at the present time. In addition to the poets, this age produced some of the most illustrious names in human history such as Jaber Ibn Hayyan (جابر إبن حيان) the acknowledged father of the science of Chemistry, Al-Khowarazmi (الخوارزمى) the mathematician founder of Algebra, the philosopher Al-Kindi (الكندى)(185-256 H 805-873) and Al-Jahez (الجاحظ) (159-255 H / 776-868) who is considered the unrivaled master of Arabic prose and a noted member of the Mo'tazalh (المعتزلة) school. As seen from the above discussion, politics and violence has afflicted Islamic life since the assassination of Othman (عثمان إبن عفان). This had profound implications for the way Islamic thought concurrently developed. Above and beyond any other considerations, the primary concern for most Muslim scholars of that era was the determination as clearly as possible of how political developments affected Jurisprudence. The liberal approach of Al-Imam Abu Hanifa Al-No'man (أبو حنيفة النعمان) was developed mostly in Kufa (الكوفة) and therefore was greatly influenced by the many merging cultures there.

On the other hand, contemporaneously others did not have any such influences and were deeply rooted in the traditions of the Arab heartland. Obviously, Al-Madinah could unequivocally boast of its being the place where Mohammad (محمد - عليه الصلاة و السلام) resided followed by his rightly guided Caliphs. It is where the foundation of Islamic values, rules and regulations took form and from which they spread to other parts of the state. It was natural that scholars of Al-Madinah would develop their own approach. Clearly, liberalism of Al-Imam Abu Hanifa Al-No'man (أبو حنيفة النعمان) did not set well with Al-Madinah school and a more conservative backlash was to be anticipated.

AL-IMAM MALIK IBN ANAS (الإمام مالك إبن أنس)
(93-179 H / 715-795)

He was born in Al-Madinah to Arab parents from Yemen. His grand father was a well know follower (التابعين). Two of his paternal uncles were well respected scholars who taught him much. He memorized the Qur'an as a young boy as well as the sayings of the Prophet (محمد - عليه الصلاة و السلام). Al-Madinah was the Prophet's (محمد - عليه الصلاة و السلام) hometown and as such his sayings were of particular interest to its scholars. Malik Ibn Anas (مالك إبن أنس) is the very first scholar to write down the sayings of the Prophet (محمد - عليه الصلاة و السلام) (الحديث) in a self contained volume. He attended the learning circles of most of the scholars in Al-Madinah since he was a very young man. His main interest focused on sayings of the Prophet (محمد - عليه الصلاة و السلام) and judicial opinions of the companions and the followers (التابعين). He studied all the prevailing schools of thought regardless of the universal low opinion concerning some of them. He is known to have attended the lessons of Ja'far Al-sadiq (جعفر الصادق) who is one of the Twelvers Shi'a (الشيعة الإثنا عشرية) Imams known for his piety and legal opinions that currently form the backbone of the group's Jurisprudence and a direct descendant of Ali (على إبن أبى طالب). The aura of holiness that Al-Madinah rightly earned from its illustrious former residents including Mohammad (محمد - عليه الصلاة و السلام) himself had a profound impression on Al-Imam Malik Ibn Anas (مالك إبن أنس) and his legal opinions. He followed closely what was narrated in Hadith (الحديث) and what the previous prominent personalities have said and did not go beyond that to offer a personal opinion in most cases. Malik Ibn Anas (مالك إبن أنس) is the very first scholar to write down the sayings of the Prophet (- محمد عليه الصلاة و السلام) in a self contained volume. He is the protégé of Ibn Hormoz (أبن هرمز) and a long time student of Naf'a (نافع) the associate (مولى) of Abd Allah Ibn Omar (عبد الله إبن عمر إبن الخطاب). When he was forty, he started having his own circles of lectures. One was at the Prophet's Mosque. The other was at his home where Abd Allah Ibn Mas'oud (عبد الله إبن مسعود) used to live earlier. He is very well known for repeatedly answering questioners "I don't know" in addition to refusing to discuss any theoretical issue. When the time came to hold his own circle, he chose to station himself where the Caliph Omar (عمر إبن الخطاب) used to sit in the Prophet's Mosque to dispense with community's affairs. When he fell sick later in life, he continued his lessons at his home. He was not a wealthy man but every now and then accepted gifts from the Caliph and no one else despite the fact that he deliberately avoided associating with them whether Omayyad or Abbasid. When Abu Ja'far Al-Mansour (أبو جعفر المنصور) forced people to accept his claim for the Caliphate for himself and his family after him, Al-Imam Malik Ibn Anas (مالك إبن أنس) dismissed giving him the oath of allegiance (which is a fundamental aspect of consent in Islam that cannot be revoked) as made under duress and consequently is invalid. That brought the wrath of the local powers that be in Al-Madinah. He was arrested and beaten and left with a dislocated shoulder. That disgraceful act roused the populace compelling Abu Ja'far Al-Mansour (أبو جعفر المنصور) to claim that he had absolutely nothing to do with it. His ordeal was all the more puzzling because he is known to have opined against revolting against the ruler even when he is a tyrant. That opinion has been used and abused by Muslim Authorities countless times throughout history. One of the common Arabic sayings till the present is traced to his stature as a scholar when people at his time said "no one can give a legal opinion as long as Malik is in town" (لا يفتى و مالك فى المدينة). He lived through the reigns of four of the Abbasid Caliphs and is known to have accepted gifts from them but used them to support the poor and the destitute. He was a very close friend of the great wealthy Egyptian scholar Al-Layth Ibn Sa'd (الليث إبن سعد) who used to correspond with him and send him gifts but unfortunately was not keen on writing down his opinions so little is known of his work except that he was one of the most knowledgeable scholars of his time. An interesting observation is due in the context of discussing the opinions of Al-Imam Malik Ibn Anas (مالك إبن أنس). He uncompromisingly opposed any figurative or allegorical interpretation to any of the words of the Qur'an despite his admission to his inability to give any specific direct meaning. This is a fundamental thesis of this endeavor when advancing the claim that for example any Ayah dealing with the physical universe can find a suitable interpretation in line with the state of the art of human knowledge so far. In the meantime, it is unequivocally asserted that universal meanings to commonly used words do exist expanding human understanding of its universe in conformity with the Qur'an and the Sunnah. This approach elevates the argument that the Qur'an and the Sunnah include every bit of knowledge till the end of time to certainty.

AL-IMAM MOHAMMAD IBN IDRIS AL-SHAFI' (الإمام محمد إبن إدريس الشافعى) (150-204 H / 767-820)

He was born in Gaza in the same year Al-Imam Abu Hanifa Al-No'man (أبو حنيفة النعمان) died. His father passed away when he was an infant leaving the family poor. Since he descended from the great uncles of the Prophet (محمد - عليه الصلاة و السلام) belonging to the line that strongly supported Islam from the very beginning, his mother decided to move him back to Makkah to secure his future. Surprisingly (or maybe not surprisingly) he is the only scholar of note who is related to the Prophet (محمد - عليه الصلاة و السلام) except obviously the Shi'a Imams descendants of Ali (على إبن أبى طالب). From his early years he dedicated his time to learning and when he heard of activities of Al-Imam Malik Ibn Anas (مالك إبن أنس) decided to travel to Al-Madinah to meet him taking a letter of introduction from the governor of Makkah. His initial meeting accompanied by governor of Al-Madinah deeply impressed him of the dignity of and the high esteem accorded Al-Imam Malik Ibn

Anas (مالك إبن أنس). Realizing the promise in the young man, Al-Imam Malik Ibn Anas (مالك إبن أنس) took him as a protégé for nine years till passing away. Having meager means to support himself, Al-Imam Mohammad Ibn Idris Al-Shafi' (محمد إبن إدريس الشافعى) decided to go back to Makkah where he accidently met the governor of Yemen, was highly recommended to him as a most qualified jurist and landed the job that took him to Yemen as a state official. Under his jurisdiction, people were used to bribe officials to get their way. That did not suit Al-Shafi'e (محمد إبن إدريس الشافعى) who severely admonished them. While there after about five years, he had to contend with another corrupt governor who decided to get rid of him by insinuating to Caliph Haroun Al-Rashid (هارون الرشيد) that he supported the revolt against the Abbasid dynasty; an accusation usually punished by death at the time. The advisor to the Caliph happened to be Mohammad Ibn Al-Hassan Al-Shaybani (محمد إبن الحسن الشيبانى) the second Protégé of Al-Imam Abu Hanifa Al-No'man (الإمام أبو حنيفة النعمان) who promptly interceded in his behalf having known of his scholarship and decided to sponsor him for a while. Providentially, Al-Imam Mohammad Ibn Idris Al-Shafi' (محمد إبن إدريس الشافعى) acquired knowledge of the two well established schools of jurisprudence from their originators first hand. After close to two years in Baghdad, Al-Imam Mohammad Ibn Idris Al-Shafi' (محمد إبن إدريس الشافعى) moved back to Makkah where he stayed for nine more years formulating his unique approach apart from the previous two schools and earning a well deserved reputation as a master teacher. Moving again to Baghdad as the center of scholarly activities, he nonetheless stayed less than two years where he published most of his work in books that found their way to all corners of the vast state. At this point in time, Al-Ma'moun (المأمون) ascended the Caliphate and it was not good to be a highly respected scholar of Arab lineage and an outspoken opponent of the Mo'tazalah (المعتزلة) philosophical approach that was dear to the heart of the Caliph. Egypt was where he thought he could be as far as possible from Baghdad's intrigue to free himself to study. One of the most fascinating facts about the scholarship of Al-Imam Mohammad Ibn Idris Al-Shafi' (محمد إبن إدريس الشافعى) is after moving to Egypt in 199 H, he realized that most universally agreed scholarly opinions emanating from Madinah and Iraq, including his own, were actually based on the strictly Arab traditions reflecting the harsh desert environment rather than being rooted in infallible rules of Islam. Surrounded by the distinctly different settled agricultural society in the Nile Valley, he drastically changed his opinions to be more accommodating to various local traditions in the meantime retracting and modifying his previous opinions. This stance has had profound influence in later centuries. It should be mentioned at this juncture that while developments related to the Arabic language are heavily emphasized in this undertaking, knowledge in other literary and physical fields were by no means stagnant. However, there is no denying that the bulk of scholarly effort in the early Islamic history was dedicated to issues related to Jurisprudence which ultimately depends on linguistic proficiency for obvious reasons. These efforts were carried out by pioneers who are giants in their respective field of expertise. That explains the unfortunate tradition among contemporary Islamic scholars to limit any interpretations of any issues, Qur'an, Sunnah, law, science, etc., to linguistic arguments.

Besides being an accomplished military general, Al-Ma'moun (المأمون) was also an astute politician. Knowing that killing of his own brother after he was captured did not set well with the populace and that the perennial Abbasid feud with their cousins and their supporters would have unwelcome consequences, he decided to designate the eighth Shi'a Imam (of the twelvers) Ali Al-Redha Ibn Musa Ibn Ja'far (علي الرضا إبن موسى إبن جعفر) who is the grandson of the revered scholar and jurist Ja'far Al-Sadiq (جعفر الصادق) as his heir apparent to follow as Caliph. In a stroke of political genius, he mollified the opposition on many fronts. Intriguingly though, Ali Al-Redha Ibn Musa Ibn Ja'far (علي الرضا إبن موسى إبن جعفر) died mysteriously soon after that. It was an open secret that Al-Ma'moun (المأمون) had him poisoned. With all the back and forth schemes, Al-Ma'moun (المأمون) consolidated his power and the Caliphate remained in Abbasid hands eliminating the most potent Shi'a figure to wit. The superior military general who lacked any interest in cultural pursuits Al-Mo'tasem (المعتصم) (179-227 H / 769-824) ; younger brother of Al-Ma'moun (المأمون) became the designated heir apparent and ruled (218-227 H / 833-842).

Translation of foreign manuscripts actually started with the Guided

Caliphs to help run the affairs of the state and accelerated during the Omayyad and early Abbasid periods venturing into many secular fields. These efforts were very much utilitarian as needed and no more. With the ascendance of Al-Ma'moun (المأمون), an explosion of sorts in these efforts took place. The tragic circumstances of his assuming the leadership position as well as his maternal heritage in addition to his personal preferences made him very interested in the non-Arab cultures at a time when the blowback of the Su'obeyah (الشعوبية) movement against anything Arabic made it easy for alien ideas to find receptive audiences. Among the most famous and appropriately appreciated names of official translators are Arab Nestorian Christian; Hunine Ibn Ishaq (حنين إبن إسحاق) (194-260 H / 809-873), Hobaysh Ibn Al-Hassan (حبيش إبن الحسن) and Thabet Ibn Qorrah (ثابت إبن قرة)(221-288 H / 836-901). Regrettably, this turned out to be high time for attributing numerous forged traditions to the Prophet (محمد - عليه الصلاة و السلام). The dishonorable misdeed was carried out by non-Muslim scholars maliciously or to ingratiate themselves to the Muslim powers that be. The most far reaching unintended consequence of acquiring perceptions of ancient foreign knowledge happened in the field of Greek Philosophy giving rise to regrettable and destabilizing effects. It was Al-Ma'moun (المأمون) fondness of this field that caused him to blindly link its rationalist influence on the Mo'tazalah (المعتزلة) school to his own power as sovereign. It was no doubt a folly of the first rate for a well versed and enlightened man to merge state politics with free exchange of ideas and then limit that to what the ruler believed. Worse yet it was the great figures of this approach who encouraged him to make such egregious act of following their ideas as a litmus test to loyalty. That created what is known in both political and intellectual Islamic history as "calamity of the issue of creation of the Qur'an" (محنة خلق القرآن)

AL-IMAM AHMAD IBN HANBAL (الإمام أحمد إبن حنبل)
(164-241 H / 780-855)

He was born in Baghdad where his father died shortly thereafter. He descends from Arabic lineage on both sides. Instead of remarrying as was customarily done for support reasons among people of that era, his mother dedicated the rest of her life to raising him. His gratitude for that dedication is clear in all his work and sayings. Hanbal (حنبل) is his grandfather's name not his father's. He belongs to one of the largest and best known Arab tribes; Shayban (شيبان) which boasted numerous military heroes. They fought and defeated the Persians before Islam and commanded some of the armies that brought about the end of the empire. They were also prominent during the successive revolts that plagued the Omayyad and Abbasid dynasties where they appeared on both sides of the conflicts. Although his father was one of those heroes, Al-Imam Ahmad Ibn Hanbal (الإمام أحمد إبن حنبل) was never interested in the military pursuits but he wholeheartedly embraced scholarship especially the collection of the Prophet's (محمد عليه الصلاة و السلام) sayings or Hadith (الحديث). He inherited a little fortune to allow him to be independent devoting his time to learning and memorizing the Qur'an. He was particularly interested in collecting the sayings of the Prophet (محمد - عليه الصلاة و السلام) and travelled widely over Muslim lands in this pursuit. In the hundred and eightieth year of Hijrah, he finished writing his collection of the sayings of the Prophet (محمد - عليه الصلاة و السلام) under the title Al-Mosnad (المسند) which became a fundamental reference in Islam. With people who were said to know something that was attributed to The Prophet (محمد - عليه الصلاة و السلام) spread all over the vast land of the state, he traveled wide and far to authenticate and record the collected information. At the urging of his mother when he was a very young lad, he attended the circle of Abu Yusuf (أبو يوسف) the propagator of the teachings of Al-Imam Abu Hanifa Al-No'man (أبو حنيفة النعمان). He later joined Al-Imam Mohammad Ibn Idris Al-Shafi' (محمد إبن إدريس الشافعى) as a student having every intention to follow him to Egypt but it was not meant to be. Although he agreed to give his opinion when requested, he refused to gather a circle around himself as long as his teachers were still alive. When he commenced his teaching after the death of Al-Imam Mohammad ibn Idris Al-Shafi' (محمد إبن إدريس الشافعى), it is said that a huge crowed attended his circle with many recording his lectures. That helped consolidate his approach as a unique school of Islamic Jurisprudence. His accuracy in what he says was clear-cut that he never narrated a Hadith (الحديث) from memory but read from what he had recorded. He is known for strictly adhering to what was narrated before his time refusing to get involved in any arguments with the philosophers dominating the political scene with the ascendance of Al-Ma'moun (المأمون) to the Caliphate. That did not bode well for his relationship with the powers that be especially after he turned down the offer of judgeship. He also adamantly refused to accept any gifts from a succession of Abbasid Caliphs. He firmly dismissed any attempt at debate which did not endear him to Al-Ma'moun (المأمون) who considered himself a Mo'tazalah (المعتزلة) scholar who enjoyed debating philosophical issues. Greek ideas and their later Christian versions permeated many translated works and most leading Mo'tazalah (المعتزلة) scholars delved into exploring their meaning. The effort although commendable intellectually, created the most regrettable episode in the history of Islamic thought when Al-Ma'moun (المأمون) egged on by the Mo'tazalah (المعتزلة) scholars misused politics and the power of the state to undermine free speech and compel others to fall in line with his notions. The Mo'tazalah (المعتزلة) somehow following Christian concepts about Jesus (عيسى – عليه السلام) came to the conclusion that since the Qur'an is the "word" of God (الله - سبحانه و تعالى), it must be created not eternal. This is the doctrine Al-Ma'moun (المأمون) advocated and demanded submission to by all others. Those who refused were subjected to inhumane treatment. It was Al-Ma'moun (المأمون) who wanted to make the Mo'tazalah's (المعتزلة) argument of the creation of the Qur'an a litmus test of the scholars' loyalty to him. Most of them avoided confronting the authority and came up with many linguistic tricks to satisfy the test. Ahmad ibn Mohammad ibn Hanbal (الإمام أحمد إبن حنبل) openly denounced such attempts and stood fast against discussing this issue or debating the Mo'tazalh's (المعتزلة) philosophers. He was promptly arrested and savagely tortured for over twenty-eight months even after the death of Al-Ma'moun (المأمون) and the ascendance of his brother Al-Mo'tasim (المعتصم); a great warrior but no scholar to the Caliphate. He decided to stop the disgraceful squabbles among the scholars. Al-Imam Ahmad ibn Hanbal (الإمام أحمد إبن حنبل) thought the whole charade was meaningless and futile and refused to participate in any discussions. In his opinion, creation of the Qur'an was a non-issue that is irrelevant to the Islamic faith and adamantly dismissed the idea.

Realizing the weight opinions of Al-Imam Ahmad ibn Hanbal (الإمام أحمد إبن حنبل) carried among the populace, Al-Ma'moun (المأمون) retorted with unwarranted retaliation imprisoning and torturing Al-Imam Ahmad ibn Hanbal (الإمام أحمد إبن حنبل) to no avail. Soon after issuing the edict to exclude all who do not submit to his doctrine from any benefits the state may offer and abuse the scholars opposing him (ending up being Al-Imam Ahmad ibn Hanbal (الإمام أحمد إبن حنبل) alone) he suddenly died advising his brother Al-Mo'tasem (المعتصم) who succeeded him to continue the unwise policy. Since Al-Mo'tasem (المعتصم) was a magnificent military leader with very little intellectual curiosity, he left the status quo intact. The ordeal lasted for more than two years till Al-Mo'tasem (المعتصم) realized the futility of the process and released Al-Imam AhmadiIbn Hanbal (الإمام أحمد إبن حنبل). When his son Al-Watheq (الواثق) succeeded as Caliph, he ordered Al-Imam Ahmad ibn Hanbal (الإمام أحمد إبن حنبل) to disappear from the public and cease his teaching. This mind numbing tribulation that swept the intellectual life of Muslims for over fourteen years pitting one great group of rationalist thinkers against another great group of judicially oriented thinkers swung to the other extreme when Al-Watheq (الواثق) passed away without designating a successor. By then, the Abbasid dynasty has lasted for a century headed by fighting men who expanded the realm. This was the end of the era of great Military Caliphs and the beginning of the rising fortunes of the Turkic military generals, whom Al-Mo'tasem (المعتصم) heavily recruited, within the empire who eventually took it over. The highest officials of the state selected the other son of Al-Mo'tasem (المعتصم) as the next Caliph. The ascendance of Al-Motawakel (المتوكل) to the Caliphate ushered the anticipated backlash. The known Mo'tazalah (المعتزلة) scholars and their writings were methodically eliminated step by step. No records or writings of theirs remained, thus all that is known about their ideas is actually derived from their opponents' refutations to the dismay of later Muslim intellectuals. It is of paramount importance to bear in mind at this juncture that all concepts and arguments either side advanced were products of mental exercises, logic and linguistic manipulation not concrete physical evidence which naturally was far beyond what was available then. As is seen from the scientific analysis in another part of this book, it turns out that there was no real reason for this intellectual split to take such ugly proportions since from the physical science point of view there was no real issue to be discussed based on relativity of time. By the time Al-Imam Ahmad ibn Hanbal (الإمام أحمد إبن حنبل) finished formulating his school of Jurisprudence, the previous three schools were already entrenched in practice and his was not actually followed anywhere. A major reason for this lack of followers is the absence of judges anywhere that adopted this approach. Appointment of judges was controlled by advocates of the other three schools. However, Al-Imam Ahmad ibn Hanbal (الإمام أحمد إبن حنبل) and his tireless efforts to gather the tradition of the Prophet (محمد - عليه الصلاة و السلام) is crowned by a major reference for the Hadith (الحديث) that has been cited by Muslims everywhere through the ages. In his arguments against the Mo'tazalah (المعتزلة), Al-Imam Ahmadi Ibn Hanbal (الإمام أحمد إبن حنبل) emphasized the concept that the Qur'an is devoid of any figurative or allegorical presentations. On the other hand, whatever is meant by some expressions or words in the Qur'an which are used in common everyday language that are at odds with the Omnipotence and Majesty of the Almighty should indicate more universal meaning when associated with God (الله - سبحانه و تعالى). This is exactly the conclusion of this undertaking reached from completely different point of view using the results of the latest scientific knowledge. It was historically clear that with the completion of the works of the four Sunni Imams and the development of the Shi'a ideals by Imam Ja'far Al-Sadiq (الإمام جعفر الصادق) the foundations of Islamic jurisprudence have been firmly established. From then on only minor additions could be made. Scholarly efforts were confined to only interpretations of what was then available. Thus, with Islam representing the most dominant global and intellectual force, Muslim scholars slowly but surely came to believe that thanks to their efforts perfection has been already accomplished.

The Islamic Illusion of the "End of History"

It was universally assumed that Islamic Jurisprudence reached its peak when the four Sunni schools of Jurisprudence ending with the work of Al-Imam Ahmad ibn Hanbal (الإمام أحمد إبن حنبل) and the Shi'a school mostly adhering to the work and teachings of Al-Imam Ja'fara Al-Sadeq (الإمام جعفر الصادق) were established. Therefore, it was concurrently assumed that nothing fundamental can be added to Islamic Jurisprudence and later scholarship should be confined to elaboration on what these approaches mean and elucidation of the subtle arguments within their framework and nothing else of any substance. However, two essential points have to be expressed at this point. Firstly, regardless of the greatness of these schools and their far reaching and sweeping contributions, they are not necessarily the most intellectually exquisite products of Islamic thought. They are rather the most complete and best recorded and the ones that had the privilege of being adopted by the vast majority of judges everywhere in the state. That means that there are many fine points that were investigated by others but were left out in what became routinely the standard Islamic Jurisprudence. Secondly, all of these schools evolved and were developed to completion during the heyday of a very powerful Islamic Empire. A natural consequence is that rules and regulations advanced by these standards were unquestionably enforced by the state and accepted by the populace as well as foreign powers. That fact obviously did not survive the decline of Islamic political power. While one would think that some modifications to accommodate the changing circumstances should take place, Muslim scholars, for good or for bad, did not give ground and Islamic Jurisprudence stayed the same albeit under very different conditions. That created the dilemma of Muslim authorities claiming to adhere to Islamic Laws and norms while at the same time not enforcing them. Or, claiming to enforce them in a manner at odds with prevailing universal attitudes albeit from a position of weakness. Therefore, the idea of perfection and the end of innovative scholarship did not actually lead to the "end of history" but rather to stagnation of Islamic thought and illustrious scholarship. For the past fourteen centuries, Islamic thought concentrated on the development of jurisprudence in a fundamental way to establish the platform of its own validity, the fundamental or primary beliefs and what defines a person as being a Muslim. That resulted in an impressive body of laws governing all aspects of the daily life of the individual and his/her relationship with others as well as the society as a whole. It also determines unequivocally how an Islamic state deals with its members and its friends and its foes. Regardless of the sometimes divergent approaches of the various schools of thought, Islam's tenets and fundamental concepts are undoubtedly remaining intact. Being a Sunni or a shi'a does not represent any differences in the basic tenets of Islam but rather a difference in some of the applicable laws descending from historical and/or political circumstances. Apart from jurisprudence, far less effort was dedicated to cultural, technical and philosophical endeavors. However, the end product of these comparatively minor ventures signified a quantum leap in human history establishing what became known as the Islamic way of life and paradoxically gave impetus to the development of what evolved as the Western civilization. Obviously Jurisprudence does not represent the whole of scholarly activities in any culture. Nevertheless, for historical reasons it formed the cornerstone of intellectual activities within the lands of Islam as they physically expanded. Jurisprudence dominated the life of Muslim scholars until it reached its apex during the first Hijrah couple of centuries where hardly anything else could be added except on the periphery. Thus, it was inevitable that if confined to Jurisprudence, Islamic thought would stagnate especially under the prevailing political climate of the time. This is not to say that knowledge in other scholarly fields did not take huge forward strides but Jurisprudence casted a long shadow over any other field. It is noteworthy that almost all great ancient Muslim Scholars studying and contributing their own theories to physical phenomena had their beginnings in the field of Jurisprudence and its related disciplines. When the tumultuous political environment settled coinciding with what was perceived as the end of the road in developing rules and regulations governing the Muslim's daily life, that represented the "End of History" and the victory of Islamic Culture for Muslim scholars leading to scholarship stagnation in all fields. Other factors doubtless contributed to this process but to a lesser extent. Nowadays any casual observer can see that apart from Jurisprudence, Islamic thought (*as opposed to Islam itself*) is inadequately prepared to deal with most aspects of modern science and technology. Therefore, a **"Paradigm Shift"** in Islamic Thought is a must in order to validate the maxim that Islam is a way of life that is adequate till the end of time.

Any thoughtful person in the twenty first century can relate to this presumed Islamic notion of the "End of History" and the victory of Islamic Culture as it turned out to be an illusion in the same way this individual can relate to the collapse of the Soviet Union and communism in late twentieth century representing the ultimate triumph of Western Civilization which was patently described by most Western scholars as the "End of History" while it turned out to be an illusion for Western democracies.

THE ERA OF RISING NON-ARAB ELEMENTS (248-656 H/861-1258)

To stake their claim to the Caliphate with their dismal Islamic credentials, the Omayyad partisans over-played the ethnic Arab card to unjustifiable extremes. It was only natural that in time the political pendulum would swing the other way. In the eastern most lands of Islam; Khorasan (خراسان) or modern day Afghanistan people were used to legitimate heirs to monarchs fighting to reclaim their usurped rights for millennia before Islam. Every Muslim reveres Mohammad (- محمد عليه الصلاة و السلام) and his descendants as a matter of fact. In these parts witnessing Omayyad persecution of the descendants of the Prophet (محمد - عليه الصلاة و السلام), it was easy for people, out of habit, to assume them being the legitimate heirs to the Caliphate. The descendants of both Ali (على إبن أبى طالب) and Abd Allah Ibn Al-Abbas (عبد الله إبن العباس) who both vehemently opposed the Omayyad dynasty found very fertile ground there to promote their cause. The political pendulum irrationally reached its maximum anti-Arab swing when they demanded of the leader of their supporters Abu Muslim Al-Khorasani (أبو مسلم الخراسانى) to rid the land of any Arabic speaking person if possible. Therefore, the Abbasid dynasty owed its existence to the ethnically non-Arab elements of the state. Consolidating their power, the Abbasids dispensed with the traditional rule of appointing military leaders from among the Arabs. By the time of Al-Mo'tasem (المعتصم), who was an impressive general in his own right, Arab soldiers were suspect and he, for the first time in Islamic history, heavily recruited among the non-Arab, non-Muslim subjects to fill the ranks of his army. He found his trophy in the resilient and war-like Turkic communities who gradually embraced Islam. A feat another formidable Muslim leader; Salah Al-Din (Saladin-صلاح الدين) put to practice approximately three centuries later to the sorrow, grief and eventual collapse of both their respective dynasties. These Turkic fighters were unqualifiedly masters of their craft against any and every enemy they encountered and scored unparalleled victories in the service of Islam in both cases. It did not take long for Al-Motawakel (المتوكل) after succeeding his father Al-Mo'tasem (المعتصم) to dislike the new militaristic cadres and tried hard to counteract their presence in Baghdad while they were taking his sons under their wings. It seems that rulers everywhere never learn the hard lessons of history prompting, in this case, Al-Motawakel (المتوكل) to make the same fatal mistake committed by his grandfather Haroun Al-Rashid (هارون الرشيد) when he divided the state among his sons and then changing the order of succession for the benefit of one over the others. However, in the case of Motawakel (المتوكل) it badly backfired when at the instigation of the Turkic generals his son Mohammad Al-Montaser (محمد المنتصر) colluded with them to assassinate his father to become the next Caliph; ironically for only six months as he died in inexplicable circumstances at the age of twenty-four. From that moment on in 248 H/862, the Turkic strong men dominated the Abbasid Caliphs and the whole state till its demise at the hands of the Mongols in 656/1258. Having killed one Caliph and installed their choice, during the reigns of the next twelve Abbasid Caliphs till 333 H/946, all state affairs were carried out, including the succession to the Caliphate, at the behest of the dominant cadre of Turkic personalities. Obviously the strong hand of the Caliph holding all provinces together was gone and no more. One by one the provinces became independent in all but allegiance to the office of the Caliphate. Only four out of these twelve Caliphs died in office with natural causes. With the overthrown of Al-Mustakfi (905-949/292-338H المستكفى) in 334/946, Baghdad itself was taken over by a succession of these once provincial leaders reducing the Abbasid Caliph to undignified miserable status without actually deposing him in deference to the very powerful symbolism of his position among Muslims. In turn, these groups fought each other for dominance and power changed hands back and forth between Shi'a and Sunni till the Mongol invasion. All in all, there were 37 Abbasid Caliph. The first eight are the ones that built, expanded and consolidated the empire followed by the twelve that were controlled by their associated strongmen when provinces started more or less ignoring the central government. The remaining seventeen were mostly humiliated undignified puppets of alien largely militaristic and practically autonomous states within the state. Many big names appeared during that period but two most formidable persons stand head and shoulder above all else. These are the Seljuk Alp Arsalan (1029-1072 ألب أرسلان) and the Kurd Salah Al-Din (Saladin 532-589 H/1138-1193 - صلاح الدين). The former decimated the mighty Byzantine army at Manzikert (منذكرت) in 463 H/1071 capturing the Emperor and sending the empire irretrievably into the slow spiral of collapse. This ignominious defeat sent shockwaves into Western Christendom prompting it to undertake the folly of the Crusades. The latter is well known in the West as the one who brought the Crusaders to heel at Hattin (حطين) in 583 H-1187 and recaptured Jerusalem later the same year. Military prowess and high culture went hand in hand during the illustrious first phase of the Abbasid dynasty ending with Al-Wathiq (الواثق) when intellectual life reached its zenith on all fronts. All schools of thought in Sunni and Shi'a Jurisprudence were firmly established, huge efforts in translating jewels of foreign cultures were undertaken and philosophical debates were raging. Poetry and prose attained their highest levels and unequal scholars were numerous. Scientific pursuits flourished as well. Then, Al-Mustakfi (المستكفى) banned all debates and persecuted Mo'tazalah (المعتزلة). He indulged in all the pleasures an awesome empire can offer and was assassinated while in a disgraceful state of drunkenness. The magnificent culture that was painstakingly erected over the years was not possible to easily destroy though. Astounding intellectual activities were still carried out in all fields during the rest of the Abbasid era nonetheless. However, they moved to the periphery in the provinces. Splendid names such as Ibn Al-Athir (إبن الأثير) d.630 H/1232), Ibn Al-Jozzi (إبن الجوزى) d.597 H/1200), Ibn Sina /Avicenna (إبن سينا) (370-428 H / 980-1037), Ibn Abd Al-Barr (إبن عبد البر) d.463 H/1070), Al-Isferayini (الإسفرايينى) d.471 H/1080), Al-Asha'ri (الأشعرى) d.330 H/934), Al-Baqlani (البقلينى) d.403

H/1012), **Al-Bukhari** (البخارى d.256 H/870), Al-Blazeri (البلاذرى d.279 H/892), Al-Bayrouni (البيرونى d.440 H/1049), Al-Juwaini (الجوينى – 419-478 H), Al-Thahabi (الذهبى d.748 H/1247), Fakhr Al-Razi (فخر الرازى d.606 H/1209), Al-Zamakhshari (الزمخشرى – 467-528 H / 1047-1143), Al-Shahrstani (الشهرستانى d.548 H/1153), Al-Tabari (الطبرى d.310 H/922), Al-Qadi Abd Al-Jabbar (القاضى عبد الجبار d.415 H/1024), Al-Ghazali (أبو حامد الغزالى) (448-505 H / 1058-1111), Al-Farabi (259-339 الفارابى H/870-950), Al-Maso'udi (المسعودى d.346 H/955), **Muslim** (مسلم d.261 H/875), Al-Waqedi (الواقدى d.207 H/822) thrived during these couple of centuries. This is but a tiny sample representing the intellectual fervent of Islamic thought at the time. It is to be noted that the revered collection of the tradition of the Prophet (محمد - عليه الصلاة و السلام) in *Sahih Al-Bukhari* (صحيح البخارى) & *Sahih Muslim* (صحيح مسلم) were accomplished during that period. The rigor, thoroughness, exactitude, precision, meticulousness, and accuracy observed in these processes far exceeds any that state-of-the art high technology research methodologies of the twenty first century can claim. As one can observe the illustrious list includes not only religious scholars and men of letters, but also some of the greatest thinkers, philosophers and scientists, humanity has produced. Nonetheless all of them began their careers studying religious Islamic topics and contributed much to them. Short biographies of a sample of these great individuals are given in the part discussing contributions of Muslim scholars to sciece.

MUSLIM SPAIN (92-892 H) (711-1031 & 1031-1492)

When the Muslim State at its beginning under the second Caliph Omar Ibn Al-Khattab (عمر إبن الخطاب) became a de facto international empire by wresting Egypt from the Romans and bringing North Africa under Islamic control, it looked essential to protect the riches of the Nile Valley in addition to the primary task of spreading Islam itself. The importance of this approach can be appreciated when one considers some of the names volunteering to join the expedition when Caliph Othman (عثمان إبن عفان) called it. They include among numerous others, Al-Hassan Ibn Ali (الحسن إبن على إبن أبى طالب), his brother al-Hussein

Ibn Ali (الحسين إبن على إبن أبى طالب), Abd Allah Ibn Al-Abbas (عبد الله إبن العباس), Abd Allah Ibn Omar (عبد الله إبن عمر إبن الخطاب), Abd Allah Ibn al-Zobair (عبد الله إبن الزبير), etc. All young and all beloved by Mohammad (محمد - عليه الصلاة و السلام) and very highly regarded among the companions (الصحابة). The fiercely independent Berber inhabitants of the region originally opposed that move but eventually embraced Islam and joined in the process of spreading its message. They culturally submitted to the more superior influence emanating from the East and quickly adopted Arabic as their language as did the Egyptians before them. The countless Roman losses in North Africa prompted them to send their legendary fleet to stop the Muslim advance and retrieve the lost territories. Muslims were new at seafaring not to mention sea battles. Nonetheless, they comprehensively routed the Romans marking the very first major sea battle in Islamic history that is known as "That Al-Sawari" (35 H/655 ذات الصوارى). Thus, Muslims waded their feet in the waters of the Atlantic in very short order and the Mediterranean was the "Roman Sea" no more. Muslims turned their attention to the land across the sea; the Iberian Peninsula where the Visigoths had their principalities. Between 711 and 716 the entire peninsula came under Muslim rule. Jewish and Christian inhabitants were left to their norms without religious molestation in accordance with the fundamental tents of Islam concerning freedom of religion and belief; a practice Muslims followed everywhere else they went. Although physically the Western reaches of the state were remote, culturally there was unbroken continuity. That did not escape the eyes of the backward Western Europeans in Spain and beyond. They had the advantage of learning securely attached to the privilege of freedom of religion; a rare concept even in today's world East and West. Westerners gathered every bit of information they could find in every field of human activity. Other points of contact between the Muslims and Europeans including the Crusading Latin Kingdoms in the midst of Muslims served the same European purpose of gaining knowledge. This process lasted in Spain for eight un-interrupted centuries. These are the *only* seeds Western Civilization sprouted from. In other words, what is now called ***Western Civilization is a natural by-product of the Islamic Civilization*** in every imagined possible way. When Muslims were finally forcibly expelled from Spain in the sixteenth century during the reigns of fanatic Christian monarchs, Spain relapsed into backwardness and is even currently lacking behind other advanced West and North European fellow states. Politically, Muslim Spain passed through three general phases. The first spanned the period till the collapse of the Omayyad dynasty in Damascus. The second phase commenced when Abd Al-Rahman Ibn Mo'aweyah Ibn Hisham Ibn Abd Al-Malik (عبد الرحمن إبن معاوية إبن هشام إبن عبد الملك) better know in history as "Abd Al-Rahman Al-Dakhil" (عبد الرحمن الداخل) escaped the Abbasid massacre of his Omayyad kin and fled West to eventually land in Spain. He gradually but persistently established an Omayyad enclave there between 138-170 H/756-788. Although forming independent entities is repeated numerous times within the frame of the universal Muslim state, the Spanish and later the Egyptian Fatimid events are unique in their declaring rival Caliphates in addition to their immense distinctive cultural contributions. The Spanish Omayyad Caliphate lasted till 422 H/1031 when it completely disintegrated under the assault of the military deposing the last ruler Hisham Ibn Mohammad Ibn Abd Al-Malek Ibn Abd Al-Rahman Al-Naser (هشام إبن محمد إبن عبد الملك إبن عبد الرحمن الناصر). The third phase covers the era when power was diffused among a succession of local dynasties known as "Tai'fas" (الطوائف) who could not always effectively resist Christian encroachment leading to interference from North Africa by Al-Moravids (المرابطون) (1085-1145) then Al-Mohads (الموحدون) (1147-1238). Despite the pathetic power of the kings of Tai'fas (ملوك الطوائف), brilliant cultural progress continued unabated till the end of Muslim Spain. That was in the year 1492 when finally, Queen Isabella felt free to commission Christopher Columbus to sail west to find another path to India avoiding all Muslim controlled lands where he inadvertently discovered the "New World". Muslim Spain's contributions to the advancement of Islamic thought are immeasurable in every conceivable vein from Jurisprudence to applied science. It produced persons such as Ibn Hazm (الإمام إبن حزم) (384-456 H / 994-1064), ibn Tofail (إبن طفيل)(502-581 H / 1110-1185), ibn Rushd/Averroes (إبن رشد الحفيد) (520-595 H / 1126-1198), ibn Al-Arabi (إبن العربى)(560-638 H / 1165-1240), Abu Al-Abbas Al-Mursi (أبو العباس المرسى d.1287), Ibn Bajjah (d.H/1138) and many others that etched their names firmly in human history.

Writings of some of them have been a must reading in European learning institutions for centuries. The great Christian theologian Thomas Aquinas who introduced rationalism into Christianity is a direct product of the Philosophical school of Ibn Rushd (إبن رشد الحفيد). So is the great Jewish Maimonides that is known as the "Second Moses" who resurrected the Hebrew language from obscurity on the general rules of the Arabic language since they are both derived from Aramaic and after a couple of millennia finally established the Jewish creed closely following concepts of Islam. Ibn Rushd's (إبن رشد الحفيد) work alone wholeheartedly adopted by the early founding fathers of the "Western Civilization" can refute the myth of its Judeo-Christian basis. The influence of Muslim Spain's scholars on Islamic jurisprudence is profound but typically along the lines of the established four prominent schools. However, one individual has had disproportionate impact; that is Imam ibn Hazm.

IMAM IBN HAZM (الإمام أبو محمد على إبن أحمد إبن سعيد إبن حزم) (384-456 H / 994-1064)

Ibn Hazm was born in 384 H / 994 in Cordova (قرطبة), Spain to a well to do family during the tumultuous years towards the end of the Omayyad dynasty in Spain. His father was an adviser/minister to the court until his arrest and consequent death because of the changing political regimes when Ibn Hazm was only 18 years old. He was loyal to the Omayyad cause and joined the fight in support of Al-Murtadha Abd Al-Rahman ibn Mohammad (المرتضى عبد الرحمن إبن محمد) who lost and ibn Hazm was imprisoned for six years as a consequence. With the rising fortunes of other Omayyad princes, he later served as adviser/minister to Abd Al-Rahman ibn Hisham and then Hisham ibn Mohammad who was overthrown by his military generals in 428 H and his death in 428 marked the end of the Omayyad rule of Spain and the beginning of the rule of the kings of twa'f (ملوك الطوائف). Ibn Hazm followed the typical career of previous scholars in excelling in memorizing the Qur'an, Sunnah (السنة) and the Arabic Language. He is known to have excelled in composing poetry but after some changes in his scholarly interests, he decided to concentrate on Fiqh (فقه). By that time, the field was well advanced and the famous four schools of Jurisprudence were accepted everywhere. The Maliki school (المذهب المالكى) was dominant in North Africa and Spain at the time and it was natural for ibn Hazm to study it. However, he was attracted by its criticism by Imam Al-Shafie' and switched to study the latter's approach. Additionally, he studied the works of Abu Hanifah and his disciples as well. Ibn Hazm's in depth understanding of all these approaches to Fiqh (فقه) gave him a commanding presence among his contemporary scholars. Unfortunately, he was harsh in talking down to them due to a serious illness that left him irritable and impatient as well as the founded suspicion that they speak ill of him to the rulers which eventually gave pretext to Al-Mu'tadhid ibn Abbad (المعتضد إبن عباد) of Saville to order the burning of all his books after ibn Hazm settled there. Ibn Hazm is unique among all Muslim scholars of Jurisprudence in that he promoted independent thinking refusing to follow any of the established schools but rather to judge strictly on what he personally finds in the Qur'an and Sunnah (السنة) regardless of which school reached the same conclusions. He is known to have been a "literalist" accepting nothing other than the Qur'an and Sunnah (السنة). His great mastery of Fiqh (فقه) and his impressive scholarship makes him one of the greatest scholars of Islamic Jurisprudence whose works are still important to the present day. He is also unique among Muslim scholars in elevating the four women highly praised by the Qur'an to the status of prophethood on the account of their communicating with angels. These are Mary (مريم – عليها السلام), Sarah (سارة), Moses's (موسى – عليه السلام) mother and Pharaoh's wife. He left a wealth of writings numbered in more than 400 volumes. His most famous books on Fiqh (فقه) is the encyclopedic Al-Mohalla b-Al-Athar (المحلى بالآثار) and Tawq Al-Hamamah (طوق الحمامة) as well as others and great poetry compositions. Ibn Haz died in 456 H / 1064.

FATIMID EGYPT (358-567 H / 969-1171)

The declining authority of the Abbasid Caliphate and its tenuous hold on the provinces encouraged the establishment of several principalities in North Africa that were for all practical purposes independent of Baghdad albeit pledging allegiance to the office of the Caliphate. It was logical for rebellious figures to go as far west as possible out of reach of the central government. Those who claimed loyalty to the Shi'a Imams and their supporters were beneficiaries of this trend. In Tunisia in 297 H/909 the partisans of the secretive Ismai'li (الطائفة الإسماعيلية) sect succeeded in overthrowing the local order and in the process establishing their own principality. They named their newly founded state after the daughter of the Prophet (محمد - عليه الصلاة و السلام) Fatimah (فاطمة) who is the ultimate mother of all the Shi'a Imams. The core idea of the Fatimid approach was to unite all Muslims under a just system as a negation to the Abbasids. After several attempts, they succeeded in invading Egypt in 358 H/969. In an unusual shift they moved their government to Egypt. Thanks to its location and agriculture wealth, Egypt has been the grand prize for all empires since ancient times. Having won Egypt, the Fatimid Al-Moi'zz (المعز) declared himself Caliph. After the Omayyad Spanish claim, this was the second time someone other than the Abbasid ruler claimed that hallowed title. However, it was a very serious matter this time around as the claim purportedly covered all Muslims delegitimizing the Abbasid ruler and system. The Fatimid state with major sea power had extensive commercial ties with the Far East and Europe generating more prosperity to enable territorial expansion at the expense of the Abbasid dynasty. However, bad management of the resources and poor governing resulted in famine and the eventual weakening of the state. Unfortunately to their discredit, the late Fatimid rulers colluded with the Latin Crusaders against their fellow Muslims. With internal political strife, the Abbasid Caliph could send an army to pacify the country of Egypt without trying to overthrow the Fatimid regime. In time, the leadership of this military force fell to the Kurd Yusuf Ibn Ayoub (يوسف إبن أيوب). Step by step, he chipped at the authority of his friend the Fatimid Caliph and in a master stroke peacefully abolished the Fatimid Caliphate altogether bringing Egypt back to the Sunni fold under the nominal moral authority of the Abbasid Caliph once more. For that reason, he was popularly given the title of "Salah Al-Din" (صلاح الدين); the legendary Saladin. The very first act Al-Moi'zz (المعز) carried out in his new land was to build the city of

Cairo as his capital. He also ordered built Al-Azhar (الأزهر الشريف) institution to function as a springboard for the spreading of Ismai'li doctrine. Ironically in time, it evolved as the greatest depositary of Sunni Jurisprudence in the entire Islamic world. There is no doubt that this magnificent place of learning has had an indelible mark on the development and evolution of Islamic thought throughout history. It produced and still produces countless high caliber scholars in all fields of Islamic studies. It would give the Fatimid dynasty enough pride to have founded this place of learning even if it contributed nothing else to the Islamic Civilization. But their dynastic era did produce towering scientists such as Ibn Al-Haytham (إبن الهيثم) the astronomer and many other noted personalities. It also did leave massive architecturally unique monuments.

THE OTTOMAN CENTURIES (699 – 1342 H / 1299-1924)

The savage Mongol assault on the lands of Islam collaborating for a while with Christendom to the West ushered the hapless Abbasid dynasty into the dustbin of history leaving the Muslim world more fragmented. The rise of the Turkic elements within the Abbasid state produced many superior military generals that led the fights and eventually defeated both Mongols and Crusaders relegating them forever to history books. This struggle sapped the energies of these fierce groups except for one that was on the sidelines of the continuous contest but stood at the cusp of making magnificent epoch in world history. When it moved swiftly to fill the vacuum, it became clear that the age of the "Ottomans" has arrived. Ottomans are unique in history as the most successful nomadic people to build an enduring empire. With their warlike character, they turned into Europe's nightmare. The glorious Ottoman victories against its Western neighbors left an everlasting impression on their collective conscience. The West might have rid itself of its Christian faith but it never forgot the Islamic nature of its mortal enemy. To this day Western intellectuals and common folks alike irrationally spew venomous attacks on Islam and the Muslims indiscriminately because of their seemingly unforgettable experiences with the Ottoman Empire. More surprisingly, the real foe to Ottoman imperial aggression was the Muslim East especially the formidable Shi'a Safavid (الدولة الصفوية) state which was actually a Sunni state but promoted Shi'a doctrines to oppose the Ottomans. It is crystal clear that the Ottoman period deeply influenced not only the Islamic thought process but also the Western Christian-based mindset. It is baffling for Westerners to notice or be told that Muslims on the other hand did not much think of the Ottoman epic clash with the West as an Islamic war against Christianity. For example, while one of the greatest Ottomans is known in the West as "Suleiman the Magnificent" (سليمان العظيم) for his numerous accomplishments, he is known among Muslims as "Suleiman the Law Giver" (900-974 سليمان القانوني H/1494-1566) because of his lasting influence and imprint on Islamic Jurisprudence; his fantastic conquests notwithstanding. It is noteworthy that since the earliest times of Islamic territorial expansion, non-Arab elements were swiftly incorporated into the community. Islamic history and culture is in a real sense written by non-Arabs. This is not surprising as the Arabs as ethnically defined almost always constituted a small minority among Muslims soon after the passing of the Prophet (محمد - عليه الصلاة و السلام). However, all great contributions to Islamic thought were nonetheless produced in the Arabic Language simply because it is the language of the Qur'an not that of the Arabs. That is until Mamluk reign of Egypt followed by Ottoman domination of the Muslim world for long centuries. With the Ottoman Caliphate imposed by brute force, all old scholarly arguments about the mandatory qualifications of persons claiming the Caliphate were gone since the consensus was that the Caliph had be at least an Arab. The major grievance of non-Turkic subjects of the Ottoman Empire is that it depleted all talent from the provinces moving it irretrievably to Istanbul and its environs resulting in stagnation and decline everywhere else. Ottoman rulers declined in the most part to switch state's administrative functions into Arabic limiting their use to what is barely necessary to the ritualistic practice of Islam. The coup-de-grace came when scholarly interpretations to keep up with new developments were banned by edict. Apart from individual brilliant contributions, Arabic backbone of Islamic thought declined sharply. That is not to underestimate contributions in other Islamic languages but for obvious reasons the entire Islamic cultural heritage was in the Arabic Language especially by non-Arabs. It is essential at this juncture to emphasize that Islam originated on a solid foundation of overwhelmingly Arabic nature but immediately with its triumph ethnicity became anathema to its principle of absolute human equality. In the final analysis, the Ottoman relentless march advanced both Islam and empire but was heavily tilted on the side of empire. That is consistent with the prevailing international milieu of the times. Decline of the imperial clout vis-à-vis its antagonist and the prevalent corruption of empire's late Sovereigns awakened Islamic fervor against both the Ottomans and the Western powers that vied to inherit its domains. Although by the early years of the twentieth century the Ottoman Empire was gone, antagonism and bitter hatred of Islam was deeply ingrained in Western psyche and consciousness. Ever since, recurring episodic spasms in the West-Islam relationship testify to that fact.

Western Pseudo-scholars view of Islam and Muslim Scholars

When one scans Western literature in Latin and most European languages since the West's awareness of and interest in Islam, one faces two main thrusts. The first is lack of empathy even by those who make a good living studying this subject. This attitude covers the entire range from out-right hatred and bigotry even if that implies forgery and falsehood at one end of the spectrum to misunderstanding and ignorance even if that implies denial of historical facts and events at the other. Either way, this is a deliberate pernicious process and a malicious attempt to denigrate Islam and the Muslims. In a general sense the "Encyclopedia Britannica" has been considered a depositary of knowledge for the Western reader and the embodiment of superior Western intellectuality for a long time. Even casual leafing through that ostensibly hallowed reference concerning Islam makes that point crystal clear showing in no uncertain terms the Western mindset and its prejudices. Most items are historically unappreciative with a somewhat better change of tone after the demise of British imperialism in the second half of the twentieth century. Noticeably, only items written by individuals with Arab and Muslim heritage show some resemblance of fairness. Ironically, it is mostly Western women that are reasonably objective and evenhanded when discussing Islam. That dovetails nicely with the fact that most recent Western converts to Islam are women as well; Western stereotype of Islamic suppression of women be damned. Nonetheless, these women are usually vilified. The case of the German orientalist "Sigrid Hunke" who was known to have asserted the major influence of Muslims over Western Civilization and values publishing a book titled "Allahs Sonne über dem Abendland" ("Allah's sun over the Occident") illustrates that point. She is condemned and promptly labled as a "Nazi" as if even if true that disqualifies the validity of her opinion on Islam! It is claimed that she was a member of a Nazi conceprcy bent on the "Islamization" of Europe and the destruction of Christianity!! The second is ravenous craving and insatiable appetite for acquisition of the latest Islamic scholars' accomplishments during the Golden Age and beyond. That should have been a laudable affair to expand human knowledge were it not for the conscious suppression of credit and acknowledgement that followed. Very few Westerners did not suffer from this malady or cured themselves of it when they knew better. By and large those who did converted to Islam afterwards. Since this book is mostly interested in science rather than politics, a description of the only contribution of Muslim scholars to Western Civilization, (through the preservation and transmission of Greek knowledge) in the opinion of the editors of that revered institution of the "Encyclopedia Britannica" in the section about education and the contributions of the uniquely great "Al-Khwarzmi" (الخوارزمى) is given as a glaring example of ignorance, bigotry and pseudo-scholarship. These self-appointed "experts" on Islam and its Civilization describe he transmission of Classical culture through Muslim channels as divided into seven basic types excluding any actual contribution by Muslim scholars since all they did was translate works directly from Greek into Arabic, translate works into Pahlavi, including Indian, Greek, Syriac, Hellenistic, Hebrew, and Zoroastrian materials (the works then being translated from Pahlavi into Arabic) and finally to translate works from Hindi into Pahlavi, then into Syriac, Hebrew, and Arabic. When original works were hard to claim Greek roots to since they were certainly works written by Muslim scholars from the 9th through the 11th century, they had to claim them borrowed from non-Muslim sources despite the fact that no line of transmission could be established though. The next type of works that they attributed to Muslim scholars were no more than summaries and commentaries of Greco-Persian materials. When works by Muslim scholars were found to be advances over pre-Islamic learning these editors had to emphasize without any proof that they (the works) might not have developed in Islam had there not been the stimulation from Hellenistic, Byzantine, Zoroastrian, and Hindu learning. Obviously the great "Al-Khwarzmi" (الخوارزمى) and his works, the subject of their discussion, could not be placed under any of these humble categories because of his tremendous and original contributions. Therefore, they created a novel category to include works that appear to have arisen from purely individual genius and national cultures. However, these works according to these "know it all" editors of the "Encyclopedia Britannica" would likely have developed independently of Islam's Classical heritage of learning!!!!! This presentation of the revered institution of the "Encyclopedia Britannica" speaks for itself and needs no comments.

ISLAMIC SCHOLARS SCIENTIFIC CONTRIBUTIONS

Ironically, the breakup over time of the Abbasid Empire into full fledged states and smaller principalities which led to the political fragmentation of Muslims created competing entities that in reality elevated the Islamic cultural life. Suddenly there were numerous patrons to support innovative ideas and ways of thinking to increase the prestige of presumed new leaders. It surprisingly also invigorated the fighting spirit among Muslims to defend the realm against Byzantium. The disintegration of the Spanish Caliphate did not stunt progress in culture and science although it eventually resulted in the expulsion of Islam altogether from the Iberian Peninsula. Pure and applied endeavors actually flourished under the new diverse patrons. In the end, paradoxically the ultimate beneficiary was the West prompting it to embark on developing its tools to launch the *"industrial revolution"* that reversed the roles. It is rather startling to notice that when Muslims were the masters of applied knowledge, they let Westerners freely delve into its minute details and use them to develop their own. The same can probably be said about all other ancient civilizations. At the present time, the West is ferociously busy erecting barriers to prevent others, particularly Muslims, from acquiring its technologies. Monopoly is the intended goal as modern technology facilitates dominance. However, this is against the laws of nature and seems in the end to breed resentment and antagonism without actually protecting Western perceived supremacy. Attributing the rise of Western Civilization to acquiring knowledge from the Islamic Civilization is not idle talk but rather a decidedly historical fact. One can draw a direct line between the invention of Algebra by Mohammad Ibn Musa Al-Khwarazmi (محمد إبن موسى الخوارزمي) (164-232 H / 781-850), mathematical astronomical calculations of Nasir Al-Din Al-Tusi (ناصر الدين الطوسى) (597-673H / 1201-1274), astronomical observations of Al-Hassan ibn Al-Haytham (الحسن إبن الهيثم) (354-440 H / 965-1039) and four centuries later Newton's (1642-1727) groundbreaking work. The same is true of the work of Jaber Ibn Hayan (721-815) in Chemistry, medical teachings of Ibn Sina (Avicenna- 370-428 إبن سينا H/980-1037), map constructions of Al-Idrisi (الإدريسى) in geography and navigation not to mention the basis of rational thinking established by Ibn Rushd (Averroe 595- إبن رشد H/1126-1198). Other examples in all fields are countless. To ilucidate and elaborate on this point, presenting brief biographies of some of the outstanding Muslim indviduals who contributed so much to human knowledge is therefore in order.

AL-KHWĀRIZMĪ-ALGORITMI (محمد بن موسى الخوارزمى)

It has been explained by many great personalities in several fields of human endeavors, especially science, that they may look like having attained the pinnacle of their disciplines only because they stand on the shoulders of many greater earlier individuals. Mohammad ibn Mūsā al-Khwārizmī (Algoritmi - محمد بن موسى الخوارزمى) is undoubtedly one of the broadest such shoulders. Humanity owes much to his work that represents an essential part of the foundation of human scientific and technical progress. Al-Khwarizmi (الخوارزمى) was born 164 H / 781 in the town of Khwarizm (thus his name) which is approximately the town of Khiva in modern day Uzbekistan. The early years of the second Hijrah century (the eighth century) witnessed the ultimate cultural and technical turn-around of the Muslim community from its humble and primitive beginnings in Arabia to its flourishing in Baghdad the newly established capital of the Abbasid dynasty. The scholarly and patron of scholars Caliph Al-Ma'moun (الخليفة المأمون) built "Dar Al-Hikmah (House of Wisdom – دار الحكمة) as a grand research and teaching institution. It served a purpose akin to modern day National Academies of the United States or the Royal Societies of the United Kingdom and other international similar institutions. Al-Khwarizmi (الخوارزمى) joined its ranks for most of his life. His most important contributions were in the fields of mathematics, astronomy, Geography and Cartography in addition to other minor works. While Al-Khwarizmi (الخوارزمى) worked and made his indelible mark in Baghdad, the characteristic Islamic notions of learning and freedom of movement carried his writings everywhere in the vast lands of Islam and crossed over to Europe. Four centuries later, the great translator of Arabic manuscripts into Latin "Gerard of Cremona (1114 - 1187)" and others took the fateful step of making his contributions available to the up and coming Western mathematicians. Al-Khwarizmi's (الخوارزمى) revolutionary conceptual addition of the number "Zero" resulted in the development of the "calculus of limits" (حساب النهايات) eventually culminating in Newton's formalizing differentiation and integration as essential components of mathematics and physics. These approaches enabled the mathematical description of natural phenomena which in turn allowed manipulation of the forces of nature. This is the underlying foundation of the "Industrial Revolution" and Western scientific and technological progress till the present day. The legacy of Al-Khwarizmi (الخوارزمى) is preserved for posterity in the terminology used by all school students as well as accomplished scientists in most fields. The term "Algorithm" is derived from his very own name after it was Latinized by the translators as "Algoritmi". His name is also the origin of the Spanish word "guarismo" and the Portuguese word "algarismo" both meaning "digit". The word "Algebra" is taken from the title of his seminal book "al-Kitab al-Mokhtasar fi Hisab al-Jabr wa al-Muqabala" (The Compendious Book of Calculation by Completion and Balancing – كتاب الجمع" (الكتاب المختصر فى حساب الجبر و المقابلة). Translating his book والتفريق بحساب الهند - Addition and Subtraction by the Method of Calculation of the Indians" into Latin as "Algoritmi de numero Indorum" introduced the Arabic numbers into European languages as well as the zero. His other major work "Sorat al-Ardh" (Shape of the Earth – صورة الأرض) was an elaboration and correction on Ptolemy's book "Geography" resulting in a map of the then known world. It is interesting to know that he calculated the circumference of the earth which means that the global nature of earth was perfectly known to at least the Muslims in the ninth century when the Church severely punished those not believing in the flatness of the earth according to its interpretation of the Bible till very late. He also compiled a set of astronomical tables. Puzzlingly, to most Westerners to the present time, even highly educated and cultured ones, all these undeniably very significant contributions to science and technology and the overall human knowledge were simply no more than copying the ancient Greeks after their works were translated into Arabic. It is mystifying why it is next to impossible for Westerners to credit Muslims with anything of value. It is understandable that human progress comes from the collective contributions of every civilization. Rising civilizations build on the achievements of the previous ones. Muslims never denied the great accomplishments of the Greeks that they absorbed and credited them for. Pondering recent scientific developments, one realizes that Americans for example are extremely proud of the success of the "Manhattan Project" that produced the A-Bomb in the 1940s while almost all the individuals involved to a person were not Americans. However, there is no disputing the fact that it was an American achievement nonetheless. Mohammad ibn Mūsā al-Khwārizmī (Algoritmi - محمد بن موسى الخوارزمى) died in Baghdad in 232 H / 850

AL-RĀZĪ-RHAZES (أبو بكر محمد ابن زكريا الرازي)

Al-Razi was born in 251 H / 854 in Al-Rayy which located outside modern day Tehran, Iran. **He is considered one of the greatest physicians that ever lived.** His early interest was in music but he later made contributions in Chemistry, Philosophy and Medicine. He is credited with the discovery of Sulfuric acid and Alcohol/methanol. He is most famous for being the director of the largest hospital in Baghdad; the capital of the Abbasid Empire during the reign of Caliph Al-Muktafi (المكتفى). When he arrived in Baghdad at the request of the Caliph, Al-Razi was charged with establishing that hospital. To choose its site, he had pieces of fresh meat placed in various areas of the city which he checked a few days later. The chosen site was the one with the least rotten piece of meat indicating the healthier and cleaner air. Even in the ninth century Muslim physicians were aware of benefits of sanitations. During his tenure at this hospital, Al-Razi established a special section for the treatment of the mentally ill where they were treated with respect and care. On their discharge, they were given a sum of money to help with their immediate needs until they can return to normal life within the community. He was also the first in history to describe "smallpox" and differentiate it from "measles" something that is not easy for trained physicians even at the present time. It is known that he did not charge his patients for his services. Al-Razi is considered the father of "Pediatrics" and his monograph on the subject contained 24 chapters covering illnesses of newborns, infants and older children. It was translated into Latin as "Practica Puerorum" and became a trusted reference for centuries. He also made very important contributions to neurology and neuroanatomy and was a pioneer in applied neuroanatomy. He stated that nerves had motor or sensory functions and described 7 cranial and 31 spinal cord nerves. Additionally, he classified the spinal nerves into 8 cervical, 12 thoracic, 5 lumbar, 3 sacral and 3 coccygeal nerves. He was a prolific writer publishing more than 224 books mostly about medicine. His book Al-Hawi (الحاوى) was considered the most significant medical reference in the medieval period where he extensively presented case studies. It was translated into Latin as "Liber Continents". Abū Bakr Muhammad ibn Zakariyyā al-Rāzī (Rhazes) died in 311 H / 925.

IBN SĪNĀ-AVICENNA (ابن الحسن إبن عبد الله إبن الحسين على أبو) (على إبن سينا)

Ibn Sina was born in 370 H / 980 in Afshana near Bukhara in modern day Uzbekistan to a high government official during a period of political turmoil due to the decline of the authority of the Abbasid Caliph in Baghdad and its replacement by that of local rulers and governors. Ibn Sina is known as Al-Sheikh Al-Rai's (الشيخ الرئيس) or the Grand Master due to his unparalleled and encyclopedic intellectual mastery of and contributions to so many fields. It is historically taken for granted that the very first step in a Muslim's learning process must be the acquaintance with the Qur'an as the central aspect of a Muslim's overall being. Therefore, all Islamic great intellectuals and thinkers of any renown began by memorizing the Qur'an and then branching into whatever field of interest they may prefer. In the typical manner of Islamic learning down the centuries till the modern era, Ibn Sina (إبن سينا) by the age of 10 has memorized the entire Qur'an. Also typically, he pursued the study of Islamic Jurisprudence for a few years as well. However, the field of medicine was his real calling that he dedicated his life to at the age of 16 and in two years of intensive self-teaching mastered the field in addition to other fields such as physics, metaphysics and philosophy, etc. becoming a well-respected local physician. While other physicians failed to properly treat the governor of Bukhara of his illness, Ibn Sina (إبن سينا) at the age of 17 cured him. His reward was the unlimited access to the governor's library that was considered among the best of its time. Ibn Sina (إبن سينا) left no book unread and acquired a vast amount of knowledge in many fields. That made him a close adviser to several rulers and patrons when he had to move around because of the changing political circumstances. In his travels, he met and collaborated with his contemporary the great Muslim scholar Abū Rayḥān Mohammad ibn Aḥmad Al-Bīrūni (أبو ريحان محمد إبن أحمد البيروني). Despite the constant interruptions in his life due to politics of the day, Ibn Sina (إبن سينا) wrote major works dedicated to some of his patrons. However, his move to Isfahan in modern day Iran offered him an uninterrupted 15 years of peace that enabled him to compose his masterpiece; the most famous "Al-Qanoun fi Al-Tibb" (القانون فى الطب)

or the "Canon of Medicine". This is an encyclopedia of medicine in 5 volumes containing over a million words. This reference book was used everywhere in the then known world to teach medicine for over 7 centuries. There are in existence approximately 87 translations of it in various European languages. It is considered to have laid out the infrastructure for modern medicine. While the Crusaders were ravaging the Muslim East in the 12th century, in the West the great translator of Arabic scientific works "Gerard of Cremona (1114 - 1187)" was busy translating this great Muslim's medical encyclopedia into Latin between 1150 and 1187 to help the West acquire the tools to reenter into the civilized world community. Ibn Sina (إبن سينا) wrote over 400 books of which about 240 have survived. In addition to the Canon, his other most influential book is "Kitab Al-Shifa' (كتاب الشفاء) or "The Book of Healing" which deals with science and philosophy and is meant to heal the ignorance of the soul. The book is divided into four parts: logic, natural sciences, mathematics (arithmetic, geometry, astronomy, and music), and metaphysics. It shows encyclopedic knowledge of works of ancient Greek philosophers, such as Aristotle, Hellenistic thinkers such as Ptolemy and Muslim scientists and philosophers such as Al-Kindi (Alkindus) (الكندى), Al-Farabi (Alfarabi) (الفارابى) as well as his contemporary Al-Bairuni (البيرونى). The influence of Ibn Sina's (إبن سينا) writings on Western learning and thinking in general is glaringly obvious. The most famous medieval Christian philosopher "Thomas Aquinas" adopted many of Ibn Sina's (إبن سينا) philosophical ideas (in addition to those of other Muslim thinkers) into his own work in his desperate attempt to introduce rationality into Christian theology. However, Renaissance Europe did not take kindly to Ibn Sina's (إبن سينا) works albeit taking full advantage of their benefits. While the Muslims during their hey-day offered every bit of knowledge they had to the West, one is disturbed to notice that modern Westerners grudgingly give credit to them if ever. And when they do, they make sure to claim against all evidence to the contrary that Muslims of that era did nothing more than transmit the ancient knowledge to Europe without contributing anything themselves!! Abū 'Alī al-Ḥusayn ibn 'Abd Allāh ibn al-Ḥasan ibn 'Alī ibn Sīnā (أبو على الحسين إبن عبد الله إبن الحسن إبن على إبن سينا) is without doubt one of the greatest personalities humanity has ever produced who as a physician did not even charge his patients for his services and left mankind with a wealth of useful knowledge. He died in Hamadan in modern day Iran in 427 H / 1037.

IBN AL-HAYTHAM-ALHAZEN (أبو علي الحسن بن الحسن بن الهيثم)

Abu Ali Al-Hassan ibn Al-Hassan ibn al-Haytham (Alhazen - أبو علي الحسن إبن الحسن إبن الهيثم) was born in Basra, Iraq in 354 H / 965. In the standard Islamic tradition, he began his learning memorizing and studying the Qur'an. This is another one of the broadest shoulders supporting the infrastructure of science and technology. He has made significant contributions to a wide range of scientific and technical fields. He is credited with establishing what became known as the "Scientific Method" based on experimental observations and deductions many centuries before the towering European researchers. He is known to have insisted in the long tradition of Islamic scholarship on referencing and giving credit to others for any arguments or statements he might have used in his work. Certain contributions of his in mathematics and optics are immortalized as "Alhazen's Problems". He always fearlessly rose to face any scientific or technical challenge to the point of claiming in one of his discussions that he could control the flow of the "River Nile" of Egypt to prevent floods and inundation of the valley as he learned that it flows from higher to lower elevations. That brought him an invitation by the Fatimid Caliph of Egypt to do just that. Clearly that was a big deal for the Egyptians and he was given a magnificent reception on arrival. Al-Hassan ibn Al-Haytham (الحسن إبن الهيثم) explored the various possibilities and chose a location for the proposed dam which is the same where now the Aswan dam stands more than ten centuries later. The great scientist realized that the available technical means would not permit him to achieve his goal and withdrew his offer. Unfortunately for him, the Fatimid Caliph at the time was none other than Al-Hakim bi Amr Allah (الحاكم بأمر الله) whose eccentricities were well known and who additionally demanded that Al-Hassan ibn Al-Haytham (الحسن إبن الهيثم) build him an observatory to use for astrological conclusions about his fortunes. The great astronomer did not succeed in satisfying the wishes of the Caliph and feared for his life; disappearing till the mysterious death of Al-Hakim bi Amr Allah (الحاكم بأمر الله). Al-Hassan ibn Al-Haytham (الحسن إبن الهيثم) left more than 237 books on a variety of subjects. Those on optics and mathematics were of much importance in their corresponding fields. His most famous writings included "Kitab Al-Manazir" (كتاب المناظر - Book of Optics) contained in seven volumes which was written between 1011 and 1021. This book became available to renaissance Europeans such as Roger Bacon, Leonardo da Vinci, Galileo Galilei, Rene Descartes, Johannes Kepler, etc. upon Gerard of Cremona's (1114 - 1187) translating it into Latin. That information influenced their ideas and made it possible for them to introduce their great contributions in the meantime unleashing the scientific challenges to the Church's teachings and doctrines approximately four centuries later. For instance, Kepler's theory of the retinal image was directly built on his conceptual framework. Another major contribution of Al-Hassan ibn Al-Haytham (الحسن إبن الهيثم) is his rejection of the age old theories propagated by Euclid, Ptolemy and Aristotle attributing vision to the eye emitting rays of light or receiving physical forms from an object. He established the fact that "from each point of every colored body, illuminated by any light, issue light and color along every straight line that can be drawn from that point". Additionally, he experimentally proved that light travels in straight lines. It is also suggested that Al-Hassan ibn Al-Haytham (الحسن إبن الهيثم) is the first to claim limits on the speed of light. Al-Hassan ibn Al-Haytham (الحسن إبن الهيثم) corrected many notions that were wrongly popular because of their Greek origins especially those significant errors of Ptolemy's about binocular vision. He wrote a book "Al-Shukuk ala Batlaymous" (Doubts Concerning Ptolemy – شكوك على بطليموس), (Corrections to Almagest – تصويبات على المجسطى) and (Resolution of Doubts Concerning Almagest – تحليل شكوك حول المجسطى). It is thus absurd for some Westerners to claim that he, among other great Muslim scholars, contributed nothing more than preserving and transmitting Greek knowledge to the West. In the number theory, he came up with the formula that every even perfect number can be described in the form $2^{n-1}(2^n-1)$ where 2^n-1 is a prime number. It was the great Swiss mathematician Leonhard Euler (1707 – 1783) who succeeded in proving that formula after close to eight centuries later. Another very significant mathematical contribution of Al-Hassan ibn Al-Haytham's (إبن الحسن الهيثم) is the establishment of the first principles of "integration" which is a fundamental step in the progress in solving equations describing natural phenomena. Interestingly, he wrote a book (نموذج الكون) on the "Configuration of the Universe" and another (درب التبانة) on the location of the "Milky Way" galaxy. As a dedicated Muslim, he wrote about the calculations to determine the direction of Makkah and the Qiblah (القبلة) necessary to performing the prayers. The appreciation of any Muslim personality especially those who influenced Western intellectual and scientific progress in a significant way can hardly be found in any Western literature till the end of the humanly disgraceful era of European imperialism in the second half of the twentieth century. Muslims were by no means an exception but they earned most abuse because they historically had the most profound influence on the Western Civilization. The Great Al-Hassan ibn Al-Haytham's (الحسن إبن الهيثم) contributions were only acknowledge by some Western scholars very late in the past century. His prominent status in the history of physics was described by the British historian of science H. J. J. Winter eloquently saying that after the death of Archimedes no really great physicist appeared until ibn Al-Haytham. He also said that if, therefore science historians confine their interest only to the history of physics, there is a long period of over twelve hundred years during which the Golden Age of Greece gave way to the era of Muslim Scholasticism, and the experimental spirit of the noblest physicist of Antiquity lived again in this great Arab Scholar from Basra" Abu Ali Al-Hassan ibn Al-Hassan ibn al-Haytham (Alhazen - أبو علي الحسن إبن الحسن إبن الهيثم). Nonetheless, Greek science had to be mentioned. ibn al-Haytham (Alhazen - أبو علي الحسن إبن الحسن إبن الهيثم) died in Cairo in 440 H / 1039.

Ibn Khaldun إبن خالدون

Abu Zayd Abd Al-Rahman ibn Mohammad ibn Khaldun Al-Hadhrami (أبو زيد عبد الرحمن إبن محمد إبن خلدون الحضرمى) is probably one of the best known Muslim personalities of old in the West because of his unparalleled genius and contributions to modern Western scholarship. In a rare and effusive tribute to his legacy, the Scottish philosopher Robert Flint (1838 – 1910) said of that as a theorist of history Ibn Khaldun (إبن خلدون) had no equal in any age or country until Vico (*the Italian philosopher Giambattista Vico 1668 – 1744*) appeared, more than three hundred years later. He also elevated his status over the greatest Greeks and Church fathers such as Plato, Aristotle, and Augustine who he described as not his peers while all others were unworthy of being even mentioned along with him. The great British historian "Arnold Toynbee" (1889 – 1975) has described Ibn Khaldun's (إبن خلدون) writings as a philosophy of history which is undoubtedly the greatest work of its kind that has ever yet been created by any mind in any time or place. Ibn Khaldun (إبن خلدون) was born to a prominent family that immigrated to Spain from Southern Arabia and settled in Seville during the early days of Islam. Members of the family held high offices in the Umayyad, Al-Moravid and Al-Mohad dynasties. When Seville fell to the Christians in 1248, the family emigrated to Tunisia where Ibn Khaldun (إبن خلدون) was born in Tunis in 732 H / 1332. As a young boy, he went through the standard Islamic routine of memorizing and studying interpretation of the Qur'an and Sunnah (السنة), Jurisprudence and the Arabic language. He distinguished himself by studying the works of the previous great Muslim scholars such as Averroes (إبن رشد), Avicenna (إبن سينا), Al-Razi (الرازى), Al-Tusi (الطوسى), etc and wrote summaries of several books of Averroes's (إبن رشد) by the age of 20. He lost both parents in the "Black Death" that hit Tunis in 1348 – 1349. Soon after this calamity, Ibn Khaldun (إبن خلدون) started his long career of public service in different places serving different rulers interrupted for a few years every now and then where was imprisoned or in seclusion concentrating on his scholastic endeavors. This service took him to Fez (فاس) in Morocco, Granada in Spain where he negotiated a peace treaty with the Christian ruler of Castile (قشتالة), Pedro the Cruel, Tlemcen, Bougie, Biskra - Algeria (تلمسان, بوجى, بسكرة) and back to Granada. In 1375, having tired of politics and intrigues, Ibn Khaldun (إبن خلدون) decided to retire in the domain of the tribe of Awlad Arif (أولاد عارف) in Western Algeria where he lived with his family for four years in Qal'at ibn Salamah (قلعة إبن سلامة) near the modern day town of Frenda. This is where he wrote his monumental work of Al-Muqaddimah (المقدمة) as an introduction to history of the world which he intended to write. Al-Muqaddimah (المقدمة) is an analytical study of the rise and fall of states and empires. It is universally considered the inaugural work of what is currently known as the discipline of "sociology". The underlying principle of the analysis is what he termed "Al-Asabiyyah" (Social Cohesion - العصبية). He introduced the concept of "cyclical history" where a state starts as a group of barbarians with strong social cohesion, evolves into a civilization with great achievements and falls when decadence takes over as other group of barbarians defeat it to begin another cycle. The originality of Al-Muqaddimah's (المقدمة) arguments are striking and could not be traced back to any Greek or Latin sources no matter how Western critics have tried. Al-Muqadimah (المقدمة) as the name implies was simply the introduction to the massive seven-books historical narrative of "Kitab al-I'bar wa Diwan al-MMubtada' wa al-Khabar fi ma'refat Ayam al-Arab wa al-A'jam wa al-Barbar wa man A'sarahum min zawi al-Sultan al-Akbar (كتاب العبر وديوان المبتدأ والخبر في معرفة أيام العرب والعجم والبربر ومن عاصرهم من ذوي السلطان الأكبر) which deals with the history of the Arabs, non-Arabs and the Berbers and their powerful contemporaries. The first book is Al-Muqadimah (المقدمة), books 2-5 deal with human history and books 6-7 narrate history of the Berbers and the Maghreb. It is very curious that when classifying the various human ethnic communities, Ibn Khaldun (إبن خلدون) *adamantly refused to include Northern Europeans among the human race due to their inconceivable savagery.* These are currently, generally speaking, the ones that racially look down on other non-white peoples claiming superior civility and social progress. Lacking the necessary references needed to complete his writing at Qal'at ibn Salamah (قلعة إبن سلامة), he returned to his hometown of Tunis in 1378 and enrolled in the service of its ruler. Finishing his massive work without dedicating it to the ruler did not endear him to the latter. On the pretext of going to the Hajj, Ibn Khaldun (إبن خلدون) left Tunis and did not continue his travel to Makkah but disembarked the ship in Alexandria and then to Cairo where he spent the remaining 24 years of his life in the service of the Mamlouk Sultans. He was intermittently appointed and dismissed as the grand Qadi of the Maliki school of thought. He was devastated when the ship carrying his wife and children sank off the coast of Alexandria drowning all of them. In 1401, he was forced to join the Mamlouk army dispatched to break the siege of Damascus organized by the murderous Mongol leader Tamerlane (تيمورلنك). However, the incompetent Mamlouk sultan turned around leaving Ibn Khaldun (إبن خلدون) trapped in the city. Tamerlane (تيمورلنك) on learning of the reputation of Ibn Khaldun (إبن خلدون) as knowledgeable about history and politics of North Africa demanded to see him. Ibn Khaldun (إبن خلدون) obliged and spent seven weeks in discussions with the despicable Mongol who apparently had hopes for conquering the whole world and was curious about the Maghreb which Ibn Khaldun (إبن خلدون) gave him an extensive report about but acquainted himself with all the resources of the Mongol army which he described in a report to his people in the Maghreb warning them. Eventually he was released to go back to Egypt. He spent the rest of his life teaching at the famed Al-Azhar University. Among his students was the future renowned Arab historian Taqi al-Din Abu al-Abbas Ahmad ibn 'Ali ibn 'Abd al-Qadir ibn Mohammad al-Maqrizi (تقى الدين أبو العباس أحمد إبن على إبن عبد القادر إبن محمد المقريزى). As always happens with the illustrious writings of Muslim scholars of the middle ages, Ibn Khaldun's (إبن خلدون) works particularly Al-Muqadimah (المقدمة) were translated into Latin and then into several European languages. The complete Arabic edition was published in Europe as late as 1858 subjecting it to extensive study. It is fascinating to learn that the present day American economist Arthur Laffer who developed the supply-side economic theory relating tax rates to the amount collected by government described by the famous "Laffer Curve" acknowledged his indebtedness to the ideas developed by Ibn Khaldun (إبن خلدون) that show that the more an economic activity is taxed, the less of it generated. It is quite remarkable that despite the political turmoil that he lived through, the constant interruption of his career and the personal tragedies he suffered, the ingenious Ibn Khaldun (إبن خلدون) was able to put all that aside and create such formidable intellectual works. For all his prodigious contributions, he deservedly occupies a place of honor in the history of civilization as he eternally lives in the collective memory of the Muslim nation. Ibn Khalun (أبو زيد عبد الرحمن إبن محمد إبن خلدون الحضرمى) died in Cairo in 808 H / 1406.

Ibn Rushd – Averroes (أبو الوليد محمد ابن احمد ابن رشد)

Abu Al-Walid Muhammad ibn Ahmad ibn Muhammad ibn Rushd أبو الوليد محمد ابن احمد ابن رشد was born in 520 H / 1126 in Córdoba (قرطبة). He is usually referred to as the "Grandson" to distinguish him from his grandfather since like the grandfather, he has the same name and illustrious career as Chief Judge. Like all great Muslim scholars before and after him, he memorized the Qur'an at an early age. He studied jurisprudence according to the Maliki school. He excelled in medical studies as well and his writings were used in all reputable European universities for several centuries. He interacted with his prominent contemporaries such as Ibn Bajjah (Avempace 1080 - 1138 إبن باجة), Ibn Tufayl (Avetophail 1105 - 1185 إبن طفيل) and Ibn Zuhr (إبن زهرAvenzoa 1072 - 1162). Ibn Rushd (Averroes – إبن رشد الحفيد) had a tumultuous career as a civil servant that landed him in prison when the political circumstances changed. Tributes to his legacy appear in naming the plant genus "Averrhoa" whose members include the starfruit and the bilimbi, the lunar crater "ibn Rushd" and the asteroid 8318 "Averroes". Ibn Rushd (Averroes – إبن رشد الحفيد) wrote at least 67 on a wide variety of topics ranging from theology to medicine. He also wrote commentaries on Aristotle's and Plato's works that were translated into Latin and Hebrew. Several of his philosophical opinions about God (الله - سبحانه و تعالى) and His attributes were not acceptable to the traditional mainstream Islamic scholarship and landed him in intellectual as well as physical trouble and contributed to overlooking his achievements for a very long time. Some of his major books are "The decisive treatise on the relationship between wisdom and jurisprudence" (فصل المقال في ما بين تهافت الحكمة والشريعة من إتصال), "The incoherence of the incoherence" (التهافت) in response to the attack on philosophy and in medicine "General Principles of Medicine – Colliget (Latin)" (الكليات فى الطب). It is ironic that ibn Rushd (Averroes – إبن رشد الحفيد) has been, till recently in the twentieth century, by far better known and appreciated in the West than among his fellow Muslims. The bizarre unrecognizability of his great accomplishments in the Muslim world over the centuries are due to conversion of several unusual circumstances. He lived and served under the relatively harsh religious intolerance of Al-Mohad (الموحدين) rulers who actually persecuted him for his opinions. Probably more importantly was the conservative wave emanating from the east that swept the entire Muslim world at the time. This wave firmly established the foundation of Sunni Islam and was spearheaded by the unequal scholarship of Imam Abu Hamid Al-Ghazali (450-505 H / 1058-1111 الإمام أبو حامد الغزالي) whose book "Tahafot Al-Falasefah" (Incoherence of the Philosophers – تهافت الفلاسفة) vehemently attacked inserting philosophical arguments into religion. On the other hand, his rational approach to religious discourse greatly influenced Jewish and Christian major thinkers. Among these in the Jewish tradition is the great Maimonides (موسى إبن ميمون 1135-1204) who is also known among Jews as the "Second Moses" for his central role in reviving Judaism and the Hebrew Language. This celebrated Jewish product of the Islamic Civilization is known to have dedicated 13 years of his life to studying ibn Rushd's (Averroes – إبن رشد الحفيد) works and then followed his methodologies to introduce rationalism into Jewish scholarship. Other prominent Jewish scholars followed suit and established what is historically known as "Jewish Averroism" that reached its zenith during the fourteenth century. It was mostly this group that very early on translated his books into Latin for the benefit of Western renaissance that blossomed in the following centuries. Ibn Rushd's (Averroes – إبن رشد الحفيد) influence has had immense impact on Christian scholars with far reaching consequences. Western Christian thinkers during the twelfth and thirteenth centuries learned about Ancient Greek philosophy only through the Latin translations of his commentaries on Aristotle's works to the extent that he was referred to in their writings simply as "The Commentator". They also collectively formed what is historically known as the "Latin Averroists" and Averroism commanded prominent position in Paris and Padua. That alarmed the Catholic Church which late in the thirteenth century issued a sequence of condemnations of his own work and his commentaries on Aristotle's. Western thinkers of the European renaissance continued to adopt his ideas till the sixteenth and seventeenth centuries nonetheless. The eminent British writer Geoffrey Chaucer (1343 -1400) mentions ibn Rushd (Averroes – إبن رشد الحفيد) as a major medical authority in Europe in his 1387 Prologue of "The Canterbury Tales". As with prominent Jewish thinkers, eminent

Christian ones studied and absorbed ibn Rushd's (Averroes – إبن رشد الحفيد) teachings. In particular, Thomas Aquinas who is considered one of the fathers of the Catholic Church employed his methodologies in rationalizing the Christian faith and doctrines claiming that Christianity and reason are compatible. However, Aquinas is known to have strongly objected to ibn Rushd's (Averroes – إبن رشد الحفيد) fundamental Islamic principle that all humans share the same intellect. Although deeply impressed and profoundly indebted to his intellectual contributions, renowned renaissance figures treated ibn Rushd's (Averroes – إبن رشد الحفيد) legacy with resentment more indicative of their envy and jealousy. The illustrious painter Raphael (Raffaello Sanzio da Urbino 1483 -1520) depicted him as a minor figure among Greek and European individuals in his 1501 fresco "The School of Athens" that is in the Vatican's Apostolic Palace. More revealing of this love-hate attitude is Dante Alighieri's (Durante degli Alighieri 1265 -1321) placing ibn Rushd (Averroes – إبن رشد الحفيد) in "limbo" in his epic poem of "The Devine Comedy". It is known that this poem is a blatantly plagiarized adaptation of the premise of the Qur'anic story of Isra' and Me'raj (الإسراء و المعراج). ibn Rushd (Averroes – إبن رشد الحفيد) died in Marrakesh in 595 H / 1198.

ORIGINS AND PROBLEMS OF ISLAMIC THOUGHT

Remarkably, Muslim scientists of old usually combined technical competence with religious scholarship as opposed to the domineering Western Church's antagonism to and suppression of scientific tendencies to explain physical phenomena. Unlike its Western counterpart since the dominance of Western Christianity, Islamic thought process was never confined to the realm of the holy but explored all aspects of existence physical as well as spiritual. Contributions to the hard sciences went hand in hand with jurisprudence and interpretation of the sacred texts. There was no conflict between those pursuing science and technology and those who explored the deeper meanings of religion. Almost all conflicts in Islamic history had political underpinnings. After its long interval of glorious days and in spite of the existence of numerous great thinkers, Islamic thought went into deep decline especially with the overwhelming successes of the up and coming militarist Ottoman Empire. This coincided with Western anti-religion enlightenment that replaced faith in a *Higher Power* with a newly acquired belief in the capabilities of science and technology to explain everything. In the enlightened Western mind, the notion of the existence of the Divine was dropped and the idea of God (الله - سبحانه و تعالى) was made obsolete. Advances and contributions by Muslim philosophers and scientists available through Spain, Sicily and other points of contact gave birth to Western Civilization. With the passage of few centuries the West made huge strides in science and technology to become physically dominant. At the present time, it is the Muslims who are trying to catch up with the West in these materialist fields. *Strictly speaking, in perfect hindsight there is no clear reason why Muslims did not continue their pursuit of material advances.* Obviously there is absolutely nothing in Islam that obstructs the acquisition of physical knowledge otherwise those glorious days would not have taken place. However, Muslims started on what seems to be a deliberate decline after furnishing the foundations of the industrial development that took place in the West instead. The concept of the purposefulness of the on-going creation process has been emphasized and elaborated on in previous books of this "Islam and the West" series. A rudimentary explanation as to why this decline had to happen was given. It suggested reaching a critical historical moment in human material development that mandated this decline in intellectual aptitude albeit in conjunction with remarkable improvements in militarist power. At that moment in time science stood at the cusp of inevitably developing evil applications of knowledge such as means of wholesale destruction. Alteration of God's (الله - سبحانه و تعالى) intentions for humanity appeared to be just over the horizon and a small step ahead. Muslims, perceived as the guardians of God's (الله - سبحانه و تعالى) final message to humanity, are implicitly disqualified to take responsibility for such developments. Hence, they had to start the march of decline. However, the situation has dramatically changed beginning with the twentieth century and its calamitous world wars and the demise of the Caliphate. Scholars began speaking about the "Decline of the West" coinciding with attempts at reviving the Muslim spirit in conjunction with other non-Western societies. As the

West brutally opposed this worldwide revival, it precipitated a valiant struggle against it among non-Westerners that resulted in eventually fanatics openly challenging Western hegemony and the world is engaged in the current turmoil of terrorism. It seems that Muslims are determined to reclaim their rightful position among the advanced nations regardless of the hurdles thrown their way by the West. What has changed? Evil applications are everywhere and that does not in itself re-qualify the Muslims. Knowledge cannot be unlearned and technological destructive means cannot be un-invented. There must be another more profound reason for the somewhat slow but sure re-emergence of the Muslims on world stage. At this juncture one has to concede the fact that the process of declining civilizations is a universal natural phenomenon that befalls all. Equally conceded is the fact that re-emergence of civilizations is a natural one as well albeit less universal. One can mention the many Middle Eastern and Asian civilizations of yester years to prove that point. China for example went through both processes and is currently becoming a world power again. Re-emerging does not in general keep the ancient foundations intact though. Theoretically speaking, ancient great personalities coming back would hardly recognize the basic ideals of the re-emerging daughter society. While Westerners claim to descend from the ancient Greek civilization, it is absurd to imagine Aristotle or Plato recognizing modern day Western democratic principles despite using the same expressions word for word. Even a very short historical time span would not alter this concept. Founding Fathers of the American Republic would have no clue about what is being promoted nowadays in their names as their legacy. However, the Islamic Civilization is the sole unique exception. While the great Muslim personalities of old would probably look down in disgust on the social and political practices of the vast majority of the current Muslims, they would perfectly understand the still intact foundations of Islam as established by themselves even when they are not followed to the letter. Therefore, by any definition the physical reemergence of the Islamic Civilization (expressed in its moral values) is an exceptional process. The question then becomes; why it declined in the first place if it is reemerging with necessarily the same ideals? And what purpose does this serve? Like any other natural phenomenon, Islamic thought passes through phases from the primitive to the sophisticated. However, after rapidly reaching its heydays Islamic thought stagnated. Uniquely among other cultures it sustained its fruits especially in jurisprudence for many centuries till the present time in the face of severe intellectual competition and ruthless political abuse. In the course of its evolution over the past fourteen centuries, there were periods of rational debates sometimes leading to social and political turmoil and others that were dominated by irrationality. But on the whole unlike other thought processes, it preserved its integrity and never lost contact with its roots in the heavenly revealed message of Islam. It is in this connection one finds its greatness and resilience. On the other hand, since the source and foundation of Islamic thought is the divinely revealed word which by definition lends itself to human interpretation, it is imperative to continuously update such interpretation in conjunction with progress of humanly acquired knowledge. This may not be apparent in the evolution of the fields of literature and other related liberal arts since the process is intimately related to the indigenous social norms and is mostly independent of any external influences. However, it can be seen unambiguously in the field of jurisprudence as it is next to impossible to find more adequate and comprehensive set of rules and regulations to organize human societies than what is included in Islamic Jurisprudence. Muslims cannot take credit for this fact however. They rightly understand that perfection of the "Shari'a" (الشريعة) is due to its divine origin rather than their own efforts to come up with a human-made system. The inadequacy of human-made systems has been illustrated numerous times in recorded history. The contemporary perfect example of this inadequacy is the turmoil in dealing with the countless problems Americans face in enacting laws according to the various conflicting interpretations of their constitution that they consider the best document human mind ever created. Paradoxically, while they reject the "Shari'a" (الشريعة) as a rigid system belonging to centuries past that does not evolve with new developments, they find it practically impossible to modify their merely 200 years old constitution to respond to modern issues. That is why the "Shari'a (الشريعة) has been continuously under constant assault mainly by Westerners down the centuries till the present. It is clear that Islamic thought reached its zenith in the collective rules and regulations of the "Shari'a" (الشريعة). Critics may find some inadequacies here and there in scholars' interpretations of Islamic Law that does not quite suit demands of modern life. This situation does not negate the fundamental aspects of the "Shari'a (الشريعة) but it forces contemporary Muslim scholars to take another look at interpretations of the ancient great works. However, by its nature this is a very limited domain of scholarship.

STAGNATION AND DECLINE OF ISLAMIC THOUGHT

Muslim scholars having developed a close to perfect body of rules and regulations that guided everyday life of the individual, Muslims could justifiably speak of Islam as a "Way of Life". However, Islam in the most universal sense is more than just how individuals deal with each other. Islam is not only the rules of the "Shari'a" but it is rather a comprehensive system explicitly expressed in the statements of the "Glorious Qur'an" that points to how everything in the created universe should begin, evolve, run and end according to the Creator's intentions. Only human self-centeredness confines that scope to what immediately affects humanity. Therefore, *perfecting the Shari'a while a necessary step in elucidating Islam, it is by no means sufficient to fulfill the task of truly appreciating God* (الله - سبحانه و تعالى). Comprehending the various statements of the Qur'an (as the complete representation of God's (الله - سبحانه و تعالى) intentions) is the ultimate task of Islam. Interestingly, jurisprudence occupies a minor portion of the text of the Qur'an. The major part elaborates the magnificence of God's (- الله سبحانه و تعالى) creation demonstrating His sovereignty and Oneness. Believing to have reached perfection, Muslim scholars, by and large, do not recognize the copious short comings in the continuous development of Islamic thought to contribute to and share in the formation of the future of mankind. They persist probably for reasons of self-preservation that scholarship (the all-encompassing Arabic word–علم) as undoubtedly extolled in the Qur'an and by the Prophet (محمد - عليه الصلاة و السلام) himself on numerous occasions is confined to religious studies prominent among them "fiqh" (فقه). Studying the Arabic language in its minute intricacies occupies the pinnacle among these studies for obvious reasons. Scholarship (علم) in their opinion is the realm of interpreting the Qur'an and the other issues of Islamic fiqh (فقه) in its broadest possible sense excluding all other activities. This is in contravention of the normal usage of the Arabic word (علم) in modern times to broadly mean science in all its endeavors including fiqh (فقه) and jurisprudence. According to their perspective, the domain of scholars (the all-encompassing Arabic word - علماء) is only wide enough for themselves. Critique of modern Muslim scholars blindly following the conclusions of the great ancient ones is not a new issue. Lamenting this attitude, the great Egyptian intellectual and philosopher Dr. Zaki Naguib Mahmoud (زكى نجيب محمود) published a sarcastic article in the 1940s under the title of (بيضة الفيل) "Elephant's Egg" and reproduced it in his 1983 book (قصة عقل) "Story of a Mind". It shows in Chapter 2, sections 4 and 5 the total disconnect between reality and potential debates of fictitious issues occupying Islamic scholars of the time. It is intriguing to recall that almost all of the great historical Islamic figures that enriched human knowledge in diverse fields such as mathematics, philosophy, medicine, astronomy, sociology, etc. began their careers studying fiqh (فقه) and jurisprudence and had some significant contributions in them. It is also fascinating to realize that the oldest and humanity's first systematically organized place of learning (what is currently called a university) is Al-Azhar (الأزهر الشريف) with history going back over a millennium. The first university in Europe was established by Muslims in the year 841 in the Italian city of Salerno. Muslims also established universities in Toledo, Seville and Granada in Spain. It is fascinating that Westerners to this day and age celebrating their great scholarly accomplishments (e.g. graduation ceremonies, conducting highly dignified academic procedures, presenting scholarly awards, etc.) make a point of being dressed in cap and gown which are the age-old Muslim garments of scholars. The highest possible level of scholarship a person can attain in order to be able to give opinion of his/her own in matters of fiqh (فقه) and jurisprudence is called (العالمية) which is exactly the title of the academic degree a person is awarded on completing all his/her studies at Al-Azhar (الأزهر الشريف) which is equivalent to doctorate. There is a major crisis in the current outlook of Islam and its interpretive scholarship leading to tragic consequences both physical and intellectual due to lack of that recognition of the shortcomings of Islamic thought. To avoid these consequences, Islamic thought has to undergo a "**Paradigm Shift**". Muslims have to sincerely, not just by paying lip service, equate scholarship (علم) with all efforts to seek knowledge whether in religious studies or material or social fields. A disconcerting and lamentable prevalent attitude among Muslims at the present time is the assumption that the expression "Islamic Thought" naturally means religious studies. Pursuits of science and technology are taken for granted to be different categories of enterprises that were mainly pioneered by the secular West. Muslims who are greatly interested in these endeavors turn their face to the West in either admiration or contempt. The term "Westernizer" was specifically coined to mostly denigrate those looking west for aspiration and material advancement. However, this attitude contradicts the historical evolution of the subject matter. Islamic thought from its inception covered both aspects albeit with religious studies dominating the field. The obvious reasons for this tilt have already been exhaustively explained. For this claim to be valid, it is inevitable to raise the question of *why after a very long and magnificent advance, Islamic thought came to a halt and actually declined particularly in the physical and applied sciences arena.* As mentioned in this undertaking there were no rivals to challenge Muslims' contributions and compete with their ideas for several centuries till the European enlightenment which owes its very existence and origins to Islamic approaches to learning. The work at hand contends that the *purposefulness of the on-going creation process* demands such decline and then after several centuries the renaissance of Islamic thought at the relevant time and space. A major contributor to the decline of Islamic thought is the natural phenomenon afflicting all historical great powers. That is the tendency after a period of civilized endeavors to favor territorial expansion and spreading global influence. When societies become overly militaristic it is inevitable that they decline and eventually fall. This is the lesson of history from the Ancient Egyptian, Persian and Roman empires to the French, British, Soviet empires and the imperial tendencies of the United States at the present time. None escape the verdict of nature. The rise of the Ottoman Empire marked the beginning of a practically exclusively proud militaristic outlook to the Islamic Civilization. The most basic reason was the adamant refusal of the new masters of Islam to adopt the language of the Qur'an as their primary tongue as did the Egyptians and the Persians before. This way they unintentionally practically cut themselves off from the long magnificent cultural and intellectual heritage of the civilization they became custodians of albeit holding fast and uncompromisingly to the religious aspects of it. That is not to say that the Ottomans represented an aberration. They existed during a period of empire building by nations east, west, north and south. They simply were the best at it. However, if militarization was the only factor in play, one would expect the civilization to vanish and for all practical purposes never come back as happened to old ones. That is not the case concerning the Islamic Civilization though. The thesis advanced in this work is that decline and the unmistakable signs of renaissance of Islamic thought follow what was called many times in this work and the previous work by the author the *purposefulness of the creation process aaccording the precepts of* "**The Grand Cosmological Divine Plan**".

Transition from Islamic to Western Thought Processes

After long centuries of coexistence on the Iberian Peninsula between Islam and Christianity that breathed life in the atrophied Western mind things took a violent and nasty turn. No sooner had Islam been expelled from Spain than the Western Church felt unrestrained and instituted the infamous Inquisition. It is instructive to learn that the institution of the Inquisition was never abolished and is still an integral part of the Church of Rome to this day albeit under a more benign name. Pope Benedict who resigned the papacy in 2013 is the last one to date to head that office before his election as pope. Because it was impossible to eradicate what Westerners conceptually absorbed over these many centuries due to Islamic tolerance, the Church doubled down on suppression of any liberal ideas. Free of any checks on its actions, tyranny of the Church had neither physical nor intellectual bounds. Heresy according to the Church was not confined to religious beliefs and practices but extended its reach to matters of science and technology. People who adopted heretical concepts were swiftly and severely punished. The sorry history of these days is well documented and there is no need to reiterate any of its examples at this juncture. However, its impact on scientific thinking is of particular interest to this endeavor as it eventually spawned atheism as an extreme intellectual reaction. In the realm of scientific pursuits, Polish/German mathematician/astronomer Nicolaus Copernicus (1473-1543) using what was learned from Muslim Astronomers centuries before formulated a new model of the universe that contradicted the one developed eighteen centuries earlier by the Greek Aristarchus of Samos as it placed the sun rather than the earth at the center. Publication of his book *"De revolutionibus orbium coelestium"* (*On the Revolutions of the Celestial Spheres*) ushered what is known in the history of science as the Copernican Revolution marking the beginning of Western methodical approach to science. That affronted the eternally held belief of the Church directly derived from its interpretation of the Bible. Death soon after publishing his book probably saved Copernicus from the wrath of the Church. However, his thoughts have had lasting impression on others. The German mathematician/astronomer Johannes Kepler (1571-1630) elaborated on the Copernican model to come up with three laws of planetary motion, based on his works *"Astronomia nova"*, *"Harmonices Mundi,* and *Epitome of Copernican Astronomy"*. These works provided impetus for Isaac Newton's theory of universal gravitation. Although his works followed the Copernican model he presented them in religious terms as God's (الله - سبحانه و تعالى) way of sustaining the universe. He linked his work to that of Aristotle. Aristotle's model was adopted by the Church from ancient times as basis for describing the universe. Kepler at some time in his career assisted the Western pioneering astronomer; the Danish Tycho Brahe (1546-1601) and was contemporary to the renowned Galileo Galilei (1564-161642) who is famous for his prosecution and condemnation by the Church which was only reversed in the second half of the twentieth century. Although Galileo was condemned by the Inquisition, he did not suffer much indignity as he was a personal friend to the pope. These towering personality at the dawn of Western scientific revolution lived during an age that saw egregious crimes committed by church authorities and the Inquisition, they did not pay a high price for their scientific opinions due to their relevant circumstances. That was not the fate of the Italian mathematician Giordano Bruno (1548-1600) who proposed a model for the cosmos based on the Copernican model that rejected the uniqueness of the solar system or its centrality. It was thus natural for him to question the fundamental precepts of Christianity based on his conclusions in the meantime undermining the fundamental dogmas of the religion itself as established by the Church's interpretation of the Bible. For this the Inquisition found him guilty, and he was burned at the stake in Rome's Campo de' Fiori in 1600. After his death, he gained considerable fame, being particularly celebrated by 19th- and early 20th-century commentators who regarded him as a martyr for science. His case is considered by historians as the most famous landmark in the history of science's struggle against the Western Church. It is remarkable to observe that historians attribute his thoughts and approaches to physical facts to the very strong influence works and opinions of the great Muslim Philosopher ibn Rushed (Averroes), who lived in the twelfth century (1126-1198), played in forming his thought process. It is thus clear that four centuries elapsing did not diminish the influence Islamic scholarship exerted on Western thought process and guided it in the right direction. It has to be mentioned here that unlike in the case of evolution of the Western Civilization, history rarely records any person being burned at the stake or simply condemned to death for their opinions within the context of the Islamic Civilization. There are countless number of Westerners males and females who met the same fate at the hands of the Inquisition which deployed the might of the Western Church to control the lives and minds of the Christian populations. These wretched souls did not even have opinions in most cases. Therefore, it was inevitable that a formidable social as well as intellectual backlash against Church's excesses would take place. Social and intellectual great personalities appeared and defied the Church calling for justice, fairness and curbing Church's interference in people's lives and even dismissing the need for Church's existence altogether. Seeds of atheism were sown at that stage. In the scientific realm two names stood head and shoulder above all else and have had an enormous influence on the evolution of modern science and institutionalizing atheist as a respectable label Western Scientists were and still are eager to acquire. These were the British Isaac Newton (1642-1727) in the Physical Sciences arena and Charles Darwin (1809-1882) in the Life Sciences arena. After the great achievements of science since the Enlightenment, it is easy to see how atheists could use it to dismantle the basic principles of Christianity one by one. One of the most celebrated results of the Pauline Christian faith by atheists that enabled them to succeed in their endeavor is the fact that Archbishop Ussher of Armagh, in his *"Annales Veteris et Novi Testamenti"* (1650-4), following the biblical narrative calculated that the date of Creation was 4004 BC, and this was widely accepted, especially in the Protestant world against all evidence to the contrary.

REFLECTIONS AND AFTERTHOUGHTS

It is universally accepted among modern scientists that Newton's Laws are the basis of physical sciences and *"Newtonian Mechanics"* is the cornerstone of the technical knowledge that helped launch the vaunted *"industrial revolution"* which eventually resulted in Western material superiority. These were the very first theories that brought acts of nature under precise mathematical formulation which made it possible to systematically manipulate the forces of nature in the service of humankind. Thus, Europe was stumbling around in ignorance when Muslim scientists were exploring the physical world. It took *four centuries* before Newton compiled and put together in clear form what his Muslim predecessors have accomplished. It is not a huge leap from their work to his though. It appears from this argument that four centuries constitute an amazingly long time when Muslim scholars had all the primary knowledge at their fingertips. They could have built on their own work to launch their own *"industrial revolution"* with fundamental consequences to humanity and its history. The inevitable question therefore is "***Why then didn't they do it?***"

The Grand Cosmological Divine Plan

INTRODUCTION

In the first book of the series "Islam and the West" titled "Why Do They Hate Us So Much?", it was forcefully argued that there is a purpose to the creation process which is still on-going. The assertion that nothing is random in the unfolding of the ongoing creation process but rather it proceeds precisely according to a divine plan has been systematically propounded and developed to its logical conclusions. The same thesis is used here to give a short answer to the vital question just posed. The detailed answer is given in the discussion of the relationship between Islam and science. Contrary to the humanity-centered idea advanced by the "Anthropic Principle", humanity does not represent that purpose but its existence is just another phase albeit a major one. Nothing in the history of the universe, let alone human history, has happened by accident. Certain cosmic events took place at the specific time they did to serve certain purpose. As for human history, certain persons/groups are associated with certain events for exactly the same purpose. In that book, that purpose was quoted from the Qur'an, which is considered in Islam as the literal word of God (الله - سبحانه و تعالى), in Surat Al-Zareyat (The Scattering Wind - الذاريات) Ayah 56 to be the worship of God (- الله سبحانه و تعالى) that is universally interpreted as getting to know and to acknowledge His sovereignty over all creation. The relevance of cosmic events to this approach is discussed in other sections of this book. On the other hand, the unfolding of human history is dealt with here. It is explicitly stated in the Qur'an in so many places that the first human being "Adam" (آدم - عليه السلام) somehow acquired all knowledge from God (الله - سبحانه و تعالى) Himself and afterwards passed it along to his immediate family. It is not important at this juncture to identify the physical environment where this process took place. However, it is clear in the Islamic tradition that when the first humans appeared on earth they had such knowledge and "Adam" (آدم - عليه السلام) narrated to his progeny the many fantastic privileges he and his wife "Eve" (حواء) had before their *departure* from Al-Jannah (الجنة - Paradise). That is to say that humanity started its existence on *earth* with full knowledge of everything there is to know. However, survival in the new environment had to obviously take precedence over contemplations of the universe and what it entails. Consequently, with the passing of the first several generations this knowledge was lost and humanity was reduced to a wild status fighting for survival. Step by step it evolved into communities that could build relatively decent lives for themselves distinguished from wild beasts. At that stage, the march of civilization and the instinctive longing to re-acquire the lost knowledge began. A captivating (albeit highly speculative) interjection at this point may be in order. The Qur'an states explicitly that "Adam" (آدم - عليه السلام) was given dominion over Al-Ardh (الأرض) as God's (الله - سبحانه و تعالى) legate where he originally lived in Al-Jannah (الجنة - Paradise). Hence, one can conclude that Al-Jannah (الجنة - Paradise) is located somewhere within this domain. This is consistent with having already established conclusively that Al-Ardh (الأرض) as used in the Qur'an and belonging to the Arabic Language is not what is conventionally referred to as the third planet from the sun in the solar system or sometimes affectionately as the "Blue Planet" or simply earth in the English Language. As intimated before, it could possibly be the Milky Way Galaxy or the entire three-dimensional universe. When he did not follow God's (الله - سبحانه و تعالى) explicit instructions, "Adam" (آدم - عليه السلام) did no longer fit in the blessed environment and had to *depart* to earth. There is no compelling reason to assume that earth was not already populated by a host of creatures that he had to compete with for survival. It is also plausible to suggest that he and his progeny afterwards inter-married with some highly developed species (Neanderthals maybe) fusing in the process both biological fundamental building blocks and creating a new genetic make-up leading to modern humans. This can crudely explain the undeniable genetic/DNA similarity between humans and other creatures and ultimately makes the process of evolution an immaterial argument for or against development of humanity. It can also interpret the affirmation in Surat Al-Tin (سورة التين) that at some point humans' status has been reduced to the "lowest of the low" (أسفل سافلين). Adopting this approach (and there is nothing in Islam that negates it regardless of its admittedly highly speculative nature) makes one puzzle over Muslims of all stripes thoughtlessly condemn any discussion of evolutionary science blindly following frivolous Christian arguments. Be that as it may, there is no question that the original humans under the new circumstances had to strive to survive and in the meantime lose their advanced knowledge concentrating instead on acquiring the primitive methods of survival. In essence, "Adam" (آدم - عليه السلام) departing his privileged existence in Al-Jannah (الجنة - Paradise) thanks to his failure to conform to God's (الله - سبحانه و تعالى) instructions had to fend for himself and his family on earth against the odds. The new conditions caused the inevitable quick deterioration in human status for him and consequently his progeny. Human history is simply the struggle to regain that lost status. This argument is introduced here because it is universally agreed that for example, Ancient Egyptians (as the oldest known builders of a civilization that left recognizable monuments) must have had very impressive knowledge about the cosmos or at least the mathematical relationships governing the solar system as is obvious from studying their monuments. With more is discovered every day about what they must have known beyond that. Many historians actually firmly believe that there were other advanced civilizations in existence before the Egyptian one and they should be searched for. The most important point emphasized here anyway is that humans must have had fantastic overall knowledge that was eventually lost and millennia later is being reacquired. Ancient civilizations established the foundation that propelled humanity into the modern age. When one civilization fell, another replaced it taking full advantage of the acquired knowledge so far. According to Islam, physical and intellectual advance was intimately coupled with warnings to stay the course in worshiping and getting to know God (الله - سبحانه و تعالى). That took the shape of *local messages* for local communities conveyed by *local prophets and messengers of God* (الله - سبحانه و تعالى) sent to every human community with no exception. This process continued until humanity reached mature stage where it was ready to receive the *universal* message of Islam. It was the burden of the Islamic Civilization to perfect the path to getting to know God (الله - سبحانه و تعالى) and acknowledge His sovereignty over all creation. It was also its burden to continue the physical evolution of humanity as God's (الله - سبحانه و تعالى) vicegerent on Al-Ardh (الأرض). It met both challenges with flying colors as demonstrated in its many acheivements in the span of few years and their following centuries. But it had to eventually succumb to nature after making its indelible mark on human history. The process involving the rise and decline of the Islamic thought and civilization taking place in the associated specific space and time is another testimony to the purposefulness of the creation process. It is shown that after reaching certain peak, Islamic Civilization *had to* decline to accommodate the physical limits of human knowledge at that time which contradicted the basic precepts of Islam. When human knowledge later on advanced to a degree consistent with Islamic fundamentals, a renaissance in Islamic thought began to take shape. This point of view is elaborated on in the following sections. Insightful investigations can easily relate to that process as an inevitable part of God's (الله - سبحانه و تعالى) plan for humanity which is explicitly detailed in the Qur'an.

The Mechanical Deterministic View of the Universe

There is general agreement that Western science transitioned from basic ideas about the physical universe espoused by great thinkers like Copernicus and Galileo (which were wholly based on the works of Muslim astronomers and mathematicians of several centuries earlier) to a more firmly established foundation with the work of Isaac Newton. Instead of Chaos, the world after Newton seemed to run according to fundamental laws expressed in stringent mathematical relationships for the first time. The transition marked an extremely important transformation in scientific pursuits. It was now possible to predict with close to certainty object's future behavior according to certain rules instead of simply observing it and recording only its current status.

From that instance on, humanity assumed that it brought nature under its control and could manipulate it for its benefit. That was a very radical departure in the Western mindset from numerous centuries of prevailing Western Christian Church intellectually tyrannical edicts describing the universe and how it runs. The most basic phenomenon that attracted Newton's attention was what every human being experienced since the beginning of their existence. It is gravity and it was the one thing that Isaac Newton tried to understand according to cause and effect rather than mythology or Christian doctrine. The result was his most famous three laws governing the motion of an inertial (an object at rest or moving with constant speed) system. According to these laws, gravitational attractive force is assumed to be inversely proportional to the square of the distance between two objects. Newton applied his assumptions to the then known motions of celestial bodies with special emphasis on members of the solar system. Astoundingly these simple relationships proved that the laws that governed gravitational interactions of objects on earth were the very same ones that controlled planetary movements. The spectacular success of these laws in explaining the celestial motion of the planets immortalized his name in science. However, this success did not obscure some contradictions apparent even to Newton himself. According to this approach, gravity acted *instantaneously* resulting in *action at a distance* which eventually demanded the creation of the esoteric medium of *aether* with its contradictory characteristics. Attraction among finite number of celestial bodies implied immediate collapse of the universe. This forced Newton to assume that the universe is infinite with infinite number of stars. Therefore, in accordance with this theory, *the universe was eternal with no beginning or end*. The resulting *static* universe required *absolute time and absolute space*. When problems one after another were solved applying Newton's laws, scientists took these *arbitrary* assumptions as the underpinnings of nature itself. Clear faults in this approach were ignored for over two centuries due to the sheer weight of Newton's name and nothing else. **What he actually established was a view of the universe as a precise mechanical system running according to explicit laws described in simple mathematical formulae**. This Newtonian description of the universe (which at the time was only represented by the solar system) was adopted by most scientists and philosophers. The natural consequence of this approach is that *the universe is a self-sustaining mechanical system. In other words, it is just "IS" and does not need any external higher power to run.* In short order that was considered the ultimate triumph of atheism over the Christian religion and its Western Church. Generations of great scientists and mathematicians adopted that attitude as learning and enlightenment became synonymous with atheism. In the enlightened Western mind of the couple of centuries that followed the establishment of the world according to the Newtonian theories, being a scientist necessarily meant being an atheist. This is so eloquently phrased by the great French mathematician "Pierre Simon Laplace" (1749-1824) responding to Napoleon's question as to why not mentioning the Creator in his seminal work "Celestial Mechanics" when he answered that he had no need for that hypothesis. Those who clung to the concept of God (الله - سبحانه و تعالى) according to Christianity (apparently including Newton himself) advanced the idea that since the universe runs just like a vast cosmic clock, it was God (الله - سبحانه و تعالى) who wound the spring and then let it run according the laws He determined without His interfering afterwards. That was a very flimsy description of the role of God (الله - سبحانه و تعالى) in the universe; *a disinterested God* (الله - سبحانه و تعالى). Additionally, what the Newtonian laws implied using the concepts of absolute time and space was inescapable *determinism*. The motion, purpose and fate of any object in the universe, regardless of its size, can in principle, albeit with hugely involved calculations, be precisely determined at any time; past present or future. Anything and everything in the whole universe is deterministic. In other words, the universe according to Newtonian viewpoint does not allow for human *free will* and its consequent accountability. Here was another more formidable blow to the concept of God (الله - سبحانه و تعالى), as described in the Bible, who rewards and punishes based on human actions. The conclusion of this very brief analysis of the history of science at that juncture of time is that when the West finally mastered what Muslim scientists developed several centuries before but never actually put together into an integrated system, the genius of Isaac Newton naturally appeared on the stage. Although only minor obvious steps were required for Muslim scientists and mathematicians to achieve what Newton magnificently did a couple of centuries later, they never finished the task. That was not accidental but rather a clear demonstration of the deliberate purposefulness of the on-going process of creation as emphatically established in this work. To reiterate, Muslims (perceived as the guardians of God's (الله - سبحانه و تعالى)) final message to humanity) are clearly disqualified as participants in, let alone being responsible for, this stage in human history. *Irrefutable scientific concepts of that era and their corresponding results were in contradiction with the primary foundations of Muslims' faith.* Therefore, it is logical that they had to decline and give way to others to fulfill that task in the on-going purposeful creation process. *The concept of God's (الله - سبحانه و تعالى) indifference to the universe and its affairs in addition to abrogation of free will are diametrically opposed to, inconsistent and incompatible with the very fabric of Islam.* Muslims could not, according to their divine message, participate in such ventures bearing in mind that these ventures are essential and necessary stages in human progress. However, it has to be absolutely clear that without Newton's laws and their shortcomings, human knowledge would not have moved into modern understanding of the universe despite acknowledging that this understanding is still incomplete. Newtonian description of the world was an inevitable step to get to modern physics. Newton's existence was thus essential and a minor part of the purposefulness of the divinely designed creation process.

Renaissance of Islamic Thought

If Islamic intellectual and technical decline was inevitable according to the advanced thesis of the purposefulness of the on-going creation process, *why is it inevitable for Muslims to reemerge with their moral values intact precisely with the dawning of the twentieth century?* The answer is again implied in the same thesis. With the twentieth century just getting underway, Physics was facing a daunting challenge in studying "black body" radiation according to well established Newtonian view of the world that proved its validity for close to three hundred years. The Newtonian approach has assumed the absolute nature of space and time succeeding in explaining most of the observable everyday phenomena. Most notably was the motion of the celestial bodies such as the planets in the solar system. The return of Haley's comet in 1758 as predicted by calculations (16 years after Haley's death) was celebrated as the ultimate triumph of this approach. Thanks to Newtonian Physics, the concept of a finite universe that was believed in by the ancients was abandoned in favor of an infinite universe.

In Newton's Wake

As explained before the natural next step to what Muslim scientists and mathematicians formulated was the creation of the physical/mathematical framework undertaken by Newton. However, this process resulted in a rigid mechanically static view of the cosmos. The inexorable corollary of this approach is a deterministic universe where given enough information about an object in space and time, its past and/or future behavior in space and time can be determined with exactness. This in a single stroke eliminates "free will". Newton and all his contemporaries understood these consequences very well and the picture of the universe running as a mechanical "clock" became a mantra among scientists and theologians alike. The whole thing is diametrically in contravention with the fundamental Islamic concept of "free will and judgment". That is not to say that Muslim scientists and mathematicians avoided taking their work to its natural conclusions in order not to transgress against the precepts of Islam. In principle, some of them would not care in the same vein that some poets plainly offended everything Islam advocated and are still considered among the greatest. But it is saying that they were not qualified within the divine plan to undertake that essential process in the path of human knowledge. In hindsight, knowing what misery and evil instruments created by the "*industrial revolution*" that took place in the West, one can easily see that Islam according to its basic principles stated over a millennium before would not be associated with such undertaking. It is nonetheless disingenuous to argue that Muslims could have advocated only the good applications. That is against human nature especially when one remembers that Muslims are not any different from any other humans where some Muslim rulers did not hesitate to kill and assassinate others to promote their interests. However, indiscriminate means of mass destruction is a whole different ballgame that is abhorrent to the norms of Islam as a divine message.

HEISENBERG, THE UNCERTAINTY PRINCIPLE AND ISLAMIC RENAISSANCE

As is previously explained, it was observed that Muslim scientists and technologists did not take the natural next step to establish the mechanical view of the universe that was put together by Isaac Newton several centuries later despite having developed and attained all the necessary knowledge to do so. It was also concluded that the essential reason was the deterministic nature of that view which negates *free will and accountability* in contravention of a fundamental precept of Islam. However, by the nineteenth century, Newtonian approach to describing the world became clearly scientifically inadequate despite its phenomenal technological success in launching the industrial revolution. The inadequacy became apparent as science moved away from strictly dealing with everyday observable facts to exploring physics of the micro world according to the atomistic view of matter. On the macro scale, the failure to accurately account for the motion of the planet Mercury was well known at the time of Newton but was considered a unique one that probably needed more study and better accurate measurements. The deterministic nature of Newtonian Mechanics meant in a nutshell that future events can be determined with absolute accuracy given enough information about their past. Scientists assuming the molecular nature of gases realized that the sheer number of randomly moving molecules makes it impossible to determine with certainty the future distribution of molecules within a gas. That is because gas molecules collide and continuously and randomly change their directions and momenta. Individual molecule's position in time is therefore impossible to determine. The statistical and probabilistic nature of the problem became obvious. That ushered the development of the new scientific discipline of "Thermodynamics". Taking a jigsaw puzzle as an example that is simple to understand, one can easily see that there are so many more ways for the jigsaw pieces to be disordered (put in any place) than ordered (put in the right place). Therefore, there are clearly far more states of disorder than state(s) of order in nature. According to this approach, any self-contained (closed) system under no external force should eventually go from the less probable state of order to the more probable state of disorder. Thus, thermodynamics most fundamental conclusion is that *"in the long run, a molecular system left to its own devices would inevitably become disordered"* This is the simplest way to put in words the *"second law of thermodynamics"* which is considered the bedrock of all physical sciences that cannot be violated. However, the entire universe and everything in it is made up of such systems. The rise of the discipline of "Thermodynamics" firmly established the *statistical* nature of the world based on the atomistic nature of matter. The leading figure in that field was the great Austrian scientist Ludwig Boltzmann (1844-1906). It is fascinating to know that although the idea of the atomistic nature of matter is an ancient one, the existence of the atom as the basic building block of matter was not proven until Einstein's work on the "Brownian Movement" in 1905. Ironically and tragically Boltzmann's suicide in 1906 was attributed to the scientific attacks he was subjected to while promoting his ideas. Poetic justice posthumously vindicated his approach to "Statistical Mechanics" rather than Newton's deterministic one as the norm in dealing with phenomena at the atomic level. During the second half of the nineteenth century, the renowned Austrian scientist Ernest Mach (1838-1916) uncompromisingly rejected the idea of the atomistic nature of matter. He used the power of his scientific standing to propagate the concept that *what cannot be seen does not exist*. Albert Einstein was one of his dedicated admirers albeit finally abandoning that position. As can be clearly seen, Mach's principle is the bedrock foundation of atheists' presumed scientific claims. However, by the close of the nineteenth century physics was in turmoil The crisis was known as the "ultraviolet catastrophe". According to the classical approach, black body radiation spectrum (the intensity versus frequency curve) should be linear as the intensity increases with frequency. However, while the relationship showed linearity at law frequencies as expected (the infrared end of the visible spectrum), it dropped at high frequencies (the ultraviolet end of the visible spectrum). There was no possible theoretical explanation for this behavior. Solution to this problem was mathematically suggested in 1900 by the great German scientist Max Planck where he tried to mathematically fit the experimental curve and came up with the assumption that the only possible explanation for this behavior was for energy to be emitted in discrete packets rather than continuously as was then believed. Energy was quantized as multiples of a minimum possible value (the quantum) proportional to frequency. The constant of proportionality is known as "Planck's constant". The absolute minimum amount of energy was named "quantum" which is that of the photon. It is equal to Planck's constant multiplied by the speed of light divided by the wavelength. Hence **"Quantum Physics"** was launched. That was the opening shot in what became one of the two most important approaches in Modern Physics. It was Einstein again in 1905 providing proof of quantization of light in his pioneering work on the "Photoelectric Effect". Although Einstein published his earth shattering great three papers (including the one on special relativity) in 1905, he remained unknown. It was Planck recognizing Einstein's work on the Photoelectric effect that proved his own quantum theory of energy who gave Einstein the first real scientific position. This was also the time when the "electron" was discovered. Experiments on electrons demonstrated its both wave and particle nature. In 1925 the Austrian Erwin Schrodinger formulated his most famous equation describing the motion of the wave that accompanied the electron in the process launching "Quantum Mechanics". It was independently arrived at by the great German physicist Werner Heisenberg as "Matrix Mechanics". Both approaches were reconciled by another great German scientist Max Born who in 1928 eventually made the startling proposal that what the solution to the equation (the wave function) represents is actually the probability distribution rather than the real value of the state (Position) of the electron. To explain the implications of the probabilistic nature of the quantum theory, Heisenberg advanced the *"Uncertainty Principle"* as a fundamental aspect of the micro world of physics. The wave/particle duality and hence the probabilistic nature of the wave function was extended to cover any object from the elementary particle to the universe itself. However, *the larger the object the less observed the effect*. A most profound implication of quantum mechanics is that **nothing in nature is deterministic anymore**. For example, object's existence at a specific point in space is not definite but is spread over the entire universe. Common everyday observations of objects represent only the extremely high probability of their state. To reconcile quantum mechanics and common experience, scientists developed what is known as the "Copenhagen" (where the physicist Niels Bohr established his institute) explanation of quantum mechanics. According to this interpretation, the wave function *"collapses"* when a measurement (observation) is made. In other words, the process of observation determines the final state of the object. Implicit in this interpretation is the assumption that an object exists in an infinite possible number of states; only a measurement (an observation) determines which one it finally exists in. Alternatively expressed, any object exists in all possible states simultaneously. Only an observation (measurement) collapses (destroy) the wave function forcing the object to go into a definite state with definite reality. Thus, the process of observation determines the final state of an object. Since the state of an object according to the uncertainty principle is undetermined until an observation is made coupled with the collapse of the wave function, *free will* is fundamentally restored. This was among the very first counter-intuitive arguments of quantum mechanics. Commenting on this argument, Schrodinger gave his famous example of "Schrodinger's cat" that exists inside an absolutely isolated box subject to certain implied quantum mechanical conditions that can lead to its death. With the arrangements of this mental experiment, the cat exists simultaneously in the states of "life" and "death" until it is observed as there is no way of determining its status until opening the box and ascertaining its condition. The issue of how an appearance of classicality might arise from a multiplicity of quantum events actually raises a number of deep questions concerning the way in which the quantum world relates to the classical one. Thus, it is interesting to explore whether or not (the appearance of) classicality comes about simply because large number of quanta are involved or from some other criteria. This is the issue of quantum reality that was never (and still is not) settled by physicists to this day. On the other frontier of science, in 1905 and then in 1916, Albert Einstein announced his **"Special and General Theories of Relativity"**. The concepts of absolute space and time were irrevocably shattered. Newtonian Physics was determined to be an excellent approximation of these theories in the observable world of everyday. The general theory merged Newton's laws of inertia and gravity into the concept of curved space-time. This in turn led to the concept of a dynamic universe. The universe is no longer inevitably infinite with no beginning as determined by Newtonian mechanics. Time is no longer absolute either while causality where cause and effect are interrelated is still preserved within the quantum realm albeit only probabilistically. Therefore, divine exclusion of Muslims from contributing to materially successful schemes that undermine divine mandates as elaborated in Islam became unwarranted. It is very interesting that for all practical purposes, Islamic renaissance movements started at the same time. It is the Muslims now who are striving to scientifically and technologically catch up with the West. Whether they succeed in their endeavor or not is not clear and only time will tell. It is not unreasonable to assume they will; bearing in mind that China, Russia, India, and others are rapidly closing the perceived gap. One has to remember that the West has had eight centuries of contact with the superior Islamic Civilization to effectively absorb its achievements.

Islam in Modern Times

The setting of the eighteenth century and the dawning of the nineteenth witnessed the unmistakable signs of a staggering Ottoman Empire that owed its survival only to Western Powers' political rivalries. It could not prevent their encroachment on its vast territories both in Europe and in the East. However, a distinctively Islamic and historically recurring phenomenon set in motion. That is, when the ruling elite fail to defend Islam, simple folks led by scholars shoulder that burden. Resistance to imperialist European movements sprang up all across Muslim lands even those not belonging to the Ottoman Empire as in the

Indian Subcontinent and South Asia. The fundamental continuity of Islamic Thought made it easy for revolutionary thinkers to move around agitating against the new scourge as well as the tyranny of the Ottoman Rulers. The early years of Islam witnessed Egypt as a haven for the persecuted. Changing its language to Arabic gave it a very special place within the worldwide Islamic community. For all practical purposes, Egypt enjoyed relative independence since the Abbasid regime. The establishment of Al-Azhar (الأزهر الشريف) elevated Cairo to become the heart of Islamic learning. Organizing and managing the pilgrimage became its responsibility till the establishment of the Kingdom of Saudi Arabia in the twentieth century. With the decline of the Ottoman Empire and the ascendance of the Mohammad Ali (محمد على) regime, its independence was obvious to all. Egypt tried twice to re-establish the Caliphate first with the demise of the Abbasid state at the hands of the Mongols and several centuries later with the collapse of the Ottoman Empire. Therefore, Cairo was the cultural center of Islam where thinkers from everywhere started agitating for an Islamic renaissance. The Pan-Islamist Persian Gamal Al-Din Al-Afghani (جمال الدين الأفغانى) (1254-1314 H / 1838-1897) moved to Egypt for his anti-imperialist efforts. He mentored and collaborated with leaders such as Imam Mohammad A'bdoh (الإمام محمد عبده) (1266-1323 H / 1849-1905), Abd Allah Al-Nadim (عبد الله النديم) (1261-1314 H / 1842-1896), Rashid Redha (رشيد رضا) (1282-1354 H / 1865-1935) and others. Abd Al-Rahman Al-Kawakebi (عبد الرحمن الكواكبى) (1271-1320 H / 1855-1902) produced his anti-tyranny polemics in Syria and travelled widely in the Arabic speaking region propagating his ideas. The Pakistani thinker Abu Al-Aa'la Al-Mawdudi (1321-1399 H/1903-1979 أبو الأعلى المودودى) exchanged ideas with Sayed Qutb (1906-1966 سيد قطب) the famous theorizer of the Muslim Brotherhood. Some great Islamic literature has been produced during that period attempting at modernization. It must have been noticeable that within the text, examples of issues pertaining to renaissance of Islamic thought are repeatedly associated with Egypt. That is not an accident since during the nineteenth and early twentieth centuries, Cairo was the center of that awakening and usually thinkers from all over the Islamic world travelled to it. Being where the fountain of Islamic thought, Al-Azhar has been continuously in action for a millennium, Cairo was a natural focus of this phenomenon. A locally flourishing printing press and publishing establishments added to its centrality as well. However, slowly but surely these efforts concentrated on political resistance. With the collapse of the Ottoman Caliphate that consequently blurred these groups' typically Islamic character. Futile attempts to resurrect the Caliphate gave birth to the "Muslim Brotherhood" movement in Egypt and spread its influence internationally with far reaching consequences into the future. With the disappearance of the Caliphate, local nationalism replaced the Pan-Islamic outlook of these efforts but could never extinguish its flame. This era saw the division of the universal Muslim land into numerous supposedly nation-states with their governments usually co-opting the learned Islamic establishments to serve their narrow purposes. One deplorable side effect of this process is the changing of the leadership vanguard of Islamic movements from the well versed scholars into the more fanatic far less competent elements. Additionally, it caused noticeable deterioration in how the public viewed Islamic scholars and scholarship. The well deserved historic hallo of respect enjoyed by such personalities vanished and is unfortunately mostly replaced by derision. The long lasting stagnation of Islamic thought caused a detrimental split in its products. Unqualified simpletons got into the act of advancing hare-brained ideas to attract a following. The arena of Islamic thought became an unguarded ground open to all sorts of distortions from within for the first time in history rather than the clearly defined and easy to refute external attacks. Valiant undertakings to revive the heritage of Islamic thought are carried out by several individuals in many places all over the Islamic World but in a manner that is more backward than forward looking adhering to re-interpreting old works rather than creating innovative new approaches. All in all, the most alarming and most detrimental to Islam is the feeling harbored by sincere maybe technically highly educated Muslims albeit with only rudimentary knowledge about the conceptual principles of Islam that they can delve into expounding on a garden variety of Islamic issues. It is not uncommon nowadays to encounter individuals who have an idea about one fact having no qualms about creating an elaborate baseless scheme to come to a pre-determined conclusion to actually defend Islam in their opinions. The situation is most-dire in the attempt to reconcile Islam with modern life in general with particular emphasis on science and technology. Currently Islamic thought is either re-interpretation of the great old works or pseudo-scholarship. Very little precious contributions can be found to augment the treasures of old. However, despite Western claims to the contrary, Islamic thought survives and can easily meet any intellectual challenges. The fundamental missing link is the present deliberate divorce between mundane aspects of scientific and technical research and scholarship in Islamic studies. This is un-natural in light of the fact that the greatest of the great ancient Muslim scholars contributing to science and technology cut their teeth originally on studying Jurisprudence and other religious issues. It is obvious that the basic tenets of Islam remained intact even with decline of its civilization. Assuming Muslims thanks to their presumed renaissance will succeed in reclaiming their scientifically and technologically distinguished place, the basic question then becomes *"Is there anything in Islam that is in direct contradiction with the findings of science?"* This is the crux of this undertaking and is discussed in detail in the later part of the book.

Islam and Science

INTRODUCTION

Atheists formulate and advance their case building on a foundation of certain arguments presumably derived from generally accepted scientific findings. The strength of these arguments is wholly dependent on that while these findings are naturally subject to modifications with the acquisition of more knowledge, the underlying principles remain valid with time. However, it has to be conceded that even these principles are based on fundamental assumptions that conform to human logic and observations rather than absolute facts. This is what modern science is all about; going from assumptions to conclusions until attaining satisfactory descriptions of nature. Science explores observations and mathematical possibilities to describe nature at its minute details rather than create it. Thus, humans are merely by-standards, as opposed to principals, in the most fundamental cosmological sense. Therefore, it behooves atheists to accept the supposedly starkly undeniable truth that their accounts are also founded on an infrastructure of carefully selected assumptions that are not necessarily self-evident or for that matter uniquely and unquestionably valid. Tragically but unsurprisingly, the potency of scientific findings crushed the Jewish and Christian religious faith of the thinking Western individual leading to modern day brand of atheism. It is bewildering that Islam is lumped in that atheistic triumph over faith while it was most determinedly not part of that Western drama. This current endeavor and countless others repeatedly do make the essential point that Islam represents a distinctly different case. Islam has its own unique approach to the process of human acquisition of knowledge in pursuit of getting to know and appreciate

God (الله - سبحانه و تعالى) the Creator and Sustainer of the universe. Its divinely revealed book, the Qur'an, is believed to contain all knowledge about everything and humans are fervently encouraged to explore its facts. In Islam there are no prohibitions on ideas and contemplations of any sorts. Muslims are explicitly told in their tradition that even reaching the clearly wrong conclusions about any issue deserves a reward if only for the act of simply trying. It goes without saying that the well-established Islamic approach and tradition precedes modern day atheism by many centuries. That should mean that the burden of proof of legitimacy falls on atheism not Islam. Islam is absolutely not bound by Judaism's and Christianity's failure vis-à-vis atheism. While atheism starts from well-defined assumptions to reach conclusions, Islam starts from divinely revealed absolute facts urging its adherents to seek their ultimate reality. Within the Islamic tradition it is implied that the sought after understanding of these realities is time and knowledge-level dependent. Since the Qur'anic revelation is not accidental as far as space and time are concerned, understanding of its facts is also time and knowledge-level dependent such that its contents are valid till the end of time and everywhere in the universe.

Confronting Atheism's Basic Thought Infrastructure

It should be transparent that discussion in this part of the book is confined to the Islamic view point with regard to atheism. Religion in this respect must be understood to be Islam where Jewish and Christian narratives are irrelevant even if they at times coincide with the Islamic one which happens occasionally. Having set the parameters of discussion thus far, there are certain concepts that are taken up by atheists in their objections to the religious notion of faith because these concepts cannot be verified by direct observation in human daily experience. It would be fruitful to explore how valid these objections are since the corresponding concepts represent essential integral parts of the belief system. It is also important to keep in mind that most of these objections are based on assumptions not proven facts and they arise late in human history against ancient firmly established rules. Nonetheless, exploring their validity should be a worthwhile undertaking in addition to the main thrust of determining whether or not Qur'anic statements of fact can be reconciled with the at the present time irrefutable findings of modern science. However, before embarking on this effort it is advisable to remind oneself that the sacred sayings within the context of Islam (the Qur'an and the Sunnah) are only understood by the rules of the Arabic Language and nothing else; a notion that is repeatedly ignored or at best overlooked by Westerners specializing in studying Islam. These rules are not invented by Muslim scholars but rather existed in many forms long before Mohammed (محمد - عليه الصلاة و السلام) announced receiving the divine message. It is historically very well known that ancient Arabic poetry was prominently celebrated all-round the year in various parts of Arabia during designated commercial exchanges culminating at Souq O'kaz (سوق عكاظ); the most famous market coinciding with the pre-Islamic Hajj season. Based on these rules and their understanding by prominent Arabs especially those at Makkah, the Qur'anic linguistic structure and Arabic authenticity were tested time and again against poetry as soon as Mohammed (محمد - عليه الصلاة و السلام) had uttered any Ayahs of the Qur'an. Qur'anic language as well as meaning and narrative were immediately and continually challenged by Islam's antagonists to promote the claim that the Qur'an is merely composed by Mohammed (محمد - عليه الصلاة و السلام) in the known Arabic tradition. The narrative is challenged as a forgery expropriated and adapted from the Jewish and Christian traditions. The seriousness of the authenticity of the Qur'an was not to be left to chance or the common sense of humans though. The Qur'an itself as the literal word of God (الله - سبحانه و تعالى) took on and directly refuted these claims. It countered with a challenge to anyone or any group till the end of time to come up with even a similar single Ayah. Being unable to meet this outstanding Qur'anic challenge and the declining mastery of the Arabic language with time, the linguistic challenge had been quashed quickly and permanently. The forgery claim had been similarly shredded to pieces countless times. However, it is curious that every now and then some ignorant bigots repeat the same claim (appallingly sometimes word for word) even at the present time as if they finally and mortally undermined Islam's foundation. This forgery claim has been briefly dealt with in another part of this book. However, the linguistic Qur'anic inadequacy is important for the main thrust of the current undertaking and is discussed in detail.

LINGUISTIC SCRUTINY OF THE QUR'AN

Quraysh (قريش) was where poets from everywhere gathered annually during the Hajj season to compete supported by their individual sponsors and tribes. Having a winner was unequaled honor bestowed on the tribe. Most prominent persons among Quraysh (قريش) in addition to their wealth were well versed in poetry and the Arabic language in general thanks to their advantageous situation. All of them either composed it or critiqued it. No historical record written or oral denied their status as the best of the best in their knowledge of the Arabic language in all its dialects. Some of the most ardent antagonists of Mohammad (محمد - عليه الصلاة و السلام) and Islam from Quraysh (قريش) gave their verdict on the veracity of the Qur'an. These persons collectively constitute the pinnacle of eloquence and knowledge of the Arabic language which reached its perfection in poetry of the time. At the head of this group were two who examined the Qur'an face-to-face with Mohammad (محمد - عليه الصلاة و السلام) and returned to the rest with a clear opinion of the Qur'an. They were Al-Walid Ibn Al-Moghirah (الوليد إبن المغيرة) and Otbah ibn Rabi'ah (عتبة إبن ربيعة). Al-Walid ibn Al-Moghirah (الوليد إبن المغيرة) being at the zenith of wealth and prominence among his people, early on denounced the idea that God (سبحانه و تعالى) would choose Mohammad (محمد - عليه الصلاة و السلام) instead of himself as His messenger to humanity. Around the third year since Mohammad (محمد - عليه الصلاة و السلام) claimed to have received revelation when despite every enticement to Mohammad (محمد - عليه الصلاة و السلام) and his clan failed to dissuade him and contain the spread of Islam especially with the approaching Hajj season, Otbah Ibn Rabi'ah (عتبة إبن ربيعة) suggested that he should personally talk to Mohammad (محمد - عليه الصلاة و السلام) to put an end to the confusion. He sought, found Mohammad (محمد - عليه الصلاة و السلام) and explained the harm his message of Islam was doing to Quraysh (قريش) among the Arabs offering in the meantime every conceivable incentive to desist. Mohammad (محمد - عليه الصلاة و السلام) listened and when Otbah ibn Rabi'ah (عتبة إبن ربيعة) was done talking, politely asked to respond. He no more than recited the beginning of Surat Fussilat (فصّلت) which is the seventy second Surat in revelation and the forty first in the Mus-haf. The Ayah (آية) describes the Qur'an as the "lucidly distinct" heavenly revelation for those who would give it heed. Otbah ibn Rabi'ah (عتبة إبن ربيعة) listened in awe and then left without uttering a word. On rejoining the others, he seemed impressed with what Mohammad (محمد - عليه الصلاة و السلام) had to say and that was not acceptable to them. Ibn Hisham (إبن هشام) (the earliest historians of Islam) narrates his comment as never hearing anything like what Mohammad (محمد - عليه الصلاة و السلام) has just recited to him. He added that it is neither poetry nor magic nor is it priestly utterance. He suggested to the heads of Quraysh (قريش) to leave Mohammad (محمد - عليه الصلاة و السلام) alone since his message may lead to great developments among the Arabs. If he prevails then it would be their win but on the other hand, if the Arabs did not like his message they would suppress him with no shame to them. They angrily dismissed Otbah ibn Rabi'ah's (عتبة إبن ربيعة) opinion as having fallen under Mohammad's (محمد - عليه الصلاة و السلام) magical spell. Arabic poetry gave the Arabic speaking people a magnificent medium for their talents and they unquestionably enjoyed that fact producing beautiful immortal compositions. However, with the revelation of the Qur'an the field has been vastly enlarged beyond calculation. Therefore, while every Arab of that era could feel the power of the Qur'anic Arabic, not very many could master the interpretation of that language. A unique category of individuals who dedicated their energies and lives to absorbing and interpreting the Qur'an began forming. The lightning speed with which Islam reached non-Arab territories east and west added huge numbers of non-Arabic speaking populations to the Islamic realm. It is amazing that in an unparalleled unique historical turn of events these populations mastered the language of the Qur'an (no longer the language of the Arabs) in less than a couple of generations and produced non-Arab Muslim scholars who quickly became the majority. It should be obvious that while the mastery of the Arabic language among common people declined with each passing generation, it immeasurably improved among the scholars' ranks. Nonetheless, centuries of interpreting the Qur'an showed that the meaning of some common everyday phrases and words used by people are less expressive than what the Qur'an conveys.

DRAWBACKS OF THE LINGUISTIC INTERPRETATION OF THE QUR'AN AT PRESENT

As is obviously clear, mastering the Arabic Language is a prerequisite for an endeavor to interpret the Qur'an. This is self-evident in case of dealing with issues of jurisprudence, rituals, history and promoting morals and good deeds. However, there are caveats attached to this rule when it comes to issues describing physical phenomena. It has been emphasized many times in this book that a paradigm-shift in the overall Islamic thought processes; particularly in understanding the terminology and lexicon used in the Qur'an is urgently needed. The pitfalls associated with interpretation of such Qur'anic narratives by religious scholars with very little if any physical sciences competence is quite apparent and does not need much elaboration. Efforts by the great Sheikh Mohammad Metwaly Al-Sha'rawy (محمد متولى الشعراوى) were mentioned in passing as an example and were elaborated on in the frequently mentioned previous works by the author. Sadly, the more appalling pitfalls in the efforts of individuals (mostly well intentioned) who may be highly qualified in the physical sciences with decent command of the Arabic Language are the real dangerous ones as they invariably impress the average person especially when sounding exotic. Any specialist can easily see their nonsensical foundations and as such, they actually discredit Islam in trying to promote it. That is the bedrock of the allegedly increase in signs of atheism among young Muslims. The widespread use of social media especially among the impressionable young enables these individuals to have undesirable impact. A sample of this ill-advised effort is illustrative of the danger involved. The Syrian Physicist Dr. Ali Mansour Kiali (على منصور كيالى) is for example active on the social media promoting his interpretations of the physical word "gate – باب" that is frequently mentioned in the Qur'an. Most old interpretations of the Qur'anic use of the word resort to figurative and allegorical descriptions. This scientist however, does not subscribe (legitimately one might say) to that approach. He insists on the existence of such doors to enter Al-Sama' (السماء) or heaven as described in the Qur'an. The common everyday usage of the word is easily related to by the average Muslim person. Dr. Ali Mansour Kiali (علی منصور کیالی) makes it his holy task to find where these gates or doors are in the world. Mixing firm religious narrative and popular knowledge with pseudoscience, he establishes with unwavering certainty the location of these points of entry on earth. The narrative of Al-Isra'wa Al-Me'raj (الإسراء و المعراج) is known to every Muslim everywhere which designates Jerusalem as the point from which Mohammad (محمد - علیه الصلاة و السلام) had to go to in order to embark on his journey of ascendance to meet God (الله - سبحانه و تعالی). Consequently, Jerusalem is one of these entry points. The fact that it is most probably where the greatest of previous Prophets resided or are buried does not get any weight in his analysis regardless of the fact that Mohammad (محمد - علیه الصلاة و السلام) and all Muslims faced in its direction for their prayers the entire period until that direction was shifted to Makkah about eighteen months after the Hijrah. Now a fantastic exotic mixture of religion and pseudoscience is formulated that appeals to the simple minded individuals. On the popular front, he explains why only certain locations on earth can be used to launch spaceships regardless of the apparent costs and technical troubles associated with these locations whereas other locations are available. It is simply because gates or doors to Al-Sama' (السماء) are readily accessible only at these locations. The implication that this reasoning designates heaven as space is overlooked and is brushed aside. Other examples are there too but it is futile to discuss them. On the other hand, cosmological *speculations* established points of entry to other universes and dimensions. These are what scientists and science fiction writers like to call "wormholes". One may be able to attach more credible connections between these scientific *speculations* and the Qur'anic narratives about gates to Al-Sama' (السماء) obviously only after accepting other dimensions or universes to designate Al-Sama' (السماء). This is diametrically in variance with what Dr. Ali

Mansour Kiali (على منصور كيالى) has done. Apart from simple minded Muslim individuals, no rational person should accept his methodology. The problem is vastly compounded when one claims to prove to non-Muslims the greatness of the Qur'an. In a bizarre way, such methodologies can actually be deployed to discredit Islam and promote atheism even among Muslims. This is the ultimate unmitigated danger in linguistically interpreting the sacred tradition of Islam; good intentions notwithstanding.

SOME WORDS HAVE DIFFERENT USES IN THE QUR'AN

At this juncture, some examples may avoid discussing abstract notions and elucidate what is meant by higher concepts intended in Qur'anic language. For example, in Surat Al-Baqarah (سورة البقرة) verse (189 آية) the Qur'an responds to inquiries about the phases of the moon (which is a very important issue for people living in the desert) by stating that they help people to establish their references in time and to fix their pilgrimage to the Ka'ba (الكعبة). There is no doubting the validity of this answer. However, it actually implies the rotation around earth and the physical principles thereof as modern science discovered without explicitly saying so or at least somehow contradicting them. As far as the Prophet (محمد - عليه الصلاة و السلام) is concerned, he for example gave a more elaborate definition for the word bankrupt (مفلس) when he initiated the inquiry about what his companions understood it to mean. The tradition is replete with such examples and here is not the proper place to discuss the Qur'anic or prophetic traditions though. It is however important to note that these answers are not proverbial but physically actual. The universal acceptance of Qur'anic interpretations as presently advocated by scholars' works rests on two deep-seated assumptions. One is the unbounded reverence accorded the original interpreters and their followers through the ages and the other is the impressive endurance of such interpretations over numerous centuries. Both contentions are without doubt meritorious. However, for example, the progress of human astronomical knowledge over these centuries did not change much for an apparent contradiction to arise and to warrant any revision and/or a change in interpretation. It goes without saying that, **that is no longer the case**. What is called for is not merely a change in semantics but rather **a paradigm shifts in the meaning and usage of the Qur'anic vocabulary and terminology.** To elucidate the relevance of the presumed absolute necessity of a paradigm shift in Islamic thought, one can introduce some examples out of countless others. The first is a simple one which interestingly enough is concerned with the definition of the term (الأرض) "**Earth**" as in the Copernican case. Another word frequently used in every day conventional language and at the same time appears often in Qur'anic text is **Al-Sama'** (السماء) which is usually taken to mean sky/heaven. Additionally, the definition of the word "**Unknowable**" as a translation of the Arabic word "الغيب" when used in the Qur'an is given. Another example is the usage of **Jonoud** (جنود) / **Soldiers**. Crucially for the discussion of Qur'anic facts about creation of the cosmos is the meaning of the word "يوم" which is taken to mean "**Day**" in the English language. These simple examples should illustrate the new approach. The outlines of these issues are given here in broad strokes while the details are dealt with somewhere else in this book in the context of the discussion of Cosmology and the process of creation of the universe.

AL-ARDH (الأرض) / "EARTH"

The fifty fifth verse of Surat Taha (سورة طه) in the Qur'an explicitly and unambiguously states: (منها خلقناكم و فيها نعيدكم و منها نخرجكم تارة أخرى). The traditional interpretation of this verse universally accepted by scholars translated into English is "We (Allah) created you (human beings) from *it* and We (Allah) shall return you (human beings) to *it* (at death) and from *it* We (Allah) shall extract (resurrect) you (human beings) another time". The *"it"* here is universally taken to be *"Earth"* what is now known to be the third planet from the Sun in the solar system where human beings live. In the English language (as in any other language developed by humans) the word "earth" has but one specific meaning since that is what generations of humans accepted as the definition of the word even if it develops later into something else.

The underlying principle is the lack of any absolute standard reference. Languages are simply human inventions. The single exception to this rule is the *Arabic language.* While it is obviously true that it was invented by humans (however, some Muslim traditionalists insist on it being eternal since the beginning of creation and not just of humans), it received an absolute reference in the form of the divinely revealed text of the Qur'an. Therefore, rules are not quite the same for the Arabic language. Common usage of a word by Arabic speakers is reflected in the Qur'an but in a more generic sense. The basic concept is the same while with progress of human knowledge other more general meanings are implied and understood. That is the power of the Arabic language and the astonishing endurance of it for millennia. The uniqueness of the Arabic language and how it developed to acquire that status is the subject of intensive discussion in the previous undertaking by the author in this series "Islam and the West" entitled "Why Do They Hate Us So Much?" and would not be repeated here. Returning human beings to *it* is taken linguistically to be an inclusive approach in the sense that burial, cremation and other methods and processes, intentional or unintentional, to dispose of the remains of a dead person indicate returning to *it* since the term "earth" implies land, water and atmosphere as well as anything else associated with the planet as a whole. In the first half of the twenty first century (approximately the first half of the fifteenth century of Hijrah) realization of inter planetary travel is merely a matter of time and effort not of feasibility and possibility. When (and probably not if) human beings colonize other planets, it is inevitable that some of them will die and be buried and/or disposed of in places other than this *earth*. It seems an apparent contradiction between reality and the statements of the Qur'an arises in this case. Thus, the Qur'anic *it* in Surat Taha (سورة طه) mentioned herewith has to have a more encompassing implication than the third planet within the solar system. It is striking that the Qur'anic word (الأرض) which is what has been taken to represent *it* in the previous discussion, and was assumed to be translated into English as **earth**, appears repeatedly (approximately 459 times) in numerous places within the Qur'anic text. In some places it certainly indicates what is understood to be earth but in others it comes with many different connotations that are indicative of more than just the planet earth itself. **It is almost certain that in a wide variety of these instances, earth, as is currently understood, is not the intended meaning.** That is not to say that earth itself is not included in this more globally encompassing expression (الأرض). One of the most intriguing instances is the vague implication in some subtle way included in the usual understanding of a number of verses in the Qur'an that (الأرض) is flattened or it takes a disc shape. Earth is therefore, certainly not what is described in these instances. Some more universal definition of the term that includes earth as a minor constituent and a limiting case should eliminate any confusion. Such possibility is discussed when dealing with the implications of the on-going development of the scientific discipline of Cosmology and the evolution and age of the universe. But tentatively, it is well known that Euclidean geometry took root in science and even popular culture for a couple of millennia where most physical phenomena were explained based on the two dimensional geometry derived from Euclid's postulates. It is quite remarkable that these postulates had no scientific bases but depended on simple logic. With the failure to find any proof for the fifth postulate and the development in the nineteenth century of the non-Euclidian geometry, curved multi-dimensional space rather than plane space entered into the scientific arena. Einstein formulated the general theory of relativity using Riemannian geometry equating gravity with curvature of the four dimensional space-time. The success of the long lasting validity of Euclidean (plane) geometry appeared naturally as an approximation to general relativity. That is because the gravitational field on the surface of the earth (due to its relatively small mass) hence the curvature of space—time is quite weak. The well entrenched plane geometry of Euclid made it rather easy for the Church of Rome (presumed God's (سبحانه و تعالى) Christian spokesmen on earth) before the age of enlightenment to interpret the direct biblical quotes as asserting the flatness of the earth regardless of ancient knowledge and measurements providing evidence to the contrary. Modern science verifies the flatness (disc shape) of the Milky-way galaxy, of which earth is but a speck. There are strong indications for the flatness (disc shape) of the entire three-dimensional universe itself. The vague hints in the Qur'an that (الأرض) is flat can be perfectly understood if one wishes to interpret it this way i.e. to refer to the Milky-Way or even to the entire three-dimensional universe where humanity lives. Contrary to the biblical expressions in any language, the Arabic usage (even by the pre-Islamic Arabs) of the word (الأرض) accommodates that interpretation with ease in addition to the common everyday use. It would also be instructive to quote the Qur'an from Surat Al-Anbia' (سورة الأنبياء) verse 105 (و لقد كتبنا) سبحانه و) Where God (فى الزبور من بعد الذكر أن الأرض يرثها عبادى الصالحون تعالى) is saying that He ordained that (الأرض) will be eventually inherited in the hereafter by His good servants among human beings as a reward. If (الأرض) refers to earth, then it is not much of a reward due to the number of humans since the beginning of time. Then (الأرض) must be understood to be something incredibly vaster than earth, bearing in mind the many other descriptions of what awaits good people in the hereafter in terms of space and enjoyments in the Islamic tradition.

AL-SAMA' (السماء) / SKY OR HEAVEN

The word Al-Sama' (السماء) occurs 120 in the Qur'an as singular and 190 times in plural form as Al-Smawat (السماوات). In Arabic, it has the generic usage as "whatever is higher". It is used in both the noun form or as an associate verb to indicate physical position, moral elevation and religiously it connotes the structure of God's (الله - سبحانه و تعالى) overall creation other than Al-Ardh (الأرض) that Mohammad (محمد - عليه الصلاة و السلام) ascended to meet God (سبحانه و تعالى) in person during the night journey of Isara' and Me'raj (الإسراء و المعراج). According to the Islamic tradition, there are seven of them where Mohammad (محمد - عليه الصلاة و السلام) encountered some of the previous messengers of God (سبحانه و تعالى) in each one and witnessed certain occurrences pertaining to rewarding the pious and punishing the unbelievers. The commonly used corresponding word in English is alternately "sky" or "heaven". While in English there is a clear distinction between the two words, there is none in Arabic which gives it a clear cut physical meaning encompassing everyday usage both physical and religious as well as cosmological application. Most Muslims raise their hands and eyes *up* when praying to God (سبحانه و تعالى). Obviously *up* in this case is meaningless physically but meaningful in the sense of higher existence. It is interesting to notice that in the Qur'an as can be seen from the above cited numbers, Al-Sama' (السماء) frequently comes in the plural form while Al-Ardh (الأرض) is always singular. It is universally understood in Islam that as of right now, Al-Ardh (الأرض) represents the lowest type of existence for humanity.

AL-SALAH (الصلاة) / PRAYER

The word "صلاة" and its related verb which is normally employed in every day's use to mean prayer or to pray occurs three consecutive times in Ayah 56 in Surat Al-Ahzab (سورة الأحزاب) associated with "God" (الله - سبحانه و تعالى), "His Angels" and the "Believers". All offer it to Mohammad (محمد - عليه الصلاة و السلام). It goes without saying that the associations are drastically different. The very same word has been interpreted in three different ways according to who is offering it. Associated with "God",(الله - سبحانه و تعالى) it is taken to mean praise of him and/or mercy conferred upon him; with "His Angels" it means asking for God's (الله - سبحانه و تعالى) forgiveness for him. As when it is associated with the "Believers" it is a command from God (الله - سبحانه و تعالى) that they ask Him to offer it to Mohammad (محمد - عليه الصلاة و السلام) in the exact sense just mentioned as He previously offered it to Ibrahim (Abraham – إبراهيم – عليه السلام). Muslims accordingly follow that divine instruction at the very end of every prayer mandatory or voluntary. Some scholars invalidate any prayer not including such statement. The same word appears in numerous other places in the Qur'an with different usage each time. Mohammad (محمد - عليه الصلاة و السلام) himself interpreted some of these incidences and scholars after him covered all these different meanings tracing them back to one companion or another. Therefore, the thesis that words of the Qur'an usually have more global meaning than what common everyday usage may suggest is not in any way off mark. However, this work extends the claim to interpretation of physical phenomena as human knowledge progresses. The fundamental issue here is that expanding the meaning of Qur'anic words (without losing the essence of everyday usage) to cover physical and cosmological phenomena is not an arbitrary sleight of hand. However, it ascertains compatibility of the Qur'an (which is the literal word spoken by God as far as Muslims are concerned) with findings of science.

JONOUD (جنود) / SOLDIERS

Words used in conjunction with every day's use are also used in conjunction with God (الله - سبحانه و تعالى). For example, the word Jonoud (جنود) occurs in the Qur'an in a variety of forms (singular, plural, etc.) 27 times. In general, it means ardent supporters and is also used to mean soldiers in the strict martial way. This word is applied in association with humans as well as with God (الله - سبحانه و تعالى). It should be obvious that its utility in both cases does not convey the same message. Therefore, it should have a universal meaning that is appropriate when employed in both cases. This is another example among innumerable ones in the Qur'an that calls for the suggested paradigm shift in the usage of Arabic words to develop better understanding of the Qur'anic text. While this example is useful for the elucidation of the suggested generalization of the meanings of Arabic words, its most important significance at that point is what it conveys when the word Jonoud (جنود) is used in conjunction with God (الله - سبحانه و تعالى). There are four places where it is employed to convey God's (الله - سبحانه و تعالى) actions; in Ayah 4 and Ayah 7 of Surat Al-Fat-h (سورة الفتح – 4 و 7 آية), Ayah 31 of Surat Al-Moddather (سورة المدثر – 31 آية) and Ayah 173 of Surat Al-Saffat (سورة الصافات – 173 آية). The most interesting incidence of this word for this discussion is the one in Ayah 31 of Surat Al-Moddather (سورة المدثر – 31 آية). It says in the context of describing Hell and its guards (Only God (الله - سبحانه و تعالى) exclusively knows who His obedient servants are). ...... «و ما يعلم جنود ربك إلا هو ......". The reason for the significance of understanding how this part of the Ayah should be interpreted is that humans are self-centered and they always give attributes to everything from their own experiences. People other than Muslims to this day give human features to their god(s). The same is true to angels, demons, etc. Islam however, considers everything and everyone at the disposal of God (الله - سبحانه و تعالى) to be employed as

He wishes. That covers the entire creation from the micro world of the elementary particles to the macro world of the whole universe(s). Awareness of this fact eliminates the illusion of the presumption of the necessity of human features to describe God's (الله - سبحانه و تعالى) obedient servants. A photon or an electron can be and is God's (الله - سبحانه و تعالى) obedient servant so is the universe as a whole and the same goes for everything else. Thus, it immediately follows that when the Qur'an for example speaks of dispatching Jonoud (جنود) to support the believers (as happened during the great battle of Badr (غزوة بدر الكبرى) for instance) that does not require humanlike persons by any stretch of the imagination. The same is true when it describes the destruction of some unbelieving community by strange means as far as humans are concerned. Comprehending this approach assuages the idea of miracles or supernatural occurrences associated with Prophets and Messengers of God (الله - سبحانه و تعالى). Abraham (إبراهيم - عليه السلام) not being burned by fire, the many blessings conferred on Moses (موسى – عليه السلام) and his people and the unusual workings of Jesus (عيسى – عليه السلام) are but a few examples among myriad others of this concept referencing God's (الله - سبحانه و تعالى) obedient servants. It is essential to understand that natural laws and norms are not broken to accomplish such acts. Rather one should think of special and rare albeit not impossible applications of these very same laws deploying whatever objects necessary whether very small (particles, atoms, molecules, etc.) or very large (planets, stars, the universe, etc.). This is the rationality of Islam at its simplest where blind faith is dismissed out of hand. While atheistic arguments against faith may sound reasonable, they are part and parcel refuted by the rationality of Islam unless one accepts the blindness of such attitude. When Muslims put faith in certain facts that they cannot understand, it is not based on blind faith but rather on the holistic nature of Islam and the Qur'an. Exactly as in science, if an approach explains numerous phenomena, one should accept it as an explanation to other yet not quite understood issues. Contrary to all other belief systems, Islam is unambiguously and unequivocally based on reason. In Islam there is a reason and an explanation for everything that takes place. Sometimes that reason is not obvious at certain point in time but should be with the passage of time till Dooms Day.

AL-GHAIB / THE UUNKNOWABLE (الغيب)

This word is a very interesting example illustrating the urgent need for a paradigm shift in Islamic thought approaches to interpreting the Qur'an called for in this study as far as the meanings of Arabic words are concerned. Limiting utilization of this word to old entrenched usages by the great ancient scholars is decidedly erroneous at the present time for a wide variety of reasons albeit it might have been more than adequate and justified in the past. This is naturally consistent with the dictum that human knowledge is continuously expanding. The common universally accepted meaning of the word among Arabic speaking people has religious connotations normally associated with God's (الله - سبحانه و تعالى) knowledge that is beyond human understanding. It is used to indicate what is "hidden" as included in the exclusive domain of God's (الله - سبحانه و تعالى) knowledge. It is usually pointing to future occurrences as in events, provisions, lifespan, etc. Historically, this is how all the greats and even the commoners used it. However, the Qur'an is replete with examples of other explicitly unexplained information that is not covered by this word. The implication is that in time humans should acquire that knowledge on their own. That should undoubtedly contradict the concept that the future is what is meant by (الغيب) Al-Ghaib since sooner or later it is perfectly understood by humans and is no longer exclusive to God (الله – سبحانه وتعالى). Within the context of the fundamental tents of Islam as implied, among many other instances, in Ayah 3 of Surat Al-Baqarah (سورة البقرة) a necessary condition to be included in the community of Muslims is to unquestionably believe in (الغيب) Al-Ghaib. The word in its various derivatives appeared in the Qur'an 53 times. In most of these occurrences the word indicates the exclusive knowledge of God (الله – سبحانه وتعالى) with no possibility for humans to ever gain such knowledge. In such cases it also explicitly asserts God's (الله - سبحانه و تعالى) ultimate undisputed and unchallenged sovereignty over all creation due to and because of this exclusivity. In the remaining cases it describes knowledge that will eventually be revealed to or acquired by humans. Obviously it goes without saying that whatever knowledge gained by humans on their own is still part of the vast unlimited reservoir of God's (الله - سبحانه و تعالى) knowledge.

ISLAM AND THE WEST

Therefore, there is no contradiction in the Qur'anic usage and hence the limited understanding of ancient scholars of the way the word (الغيب) Al-Ghaib is used in all cases. How would one differentiate between the information that is open to investigation, eventually acquired and ultimately understood by humans and that is *unknowable* no matter what as it is within the exclusive domain of God's (الله - سبحانه و تعالى) knowledge? This issue may sound academic with no practical utility but it is of paramount importance nonetheless. This becomes obvious when one bears in mind that lumping all knowledge to God's (الله - سبحانه و تعالى) exclusive domain is basically at its core a call to utter ignorance and intellectual laziness. It is easy to detect traces of that attitude in the history of Islamic thought that linger to the present day despite the unmistakable and explicit Qur'anic summons to Muslims to ponder all of God's (الله - سبحانه و تعالى) creation. Additionally, one must not ever forget the Qur'anic fact that it was God (الله – سبحانه وتعالى) who taught humans all they know in the first place. This differentiation when adequately figured out would clarify many misapprehensions of the Qur'anic narrative. As mentioned before, the future is not a proper equivalent to the word (الغيب) Al-Ghaib when interpreting the Qur'an albeit it can be and is a minor part of it. The *unknowable* is a more appropriate general definition. Now what does this definition and its relevant Qur'anic usage look like from a modern science point of view? When discussing cosmology, it was stated that matter (that is what humans are made of and can directly sense) constitute a ridiculously small part of the mass of the universe. The rest is attributed to things that are *indirectly observed* or in most cases simply *assumed*. Whether these things are *unknowable* or not remains to be seen. What is beyond doubt as illustrated before is the fact that the objects of the universe are separating by speeds unrestricted by relativity. Eventually, that speed must exceed that of light. It is impossible to tell by human observation whether this has already happened or not since information is exchanged between earth and the rest of the universe by the speed of light. It is therefore obvious that there are parts of existence which are out of reach to human observation and are therefore *unknowable*. Humans can speculate to their hearts' content about these parts without any hope of actually under the existing physical laws learning anything about them.

What humans cannot ascertain is that these parts do not exist. If one subscribes to the concept of multiverse, the same holds true since other universes are governed by different physical laws. One way or the other, there are *unknowables* out there which the Qur'an fourteen centuries ago described in one single word; that is (الغيب) Al-Ghaib. The Qur'an also unequivocally amalgamated this information with the exclusive domain of God's (الله - سبحانه و تعالى) knowledge. It is simple minded to both hold on tightly to ancient scholars' interpretations or to fault these great personalities for their incomplete understanding. This is the undisputed nature of human intellectual activity.

يوم / Day

The Qur'an explicitly states that God (سبحانه و تعالى) created heavens and earth(s) in seven "يوم" with people translating its meaning to English for example would say seven days. However, one should never lose sight of the fact that the Qur'an is only in Arabic. The Bible mentions the same idea. While neither the Hebrew Bible/Old Testament nor the New Testament as they existed after Moses (موسى عليه السلام) and Jesus (عيسى عليه السلام) are in their original language(s), the Qur'an is beyond the shadow of a doubt is in its original form as revealed. This fact allowed devotees of both Judaism and Christianity to interpret the creation process as taking seven days as measured by humans and calculate the age of the universe to be ridiculously young against available evidence. On the other hand, how long the creation process did take (and whether it is still in progress) derived from the Qur'anic statements depends entirely on the meaning of the Arabic word "يوم". The everyday usage of the word by Arabic speakers is naturally connected to their earthly experiences too. However, the Qur'anic usage differs from one place to the other including, but not limited to, that conventional usage by humans. These examples are discussed in much more detail in other parts of this work. However, the same approach of proving the compatibility of interpretation of the Qur'an using more global meanings for its vocabulary with the latest proven scientific findings is followed.

It is of the utmost importance to bear in mind that interpreting the

Qur'an according to new scientific knowledge is not the aim of subjecting its fundamental precepts to a rigorous scrutiny as has been done to the Bible for example. *The ultimate aim of such endeavors should be to ascertain or refute the compatibility and reconcilability of the Qur'anic text with the latest irrefutably established scientific facts.* These facts are obviously not absolute in nature but evolve with gaining more knowledge. However, their fundamental roots remain intact which is basically what is needed to be inspected against the Qur'anic text. *Restricting the effort to compatibility and reconcilability is what the inspection should be after.* This stems from the unique fact that most Arabic expressions can have numerous connotations belonging to the same root meaning. Some of these are more universal than others and some are limiting cases used for millennia in the daily life of the individual. This is a distinctive and exclusive characteristic of the Qur'an due to its Arabic nature. And that is the vital difference between Islam and other religions, monotheistic or otherwise. *Qur'an as the literal word of the Almighty is in (and only in) Arabic.* This is an explicit mandate spelled out in the Qur'an itself several times. That is to say that God (سبحانه و تعالى) requires the interpretation of His words, the Qur'an, done according to the rules of the Arabic language and nothing else. However, meaning of Arabic words and expressions are mostly abstract because of their development within the bleak environment of Arabia as the Arabs came into being as a separate people.

THE NECESSITY OF A "PARADIGM SHIFT" IN ISLAMIC THOUGHT

By its very nature, jurisprudence does not depend in a strong way on progress of human knowledge. The impressive huge strides taken by humanity over the past fourteen centuries in acquiring knowledge necessitate very little change in Islamic jurisprudence. While new facts such as the factors controlling human (and other living things) genetics dictate new insightful rules to be addressed within the body of the Shari'a (الشريعة), they do not preclude any existing ones. The same is true concerning other social or material developments. That is definitely not the case with understanding the magnificence of God's (الله - سبحانه و تعالى) creation. Therefore, one can proceed comfortably with adopting the inevitability of a "**Paradigm Shift**" in Islamic thought for the preservation of Islam itself without fear of trespassing on the realm of the Shari'a (الشريعة). The Qur'an being the most fundamental aspect of Islam and the core subject of Islamic thought in its historical interpretation of scholarship should be the main topic of that shift. Among other things, the anticipated "**Paradigm Shift**" requires an elementary change in the usage of the Qur'anic vocabulary and terminology. *This does not imply, by any stretch of the imagination, a change in the text of the Holy Qur'an itself but rather the way it is interpreted.* It has to be recognized by Muslims that Islamic thought even in its narrowest definitions is nothing more than scholarly human activity, albeit interpreting divine texts that should be subjected to all the constraints, rules and regulations of scientific pursuits as established through the centuries.

For the past fourteen centuries, Islamic thought concentrated on the development of jurisprudence in a fundamental way to establish the platform of its own validity, the fundamental or primary beliefs and what defines a person as being a Muslim. That resulted in an impressive body of laws governing all aspects of the daily life of the individual and his/her relationship with others as well as the society as a whole. It also determines unequivocally how an Islamic state deals with its members and its friends and its foes. Regardless of the sometimes divergent approaches of the various schools of thought, Islam's tenets and fundamental concepts are undoubtedly remaining intact. Being a Sunni or a shi'a does not represent any differences in the basic tenets of Islam but rather a difference in some of the applicable laws descending from historical and/or political circumstances. Apart from jurisprudence, far less effort was dedicated to cultural, technical and philosophical endeavors. However, the end product of these comparatively minor ventures signified a quantum leap in human history establishing what became known as the Islamic way of life and paradoxically gave impetus to the development of what evolved as the Western civilization. Obviously Jurisprudence does not represent the whole of scholarly activities in any culture. Nevertheless, for historical reasons it formed the cornerstone of intellectual activities within the lands of Islam as they physically expanded. Jurisprudence dominated the life of Muslim scholars until it reached its apex during the first Hijrah couple of centuries where hardly anything else could be added except on the periphery. Thus, it was inevitable that if confined to Jurisprudence, Islamic thought would stagnate especially under the prevailing political climate of the time. This is not to say that knowledge in other scholarly fields did not take huge forward strides but Jurisprudence casted a long shadow over any other field. It is noteworthy that almost all great ancient Muslim Scholars studying and contributing their own theories to physical phenomena had their beginnings in the field of Jurisprudence and its related disciplines. When the tumultuous political environment settled coinciding with what was perceived as the end of the road in developing rules and regulations governing the Muslim's daily life, that represented the "End of History" and the victory of Islamic Culture for Muslim scholars leading to scholarship stagnation in all fields. Other factors doubtless contributed to this process but to a lesser extent. Nowadays any casual observer can see that apart from Jurisprudence, Islamic thought (*as opposed to Islam itself*) is inadequately prepared to deal with most aspects of modern science and technology. Therefore, a **"Paradigm Shift"** in Islamic Thought is a must in order to validate the maxim that Islam is a way of life that is adequate till the end of time. Any thoughtful person in the twenty first century can relate to this presumed Islamic notion of the "End of History" and the victory of Islamic Culture as it turned out to be an illusion in the same way this individual can relate to the collapse of the Soviet Union and communism in late twentieth century representing the ultimate triumph of Western Civilization which was patently described by most Western scholars as the "End of History" while it turned out to be an illusion for Western democracies. Confining the Islamic intellectual enterprise to the development of jurisprudence and a distinct way of life, Islamic scholars assumed over time that Islamic thought reached a state of perfection. There are legitimate reasons for that attitude. They can be seen in the success and survival of the Islamic Civilization and the Islamic way of life, more or less intact, despite the enormous physical, political, cultural, etc. challenges they faced over the centuries. The flip side of these attitudes is stagnation which is obviously the clearest malady of Islamic thought at the present time. There is an unyielding conviction among Islamic scholars and to a very large extent among all Muslims that once a consensus is reached among scholars on certain subject(s) or an idea(s), it becomes part of the religion itself and deviating from it is strictly prohibited. In return this subject(s) or idea(s) acquires some kind of sacredness and challenging it is anathema. This may be valid for rules and regulations (which is what jurisprudence is all about) but human history in all its scientific, political, social, etc., details demonstrates unmistakably that even unanimous approval of an idea or a principle does not necessarily make it right or guarantees its validity. The near sanctity of Newton's Laws that were accepted for centuries without serious challenges is a case in point that is continuously cited in modern physics. They acquired their holiness because they explained almost all (but not strictly all) the physical phenomena of the motion of the planets that puzzled science for a very long time. The very minor deviation in the motion of Mercury proved to be the undoing of this aura of sanctity. Once a new theory (General Relativity) explained all these phenomena including the behavior of Mercury, scientists became believers in the new theory. However, it created new problems of its own. This is the nature of things in human thought. When a larger scope is dealt with, the shortcomings of the solutions to the narrower scope become apparent. Newton's Laws discussed the motion of the planets within the solar system whereas General Relativity deals with the entire universe. In Newton's time science did not deal with the universe but only with phenomena it could manage using available knowledge. *When knowledge expands, so does the number of problems to deal with mandating fresh thinking and new approaches.* On the other hand, history of science (and Islamic thought in general is nothing more than assumptions that are subject to scientific verification) teaches that progress comes only from challenging what is considered to be the consensus among the community of scientists (of the hard as well as social sciences). If Islamic jurisprudence is considered at a state of perfection (a hard assumption to defend), other Islamic thought endeavors have to be subjected to challenges continuously if there is any hope for progress despite any consensus at any age. It is essential to recognize the impossibility of continuing the status quo in Islamic thought. It should be crystal clear that thought stagnation has social, political, etc. repercussions.

With the end of the Second World War the phenomenon of nationalism swept the Islamic world as a backlash to Western Imperialism. It replaced the Pan-Islamic movements that ended in failure with the collapse of the Ottoman Empire in the aftermath of the First World War. The Arabic speaking part of the Islamic World went through these convulsions in a more profound way. The rise of Arab Nationalism particularly during the 1960s caused the split of the Arab world into two camps; a revolutionary (mostly republican) and a reactionary (mostly monarchical). As with the Turkish republicans, the Arab counterparts claimed being progressive reaching out to share in the exploitation of the advances in science and technology while the monarchies were traditional claiming the mantle of Islam and its glorious age's old traditions. It was natural that this division would extend to the realm of religion and its role within the society. Progressives in various degrees relegated Islam to the back stage while Traditionalists adopted the enduring old definitions of almost everything. The most intriguing incident relevant to this book endeavor was the fanatic insistence of the highest Muslim authority in the Traditional grouping on the "*flatness of earth*" contrary to all wide spread existing evidence even after photographing earth from space. What prompted him to valiantly stick to this argument was the presumably clear mention of this fact in the Qur'an as traditionally interpreted. This was a golden opportunity for Progressives to denounce the Traditionalists as irrational and backward. Although obviously this intellectually awkward incident was not the reason for the split, it nonetheless contributed to the durability of the dichotomy and made it clear cut. The lines of confrontation were drawn and each side sought help from the then dominant world great powers. The struggle eventually led to the trauma of the humiliating debacle of 1967 when realistically both camps lost and the consequent convulsions are still felt in the twenty first century with no relief to either group in sight. Mundane interpretations of Islam on one hand and their dismissal on the other made losers of all of them and still do. Lamentably the basic conflict can be reduced to the argument; do Muslims adhere to the facts as mentioned in their tradition (**bearing in mind here that they are** *oblivious to the fact that everything in this tradition is their OWN INTERPRETATION)*? Or should they search for accommodations with the tried and proven advances in science and technology (*bearing in mind here that this is exactly what Judaism and Christianity had done)?* Neither approach serves Islamic thought well. Fascinatingly in contrast, during the very same period cosmologists were facing what is known as the "flatness problem" but this time the entire universe was the subject. The basic idea in extremely simple terms was that the shape of the universe emerging from solving equations of the theory of general relativity heavily depended on what scientists call the parameter "Omega" (detailed in another part of the book) which is the ratio between the actual mass density of the universe and the critical density (which is estimated to be somewhat between 2 and 8 hydrogen atoms per cubic yard) resulting from its expansion that would put the universe just on the borderline between eternal expansion and collapse. If omega is greater than one, the universe is a closed one and will eventually collapse in a "Big Crunch"; if omega is less than one, the universe is an open one and will expand forever. If omega is exactly equal to one, then the universe is flat and expanding forever. Omega's value depends on several parameters that cannot be accurately measured at the present time due to lack of adequate technological capabilities. However, it is estimated to currently lie between 0.1 and 2 which means the universe is very close to being flat at the present time that is estimated to be 10-20 billion years after its creation in the "Big Bang". Whether it is open or closed is still beyond Cosmology's affirmation.

It would be crystal clear in the discussions to follow that all the heated intellectual arguments that took place (and still do) in Islam and their associated misery are non-issues emanating from the definitions Muslim scholars, especially the revered great ancient ones, gave to the meaning of words of the Qur'an. It is also clear that both sides are at fault where their conflicts sapped the energy and power of Muslims to no discernable useful purpose and availed unfriendly foreign powers that harbor only ill will to Islam and the Muslims with in-roads to engage in what is essentially an exclusively Islamic debate. If one in the above example substitutes "earth" (which is what scholars think the Qur'anic word means) with "universe" (which is what science irrefutably establishes), the problem goes away. This is ominously similar to the millennium old anguish generated by the issue of the "Creation of the Qur'an" (محنة خلق القرآن). The inescapable implication is that a paradigm shift in Muslims' understanding of the lexicon of the Holy Qur'an is urgently required. When the West injected itself in this uniquely Islamic phenomenon by trying to manipulate Islamic societies' social norms and forcing changes in Muslims' conceptions of their fundamental ideas like Jihad and insulting their revered symbols and personalities, the problem gets aggravated most probably beyond repair. What the West instigated as "Islamic Terrorism" is very much based on the associated fanatical groups' interpretation of ancient opinions rendered under absolutely different circumstances. This interpretive debate has been going on within Muslim societies for considerably long time with little or no violence at all. It is only when the West got into the act that unprecedented crimes of violence are committed. In a real sense, the West enters the mêlée at its own peril justifying in the minds of fanatics their deplorable actions. Muslims should be able and willing to solve this problem by peaceful dialog and respect for the others' strongly held opinions. However, the added benefit of adopting the proposed paradigm shift makes such arguments meaningless to the benefit of all.

As mentioned before, comprehending the various statements of the Qur'an (as the complete representation of God's (الله - سبحانه و تعالى) intentions) is the ultimate task of Islam. That is to say that the advocated *paradigm shift in Islamic thought* has to primarily focus on the universal meanings of Qur'an's words. The Arabic Language as evolved before Islam clearly accommodates such endeavor. This explains why God's (الله - سبحانه و تعالى) final message to humans was uttered in the Arabic Language and why it is mandatory to understand that the Arabic text of the Qur'an is the only text worthy of that description. Qur'anic words used in passing rules and regulations do not in aggregate require any expansion of meaning since human basic needs which are the subjects of jurisprudence did not change in any perceptible manner over time. However, for example those dealing with the creation process have and do urgently stipulate such expansion. The urgent call for a paradigm shift in Islamic usage of Arabic words (among other things) from common every day to more encompassing universal meanings to accommodate reconciliation of scientific findings with Qur'anic narrative should not be a reason for concern. The new paradigm should help illuminate new meanings while preserving the integrity of centuries old conclusions on rules and regulations of Islamic Jurisprudence. Jurisprudence by definition deals with humans and their interactions at the various levels. Thus, familiar common daily expressions used by humans as universally understood are what is addressed by necessity. This does not change with new findings of science and as such are not subject to any change except on rare occasions when medicine for example uncovers possibilities for organ transplants or cures to deceases. What is usually needed in these cases is simply extrapolation of the existing rules based on similarity as opposed to changing or even cancellation of these rules and regulations. The ultimate goal for this called for paradigm shift is preserving the body of the Islamic Shari'a (الشريعة) while providing updated interpretation of Islamic tradition that stands the scrutiny of science. This way one can adhere to the maxim that Islam and in particular the Qur'an is valid till the end of time.

ISLAM AND RELIGIOUS AUTHORITY

As is extremely well known by all Muslims, the Qur'an unequivocally places accountability for one's actions squarely on the individual. Simplicity of the tenets of Islam allows every human being no matter how unsophisticated they might be to understand what is right and what is wrong regarding the fundamentals. The young and mentally unbalanced are explicitly exempt in Islam from culpability or even performing mandatory rituals. However, not every issue is fundamental and obvious not just to the simple folks but more often than not to highly cultured and highly educated persons. That is for example the origin of the various schools of judicial arguments. It is customary for the worldwide Muslim communities to follow one or the other of these schools of thought. The prevailing school however, does not ban the practices of the others. When this happens, it is always for political reasons. Nonetheless, no institutional authority can shield a person from accountability before God (الله - سبحانه و تعالى). Not even Prophet

Mohammad (محمد - عليه الصلاة و السلام), as is well documented in the tradition, would be able to intercede with God (الله - سبحانه و تعالى) on behalf of his own beloved daughter Fatima (فاطمة). This is ABC of Islam that even little children know by heart. Consequently, the idea that some weird practices endorsed by some institution carrying the label of Islam or some fanatical spokesperson should automatically be held against Islam is groundless. It is not true that something is patently Islamic simply because some Muslims somewhere advocate it. This is not a convoluted way to avoid a stigma. It is an axiom in Islam that if something is logical and makes sense, especially if it is beneficial, then it is automatically acceptable. This is uniquely Islamic. It is a rule in Islam that emergencies overtake prohibitions and the rights of humans which can be lost take precedence over the rights of God (الله - سبحانه و تعالى) which can never be lost.

DIFFERENTIATING BETWEEN THE QUR'AN AND ITS INTERPRETATIONS

Islam is primarily based on the veracity of the assertion that the Qur'an is the literal word of Almighty God (سبحانه و تعالى) the Creator. How would one then reconcile reality with the universally accepted interpretation of the abovementioned statements of the Qur'an? At this juncture, it must be emphasized that the arguments thus presented, speak of *the universally accepted interpretation of the Qur'an rather than the Qur'an itself*. Interpretation is compellingly associated with the interpreter and is firmly a human activity with no actual connection to divine meaning. As such, one has to assume, if one embraces the veracity of the Qur'an as the divine word of God (سبحانه و تعالى), that **the universally accepted interpretations of the verses of the Qur'an are not universal after all.** Therefore, there are no allegorical presentations in the Qur'an except when God (سبحانه و تعالى) says explicitly He is only giving an example which happens frequently in the Qur'an. Everything else is a divine statement of fact. This makes the interpretation of numerous verses extremely difficult such as the most hallowed verse in the Qur'an which is the two hundred fifty fifth (آية

(الكرسى) of Surat Al-Baqarah (سورة البقرة). The word "الكرسى" has been given many meanings allegorical and otherwise. However, no scholar ever assumed similarity between it and the common usage (equivalent in English to "chair"). The Qur'an being devoid of any allegorical statements in the opinion of Arabic scholars seems to undermine Arabic literature whose greatness especially in poetry hinges on such usage. This is a false statement since it does not differentiate between what God (سبحانه و تعالى) says and what one understands He says. In Islam one is legitimately allowed to associate allegorical meanings to statements of the Qur'an when one does not know better due to incomplete knowledge which will always be the case till the end of time. So, there is no contradiction between saying that the Qur'an is devoid of allegory and at the same time understanding it allegorically in many instances. This is not a modern day secular argument. This general approach to interpreting the Qur'an was perfectly understood by Muslims even at the time of the Prophet (محمد - عليه الصلاة و السلام) and the immediate era after his death. There are numerous statements to that effect especially during the political crisis after the assassination of the third Caliph Othman ibn Affan (عثمان إبن عفان). Imam Ali ibn Abi Taleb (على إبن أبى طالب) is quoted several times to indicate that various interpretations of the various verses of the Qur'an presented by his followers as well as his opponents were merely their own understanding of these verses and carry no special sanctified meaning to be imposed on the rest of the community. Concepts, rules and ideas advocated by Imam Ali ibn Abi Taleb (على إبن أبى طالب) are highly regarded and accepted as references by all Muslims, their personal affiliations as Sunnis or Shi'a notwithstanding. A more compelling argument in support of the need to reevaluate age old usage of some of the Qur'anic expressions can be found in the Prophet's (محمد - عليه الصلاة و السلام) own utterances. It is well known in the Islamic tradition that the Prophet (محمد - عليه الصلاة و السلام) in numerous occasions asked his companions (and especially his wife A'isha (عائشة – أم المؤمنين) about what they understood certain words, frequently from the Qur'an itself, to mean. On answering him giving their universally accepted meaning, he emphasized the fact that although their usage is correct, the meaning is more encompassing and gives a higher level of interpretation. That is to say that *the absolute meaning of Qur'anic and prophetic expressions while imply the common usages, intend to convey much wider concepts.*

ISLAM AND THE FUNDAMENTAL ASPECTS OF ATHEISM

With the unfolding of the creation process with time and humans acquisition of more knowledge about their universe(s), all will be known. Nonetheless Mohammad's (محمد - عليه الصلاة و السلام) career as God's (الله - سبحانه و تعالى) Messenger spanning the last twenty-three years of his life was basically spent in physically establishing the framework of the faith and his word was more than enough to satisfy the believers. That has changed with his passing and his companions were repeatedly questioned about many such issues. Common folks wondered aloud about their submitting to God's (الله - سبحانه و تعالى) will in all their actions and His holding them accountable for them at the same time. The companions, their followers and scholars for the past fourteen centuries held on tightly to the well-established fact that what is hard to relate to in Islam is validated by the abundance of what is easy to understand. Generation after generation of the great interpreters of Islamic faith explained the apparent paradox by the unqualified belief in the unbounded justice of God (الله - سبحانه و تعالى). God (الله - سبحانه و تعالى) would not compel people to act in some way and then reward or punish them accordingly. That is a rationally unacceptable contradiction. For all these centuries human logic provided the answer to this apparent paradox. However, logic is not singular and antagonists could and did come up with also logically formulated opposing arguments to dismiss this fundamental concept to undermine Islam itself. There is no much harm done when arguments and counter arguments are confined to intellectual debates since those who wish not to believe in Islam will always find some way based on language and mental exercises to advance their cause and no debate will ever convince them. However, when undermining and defending Islam moved into the political arena, there was blood letting and confusion. Islam is the poorer for this. By the time hard sciences took center stage in human knowledge, shortcomings of language and purely logically based arguments explaining the universe(s) became obvious. While logic is not singular, science is. Individuals can disagree with each other's logic but universally verifiable scientific approaches are not subject to debate; they are one and the same everywhere at all times until new paradigms replace old ones and the cycle is repeated. In the meantime, any concept has to lend itself to falsification by practical means. In the final analysis, for an object to exist it has to be detected by human senses. That is the crux of the so called Western science based atheism. While belief in the "Supreme Being" is as old as human existence, science is a brand new phenomenon in human history though. So called science based Western atheism dismisses religion and the concept of God (الله - سبحانه و تعالى) as an immature human creation to cope with the ravages of nature. When humanity recently reached a semblance of maturity, it should follow Western atheists' lead and get rid of backward perceptions of faith to be able to progress and enjoy its nature endowed privileges free of the shackles of religious restrictions. Atheists reverse the order of things and demand that religious individuals prove the existence of God (الله - سبحانه و تعالى) (a concept as ancient as humanity itself) rather than the upstart atheism proving His non-existence. They take comfort in the irrationality of trying to prove the absence of something rather than its existence. Conceptual principles of Islam are self-consistent and flow harmoniously with the progress of human knowledge. Recently Muslim scholars with scientific and technical competence defend Islam by striving to prove that modern findings of science have their corresponding narrations in the Qur'an. It is avowed in this endeavor that this is a faulty approach since findings of science may change with time. The appropriate approach is to show that atheism's basic assumptions are patently wrong and at the same time proving that whatever verified science shows, albeit essentially temporary, is not contradicted by the Qur'an and that there exists an interpretation that accommodates such findings. This is the intended methodology of this undertaking and the launching pad for the call for a paradigm shift in Islamic thought and its verbiage.

Among the arguments underpinning atheism and its objections to religion and faith one can find no more than a paltry few issues regardless of atheism's claiming the mantle of science and presenting a façade of presumably scientific facts. It is indispensable for this endeavor to methodically, meticulously and carefully dissect these few arguments scientifically as the only way to show their inadequacy before embarking on scrutinizing the Qur'anic text vis-à-vis modern scientific findings. For the most part, these arguments are:

GOD

Obviously and without much arguing, this concept is the most fundamental of all. As such, questioning and doubting His existence is not a new issue. In Islamic tradition, Mohammed's (محمد - عليه الصلاة و السلام) antagonists although familiar with the word "Allah" in their heritage, refused to accept Islam's conceptions of His attributes. Like today's atheists, they demanded a physical description which in the context of the Islamic faith is beyond human knowledgeability. It takes only God (الله - سبحانه و تعالى) to properly describe Himself. Immediately, the Archangel Gabriel conveyed to Mohammed (محمد - عليه الصلاة و السلام) God's (الله - سبحانه و تعالى) answer as Surat Al-Ikhlas / Pure Sincerity (سورة الإخلاص).

«قل هو الله أحد * الله الصمد * لم يلد و لم يولد * و لم يكن له كفواً أحد»
Surat Al-Ikhlas (Pure Sincerity) سورة الإخلاص
"Say: He is God (الله - سبحانه و تعالى). One. God (الله - سبحانه و تعالى) the Everlasting Refuge. He does not beget. Nor is He begotten. And comparable to Him, there is none."

This is one of the shortest and earliest Surats in the Qur'an that was the twenty second in order of revelation in Makkah. However, it is of a profound status. Tradition has it that blessings invoked by reciting it are equivalent to those commensurate with reciting one third of the entire Qur'an. The reason is that in it God (الله - سبحانه و تعالى) describes His nature in very simple terms. The essence of this narrative is that *God (الله - سبحانه و تعالى) is exclusively the only being with intrinsic value*. Everything else in existence (whether known, unknown, knowable or unknowable to humans) is dependent on other factors to exist. *This is the crux of Muslims' faith* with everything else in Islam including the

Qur'an itself is an elucidation of this paramount fact. It is noticeable that other than Al-Fatiha (the Opening - الفاتحة) that states the obvious order of the Qur'an, this Surat is unique in that it is the only one that derives its name from a description of its content rather than a word in its text. There is practically no Muslim who ever existed that does not memorize this Surat and solicits its blessings. It is fascinating that the collective brain power of humanity since its creation describing the universe that it inhabits is expressed by modern day cosmologists in an identical way. The indisputable conclusion of all scientists, each in their own field, particularly cosmologists dealing with the overall universe is that for the observable universe to be stable, its constituents have to balance out to zero. That is to say the number of elementary particles has to exactly equal the number of their anti-particles. The number of electrons has to exactly equal the number of positrons in the entire universe, the number of neutrinos, has to exactly equal the number of anti-neutrinos in the entire universe, the number of protons has to equal the number of anti-protons in the entire universe, etc. These are staggeringly huge numbers but nonetheless, they have to be exactly equal otherwise the familiar universe becomes unstable and collapses. That is why physicists and cosmologists came up with the concept of "Black Holes" which are by definition unseen as well as unobservable in a direct manner to account for the shortage of matter in their calculations. They are currently desperately looking for what is termed as "Negative Energy" for the same balancing reasons. It is firmly believed that the material universe resulted from the imbalance in these numbers under unusual circumstances but it is evolving to correct that imbalance in a rush to reach oblivion. The upshot of all these conclusions is that *the entire universe with the fascinating phenomena that it encompasses, adds up to exactly big fat zero.* That is why a powerful mind like that of Steven Weinberg's finds the universe pointless. Put another way, *the vast observable universe has absolutely no intrinsic value of its own.* That is why atheists ascribe chance to its existence. On the other hand, the Qur'an over fourteen hundred years ago revealed that same fact in four short sentences in Surat Al-Ikhlas (سورة الإخلاص) to an illiterate Arab who conveyed it to humanity attributable to God (الله - سبحانه و تعالى) while claiming no credit for himself. This is worth pondering and contemplation before passing sweeping judgments. It is ludicrous to perfectly acknowledge limitations on current levels of human knowledge and at the same time utter blanket statements to explain the universe. The celebrated British scientist Stephen Hawking condescendingly said in an interview with the American cable news network CNN in 2010 "God (الله - سبحانه و تعالى) may exist but science can explain the universe without the need for a creator". He also mentioned that "The scientific account is complete. Theology is unnecessary". The fist quote is a manifestation of the habitual scientists' arrogance. The second can be insightful if "Theology" is defined in terms of explaining the concept of God (الله - سبحانه و تعالى) through philosophical mental exercises as done in all religions other than Islam. In Islam God (الله - سبحانه و تعالى) describes Himself in the Qur'an and no other definition is warranted or remotely acceptable. This is a foundational cornerstone of Islamic thought since Mohammad (محمد - عليه الصلاة و السلام) started receiving God's (الله - سبحانه و تعالى) revelation in the seventh century.

REVELATION

Since this is a personal rather than a communal experience, atheists dismiss its reality as hallucinatory. They advance the opinion that individual's background and experiences dictate what he/she utters or claims. Consequently, the disciplines of sociology, psychology and similar fields can easily explain such phenomenon. That is generally true of individuals with observed unusual behavior. On the other hand, prophets and messengers of God (الله - سبحانه و تعالى) are no ordinary individuals since they consistently over the entire human history advance the very same single idea which is the Oneness of God (- الله سبحانه و تعالى). They also make predictions (demise and punishment for the unbelieving masses) that come to pass. Here atheists assert that these predictions are only myths that do not actually appear in recorded history and hence never happened with Noah's (نوح – عليه السلام) flood as an example. One will never reach a conclusion debating these points concerning all heavenly religions and messengers of God (الله - سبحانه و تعالى) with the sole exception of Islam and Mohammed (محمد - عليه الصلاة).

و السلام) as they occupy a distinct undeniable position in the history of humanity regardless of one's personal agreement or disagreement with them. These are no myths and some of their predictions (mentioned in other parts of this study) are very well known which came to pass while others are temporarily waiting to take place. Additionally, the physical condition of Mohammed (محمد - عليه الصلاة و السلام) when receiving the Qur'an over twenty-three years were extremely consistent and witnessed by all members of the community. That may suffice as a compelling proof of the reality of revelations to Mohammed (محمد - عليه الصلاة و السلام) in this logical mental argument and counter argument debate with atheists. On the other hand, there is a very scientific plausibility to the phenomenon of revelation. Neurology and its researchers are convinced of the electro-chemical nature of the processes taking place within the brain leading to human thoughts, speech, vision and other senses that make humans what they are. Mohammed (محمد - عليه الصلاة و السلام) claimed to have received the Qur'an from the Archangel Gabriel whom he saw and conversed with. He also described angels as made of light. It would be absurd to dismiss the plausibility (which is all that is needed to refute atheistic presumably scientific arguments) of Gabriel who is made of light inserting photons (light) by some yet unknown but eventually knowable mechanism into Mohammed's (محمد - عليه الصلاة و السلام) brain cells to convey a message that would be processed according to the normally observed electro-chemical functions of the human brain. That also provides ways to explaining conversations with Gabriel as well. Once the message is imprinted in Mohammed's (محمد - عليه الصلاة و السلام) brain and memory, he is able to communicate it to his companions. This is how human brain and memory work according to modern science. Obviously one can still claim that the whole process hinges on the initial triggering mechanism provided by Gabriel whose existence is merely an assumption. Even so, this is not any different from the universally accepted scientific assumption of the existence of elementary particles that explain and sustain the universe which is the bedrock of all modern science without any direct observation nonetheless. One has to remember that proving the plausibility of an action, extremely rare as it might be, counter to what atheists claim to be the finite truth is sufficient to degrade their arguments and present them as what they really are; mere assumptions that need proving themselves.

ANGELS AND JINN

Humans from ancient times believed in the existence of powerful creatures that cannot be seen. They attributed anything abnormal they may experience, good or bad, to these unseen beings. One can assume that humanity had legitimate knowledge about these creatures' existence but with degradation of its status with the harsh conditions it went through in pursuit of survival, myth replaced reality. However, the fact is that till the age of rigorous science, most of humanity's ideas about causes of natural phenomena, especially disasters, turned out to be absurd. That was a potent argument for atheists in service of the scientific method which they based on the concept that *what cannot be detected by the physical senses and their extended tools does not exist.* Thus, atheism dismisses the religious belief in Angels and Jinn as mythology proving the falsity of faith altogether. Ironically, with the triumph of science in landing a man on the moon in 1969, the age old interest in extraterrestrial aliens exploded in the second half of the twentieth century to the point that no less a person than a president of the United States claimed to have encountered one! With the impressive advances in astronomy during the past few decades, it was natural to enquire about the possibility of other life forms developing in this vast universe given the realization of the utterly insignificant position of humanity and its earthly habitat within the cosmos. That scientific fact emboldened the atheistic allegation of the non-existence of a super-being wastefully creating such a vast universe for humans as asserted by the sacred Western Judeo-Christian tradition of the Bible. It was also natural to start by looking for conditions similar to what is thought to have happened on earth to evolve life anywhere else in the universe. It is assumed that some sort of a solar system is necessary and numerous such planetary arrangements were discovered with no signs of life even in its very primitive form detected yet. However, special attention is currently paid to conditions within the familiar solar system with emphasis on exploring Mars and the moons of Jupiter since they seem to offer environments somehow close to that of earth. If nothing else, these places can be employed as launching pads for human exodus into the universe to guarantee survival of the human species if/when humans destroy their earthly habitat. Very prominent scientists spearhead calls for such efforts and support for the institute in Mountain View, California that conducts "Searches for Extraterrestrial Intelligence – SETI". Other very prominent scientists find this effort absurd. The great Enrico Fermi for one posed the compelling question that if aliens do exist "Why aren't the aliens here?" That obviously does not negate their existence in less capable intelligent forms or their disinterest in humans in case of a superior intelligent beings. Those who firmly believe in the evolutionary scenario for the emergence of life on earth to extremes contend that even humans will inevitably evolve into other advanced completely different forms with time since the Sun presently is only at less than half of its presumed life. That kicks the door wide open to the possibility of far more intelligent life forms in the universe bearing in mind that there are other star-systems that are billions of years older than the familiar solar system. Elapsed time is considered the most important factor in evolution of life and intelligence. Additionally, there is no reason whatsoever to assume that life has to follow the human path especially with the current development and evolution of *artificial intelligence*. Therefore, atheistic arguments against the existence of Angles and Jinn seem very persuasive especially in the Western Judeo-Christian context. How does the Islamic tradition differ? Intellectually speaking, the Islamic tradition is diametrically at odds with that of the Western Judeo-Christian one. While that tradition persecuted to extremes the mere idea of thinking, Islam in its sacred tradition not only opposed persecution in all its forms especially that of ideas, it also encouraged pursuing such contemplations and free thinking and rewarded it both here and in the hereafter. Islamic history records exceptionally few incidents of thought suppression of rather intellectually minor individuals which always coincided with periods of decay and backwardness. That is the exception that proves the rule. On the other hand, history of the Western Church is a continuous record of such persecution of major thinkers, philosophers and scientists. For example, the Dominican monk Giordano Bruno fled from his

Naples monastery because of threats on his life due to his "radical opinions" and in 1584 published a pamphlet "On the infinite Universe and Worlds" where he stated "There are countless Constellations, suns and planets; we see only the suns because they give light; the planets remain invisible, for they are small and dark. There are also numberless earths circling around their suns, no worse and no less than this globe of ours." While currently even the least educated would find nothing unusual in this statement, Bruno was imprisoned in Rome by the inquisition for his "obstinate and pertinacious heresies" when he was lured to return to Italy in 1592 to gain a professorship at Padua. He was *burned at the stake* for his inadmissible thoughts in February 1600 in the Campo de Fiori. Ironically, that professorship went to Galileo who was also persecuted by the same Church for his ideas about the physical world. Scientifically speaking, there is nothing one can find in the Western Judeo-Christian tradition that would promote methodical scientific investigation of the universe even at the level of the physically observed phenomena let alone the unobserved ones. Now the question is "How does Islam deal with the notion of Angels and Jinn? And does that reconcile with the confirmed findings of modern science?" The well recorded Islamic tradition of the Qur'an, Glorious Divine Sayings (الأحاديث القدسية), Sunnah (السنه) and the biographies of the companions unequivocally assert the existence of Angels and Jinn. These beings are explicitly mentioned and some are even given specific names in the Qur'an and the Glorious Divine Sayings (الأحاديث القدسية). Mohammed (محمد - عليه الصلاة و السلام) repeatedly spoke of meeting, receiving revelations from and conversing with the Archangel Gabriel numberless times over the 23 years of his prophethood. He has seen countless numbers of Angels during his daily activities particularly during the prayers and occasionally crossed path with Jinn particularly Iblis (إبليس). On his "night journey" or Al-Isra' wa Al-Me'raj (الإسراء و المعراج) he encountered and dealt with numerous numbers of Angels and Jinn. These are all matters of fact for Muslims due to their association with Mohammed (محمد - عليه الصلاة و السلام) himself and the utmost trust Muslims place in him. Obviously Westerners and particularly the atheists among them do not accept that. The question of trust in Mohammed's (محمد - عليه الصلاة و السلام) integrity has been dealt with before on several occasions and there is no reason to litigate it here anymore. However, the most important issue at this juncture is what the tradition tells about the companions' experiences in this regard. It is quite well known that Abd Allah ibn Abass (عبد الله إبن العباس) who is among the great companions reported physically seeing the Archangel Gabriel twice when he was in the company of his father who did not have the same experience since he is of a lesser status as a late convert to Islam. More importantly, dozens of the companions reported seeing individuals they do not recognize fighting side-by-side with them against the unbelievers in several battles over many years. This is of particular importance since Arabs of that era constituted a very tightly closed community who knew each other (friends and foes alike) fairly well and there is no margin for error. It is also of particular importance that these companions never saw these individuals afterwards but were told by Mohammed (محمد - عليه الصلاة و السلام) that those were Angels sent by God (الله - سبحانه و تعالى) to support them. Keeping in mind that the entire structure of Christianity is built on Paul's claimed vision (one single individual) on the "Road to Damascus", it is ludicrous for Westerners to dismiss the testimony of numerous eye witnesses concerning this matter. Still that does not counter atheists' arguments nonetheless. To counter atheists' protestations, one has to resort to scientific arguments since these are understandably and convincingly the only ones acceptable to them. As all scientists know, cosmologists found a huge deficit in calculating the mass of the universe according to its matter and attributed the missing mass to black holes and dark matter for which they found some theoretical proof at best. That is not satisfactorily enough to solve that problem. Dark energy was the answer albeit with absolutely no sensible proof so far. Modern physics has no quarrel with energy materializing or matter becoming energy. Islam defined Angels as creatures made of light and Jinn as creatures made of fire or in scientific terms infrared radiation. This was done more than fourteen hundred years ago. The plausibility of this unknown dark energy representing what Islam refers to as Angels and Jinn is not farfetched at all. According to Islamic tradition as witnessed by many companions on many occasions, this energy could have materialized in terms of individuals they could see. The mechanism as how this happened is irrelevant since science does not know how energy becomes electrons and positrons either but it does. Angels (or Jinn for that matter) materializing does not necessarily imply becoming humans but rather taking human shape for recognition. The incident of some Angels interacting with Ibrahim (Abraham – عليه السلام – إبراهيم) and declining his food is a case in point. It is clear that the Qur'anic narrative is self-consistent in all its presentations of Angels. They could arrive in this three dimensional universe from higher-dimensions universes for example where they exist under completely different physical laws. It has to be emphasized here that these arguments advance mere plausibility rather than actual fact in an attempt to refute the alleged uniqueness of the atheistic arguments. It remains to mention that the Qur'an is explicit in stating that Angels and Jinn are aware of humans, observe them and are in constant contact with them despite most humans' unawareness of this fact. This is unequivocal in the Qur'an and there is a whole Surah named after Jinn (سورة الجن) detailing an incident of such interaction. The Qur'an also makes it clear in Surat Al-Baqarah (سورة البقرة) that Angels and Jinn in terms of at least Iblis (إبليس) were there receiving God's (الله - سبحانه و تعالى) news concerning the process of creating Adam (آدم – عليه السلام) and most importantly they could independently construe his and his progeny's shortcomings before he was even created. This is superior intelligence at its most obvious. Fair scientific minds can see that the Islamic tradition n its various aspects consistently gives integrated and complete answers to questions raised by the most accomplished humans more than fourteen centuries later. It would be foolhardy though to think that such arguments like the ones just given inhibit human quest for knowledge under the pretext that according to Islam all questions have been already answered and there is nothing left to learn. On the contrary, Islam not only encourages humans to acquire knowledge but it actually exhorts them to do so as it brings them closer to appreciating the sovereignty and magnificence of God (الله - سبحانه و تعالى) and His creation which is the cornerstone of the religion of Islam. Obviously learning will continue till the end of time and so will Islam's interpretations.

Resurrection

Atheists scuff at the notion of resurrection as a means to manipulate and exploit the underclasses with a vague promise of a better life in the hereafter. There is no denying that this immoral approach has been the hallmark of human socio-economic history. While Islam promises a vastly better life in the hereafter for the believers, it unequivocally condemns the exploiters and promises them severe punishment. One can plausibly argue that this does not prove anything in light of the historical reality of exploitation even within Muslim communities in addition to the lack of material proof of these rewards and/or punishments. As is always the case in this undertaking, logical arguments do not settle any debate as opposed to irrefutable scientific arguments acceptable to all. Life scientists unlocked its secrets by attributing its success in creating the vast number of species in the world to the cell self-replicating process. All information required for that process is included in the DNA structure responsible for heredity. This is how species progress from one generation to the next. Every now and then mutations take place for a variety of reasons leading to new species. This is how all creations, including humans, are related going back to a single origin according to evolutionists. Thanks to the DNA information included in the YY (female) and XY (male) chromosomes in human cells that are randomly fertilized, individual human beings are unique in their DNA structure which means each individual is constructed according to a unique set of information. An insightful look at the tree of life as convincingly determined by science leads to the persuasive unavoidable conclusion that one can reconstruct the path taken by a species to reach its current status. By the same token, one can theoretically trace the lineage of an individual human being with certainty as far back as desired. In other words, the path taken to bring about a single unique human being (the same should be true for any other creature) is unambiguously and uniquely well defined. Given that absolutely valid scientific fact, it is easy to see that every single human being can be *reconstructed* knowing its DNA structure. The only problem is the amount of information needed to be processed to accomplish that reconstruction. Therefore, in a *material* sense, it is theoretically possible to reconstruct all human beings given the required information. The principle is valid while the means are presumably lacking. Spanning the entire human history, humans in their beliefs bestowed infinite knowledge to God (الله - سبحانه و تعالى) with absolutely no hesitation or second thoughts. Some anthropologists think that was the reason for polytheism to make gods available to human comprehension and interaction. Atheists cannot refute that fact even if they question the existence of God (الله - سبحانه و تعالى). Belief in God's (الله - سبحانه و تعالى) infinite knowledge and wisdom is as old as human existence. Islam subscribes to that verity as well. Now it is trivial to push the argument one step further and state that God (الله - سبحانه و تعالى) has the required capacity to store all the information needed to *reconstruct* every single human being (creature) that has ever lived. That is simply what resurrection is about. There is nothing in science that make this process unattainable. There is another dimension to this approach following the Islamic concept of resurrection. The Qur'an is not explicit on the status of humans during this process of resurrection whether they will be materially resurrected or in some other form which is generally referred to as Al-Ruh (الروح) or the soul. The material case is so far thoroughly solved. If resurrection is for Al-Ruh (الروح) or the soul, then one should accept it wholeheartedly since the nature of Al-Ruh (الروح) or the soul itself is not known. In other words, one should accept what one cannot understand based on the validity of what one does understand perfectly well which is the material case of the same problem. Now why would God (الله - سبحانه و تعالى) want or care to resurrect human beings anyway? The answer should be self-explanatory as humans were divinely chosen (as explicitly mentioned in the Qur'an several times) to be God's (الله - سبحانه و تعالى) vicegerent in Al-Ardh (الأرض) after being fully prepared for that task and given the required tools of knowledge and free will. It is just natural for them to come back and account for what they have done and bear the consequences. Therefore, there is consistency and logical flow in the Qur'anic narrative concerning human duty towards God (الله - سبحانه و تعالى) and its repercussions. In a different approach one can give a very brief cosmological hint (cosmology is detailed in another part of this study) about the reasonableness of the concept of resurrection. Solving general relativity's equations, the Austrian logician

Kurt Godel (1906 – 1978) found out that the theory did not in itself preclude time travel to the past. In simple terms it is plausible according to the general theory of relativity for individuals to come back to face everything they have done; this is the definition of resurrection. However, this valid solution is dismissed by physicists as unrealistic because it describes a universe that is not expanding while it is also rotating. It is observationally established that the universe is expanding but that is not an absolute intrinsic fact since the rate is not exactly fixed and it may eventually come to a halt and even contracts. Rotation of the universe is a tricky question since no one can get out of the universe to observe such rotation and it has to be determined rotation with respect to what. Nonetheless the issue cannot be scientifically dismissed out of hand as atheists do.

PREDESTINATION AND FREE WILL

Islamic stand in affirming free will and consequently accountability and judgment is emphatic. A fundamental principle of Islam undergirding the faith is individual's accountability for any and every undertaken act regardless of how small or how large it is. This is a clear cut criterion in the description of what qualifies a person to be a believer and a member of the community of Muslims past, present and future till the end of time. Qur'anic revelations are numerous and explicit in emphasizing this concept. This is a uniquely Islamic approach differentiating Islam from any and all other belief systems heavenly or human made. Mohammad (محمد - عليه الصلاة و السلام) was questioned about the apparent paradox of simultaneously asserting, beyond the shadow of a doubt, humans' (and Jinn's) free will and God's (الله - سبحانه و تعالى) ultimate power over their actions when he first started inviting people to Islam. Since Islam is an integral whole, the ultimate validity of its easy to relate to aspects, underwrite the validity of those difficult to understand ones. However, there are no mysteries in Islam where the difficulty in understanding some of its assertions stems from humanity's transient lack of knowledge. Predestination and fate are time dependent by nature. That means that a determination has to be made *before* the act is carried out; there is a definite *before* and *after* events. It is also a distinct consequence of absolute determinism. Ironically, twentieth century science unequivocally concluded the relative as opposed to the absolute nature of time. The implication is that what seems to occur *before* (in time) to one observer can be observed as happening *after* to another observer according to their relative speed with causality still holding nonetheless. It also relegated determinism to the nonsensical batch of absurd notions replacing it at the most fundamental level with uncertainty. These are unquestionably counter intuitive results but they are equally unquestionably incontrovertible truths of modern science. They have been validated in numerous applications sometime with grief as in the development of nuclear weapons as an unpredicted consequence of the relativity of time. It is fascinating to learn that quantum theory is the only scientific theory that was never proven wrong despite the countless tests it was subjected to. The upshot of this discussion is that atheistic arguments (by definition are built on the concept of absolute time and determinism) writing off free will as an anachronistic human idea are scientifically baseless. Moreover, topics of fierce debate on fate, free will, God's (الله - سبحانه و تعالى) justice and reward and punishment in the hereafter are actually non-issues for all practical purposes. Antagonists involved in these debates are fighting over non-existing problems with no defined criteria. While human intellect will never advance without such abstract endeavors, it is utterly contemptible for their practitioners to descend into physical abuse of each other. This is more appalling when the abuse takes place within the boundaries of Islamic culture that is fundamentally tolerant. The reference here is to incidents in Islamic history when opposing sides resort to physical abuse sometimes deteriorating into murder of their opponents. Unfortunately, Muslims exhibited such deplorable behavior several times with the worst example taking place over the issue of the "creation of the Qur'an" (محنة خلق القرآن). It is hoped that they come to the conclusion that all logical arguments are temporarily acceptable as long as the laws of nature are not breached and that no argument has monopoly on Qur'anic interpretation regardless of the status of its advocate. Muslim thinkers deduce the concept of predestination from the explicit mention in the Qur'an in Ayah 22 of Surat Al-Borouj (سورة البروج) of the existence of a divinely guarded record of the Qur'an. As is well known (to Muslims), the Qur'an encompasses descriptions of all that is taking place from the beginning of the creation process till the end of time. Looking at it at any instant of time it represents the past, the present and the future all at once. While modern science tends to consider time as a convenient humanly introduced parameter of nature, it strips time of its absolute value in the observable three-dimensional world. However, it does not violate the principle of causality. Additionally, it strongly hints at the existence of as yet undetected higher dimensions where time may not play any role. Islam, since the first revelation in excess of fourteen centuries ago, unequivocally stipulates the irrelevance of time to God (الله - سبحانه و تعالى) where the expressions past, present and future are nonsensical. One can therefore plausibly assume that God's (الله سبحانه و تعالى) will (as well as the Qur'an) is recorded in such a way that time plays no role whatsoever. Taking another step consistent with modern scientific speculation, one can plausibly assume again that such record is preserved in a higher dimension of the universe controlling and managing happenings in the familiar three dimensional world humanity inhabits. In this manner predestination and fate are reconciled with accountability, reward and punishment. The idea that happenings in the three dimensional world are controlled and managed from a higher dimension of the universe needs some explanation. Until the end of the nineteenth century scientists studied all natural phenomena as taking place in two dimensional spaces regardless of the fact that nothing in nature is actually two dimensional. That was a very convenient approach since the extremely slow variation in the curvature of the three dimensional earth allowed approximating limited areas, where researched phenomena take place, to two dimensional planes. Additionally, well understood Euclidean geometry provided excellent tools for such studies. A three dimensional world presumably consists of an infinite number of two dimensional planes. The relationship between observers located in a three dimensional space and their counterparts living in one of these two dimensional planes can be instructive. While the three dimensional persons can easily observe (detect), control and manage everything that goes on in the two dimensional world, the two dimensional creatures are incapable of observing (detecting) the three dimensional individuals or their activities. But some of them being brilliantly educated scientists can find clues such as shadows for example ascertaining the existence of such three dimensional objects albeit they obviously cannot influence their actions. Continuous regular movement of the plane in one direction can be conventionally attributed to the passage of time in this world. Clearly the three dimensional being is not bound by the passing of time in the two dimensional world. The global view of the three dimensional observer covers all at once what would be considered the past, the present and the future from the point of view of the two dimensional creature. Additionally, and most importantly, all information about the plane person(s) is readily available to the three dimensional individual who consequently has the ability to shape, control and manage events in the life of the two dimensional creature(s) even when not exercising that option. It is trivial to imagine a methodology by which this information is recorded and stored in the three dimensional world encompassing the entire history of the two dimensional world and its inhabitants from beginning to end. If all deeds are recorded it is not difficult to consider holding individuals accountable for them and the consequent reward or punishment. One can also observe that almost every person may possibly experience what is called an event of "déjà vu" where individuals are certain that what they experience at one moment in the present has been already experienced before in the past. If one accepts the actuality of the existence of a record of deeds as described here, then the "déjà vu" incident is simply the exposure to that record somehow without constraints of time. That is to say obtaining knowledge of something before it actually happens according to time flow in the two dimensional creatures' sense. These are obviously merely interesting mental exercises that break no physical laws. On the other hand, it is so far believed by scientists that three dimensions represent the lowest possible existence to sustain life. It is also believed that higher dimensional worlds do exist enveloping this three dimensional one. The same is true of the ability of brilliant human scientists to discover clues to their existence. Therefore, one can proceed from the theoretical example of two dimensional/three dimensional worlds and take the intellectually logical but bold step of considering the reality of the three dimensional/higher dimensional worlds. It is not difficult to see that predestination, fate and the existence of a guarded record of all events taking place in this world make perfect sense if viewed from a higher dimensional world the existence of which is not a farfetched concept but rather a very scientific one. Implications of individuals' free will in this world are obvious. Therefore, Islamic perceptions of the accountability of humans and predestination do *not* constitute any physical contradictions or paradoxes. The same argument can be used to refute the atheistic denial of God (الله - سبحانه و تعالى). God (الله - سبحانه و تعالى) described Himself in the Qur'an as existing in a state of being beyond human imagination because theirs is limited by the laws governing this three dimensional world. He is not bound by these laws since He is their Creator. He has all information about this world readily available to Him including deeds of every individual determined by their free will, He interacts with all its events without forcing humans to act in a certain way, He keeps a record of these events and deeds through whatever mechanism He creates, He holds humans accountable and He will reward and punish them accordingly. There is absolutely nothing in these statements that is inconsistent with scientific cosmological discoveries. These according to Islam are facts given to humanity at its creation. When it lost knowledge of them during its pursuit of survival on earth, the Merciful God (الله - سبحانه و تعالى) sent messengers partially conveying them anew to every community. Finally, they were explicitly revealed to Mohammad (محمد - عليه الصلاة و السلام) to convey them in their entirety once and for all to all humanity. On the other hand, atheists promote the frivolous ideas of randomness and chance concerning the creation process and sustainability of the world without external omnipotent guidance simply because they think so. At this point, it may be intellectually stimulating to investigate what it would take in terms of space and time to move from a lower dimensional space to a higher dimensional one and vice versa. In the theoretical example of two dimensional/three dimensional case it is easy to see that the three dimensional world consists of a multitude (up to infinite number) of the two dimensional worlds. Thus, the spacing between a plane and the three dimensional world above (or below) is zero. Assuming the possibility of moving from one to the other exists then, the distance to be travelled in going from one to the other is zero. Additionally, that implies the elapsing of zero time to perform such transition. In Ayah 16 of Surat Qaf (سورة ق), God (الله - سبحانه و تعالى) describes Himself as closer to a human being than this person's vital organs needed for survival. Historically this was linguistically interpreted to confirm God's (الله - سبحانه و تعالى) knowledge of the deep thoughts and even unspoken ponderings of every human being. Looked at from a scientifically deduced point of view, one can interpret this Ayah in a physical way. In other words, the space (distance) separating God (الله - سبحانه و تعالى) from His creation is negligible amounting to zero. In this case no figurative or allegorical speech is needed. Obviously stating that, one is not conferring certain physical nature to God (الله - سبحانه و تعالى) but rather confirming the absolute higher existence of God (الله - سبحانه و تعالى) outside space and time albeit unfathomable and beyond human knowledge. The same argument can be used to explain the Archangel Gabriel's frequent encounters with Mohammad (محمد - عليه الصلاة و السلام) and all other Messengers of God (الله - سبحانه و تعالى) to convey revelations. Therefore, all events and incidents that atheists consider supernatural are given physical scientific explanations in Islam. As a matter of fact, Surat Qaf (سورة ق) in 45 Ayahs gives a point by point explanations of the creation process, the hereafter and the nature of the relationship between human beings and God (الله - سبحانه و تعالى).

AL-JANNAH (PARADISE - الجنة) AND JAHANNAM (HELL – الجحيم/النار/جهنم)

With free will comes accountability and then judgment in a very logical natural flow of events. Judgment is immediately followed by reward and/or punishment but preceded by resurrection of human beings. The Qur'an is very clear about a second creation or a return for such judgment to take place and reward or punishment to be meted out. But why could not this happen during the lives of humans rather than going through the process of dying to be then resurrected to face the consequences of their deeds? That may sound like a round-about way to do something even when one believes in the process itself. The Qur'an is explicit about the first human Adam (آدم - عليه السلام) residing in Al-Jannah (paradise – الجنة) right after his creation but having to *descend* to earth as a result of disobeying God's (الله - سبحانه و تعالى) instructions as opposed to being God's (الله - سبحانه و تعالى) vicegerent in Al-Ardh (الأرض). One may assume that Al-Jannah (paradise – الجنة) is part of Al-Ardh (الأرض) where earth is not a very distinctive part but where life in its familiar form began and where Adam (آدم - عليه السلام) ended up. One may equally assume that Al-Jannah (paradise – الجنة) is an exulted higher existence (e. g. dimension) different from earthly existence. One can come up with other speculative options not inconsistent with the Qur'anic narrative as well. Be that as it may, all speculations agree on humans having to go through a *reverse* transition to the one they experienced out of Al-Jannah (paradise – الجنة) to be judged before God (الله - سبحانه و تعالى). That is simply the processes of death and resurrection. Whether or not the familiar three dimensional Universe is what humans return to with all the speculations involved is immaterial to the ultimate necessity of the processes of resurrection and judgment. Nonetheless, these are intellectually and scientifically exciting issues that are worth pondering. Another universally accepted matter is that unlike life on earth, life in the hereafter that humans return to is eternal and they are immortal within its framework. The upshot of this discussion is that a *reverse* transition has to take place to allow for judgment. That transition is termed "doomsday" (القيامة) when all life comes to an end. There is nothing in the Qur'an that links the end of life to the end of the physical universe or the end of time. Obviously eternity of life in the hereafter necessarily implies the meaningless of the notion of time. Therefore, the answer to the above mentioned question is that death, resurrection and judgment sequence is not a round-about process but rather a mandatory one. Thus far, the logical consistency of the flow of events in the unfolding of the divinely enacted creation process in all its components in the Islamic narrative is crystal clear. There is nothing arbitrary, magical or unnatural in this narrative. When one understands that all these issues were deeply engraved in the human psyche since ancient times, one wonders how atheism could claim otherwise. However, it is true that major corruption of these ideals seeped into the human soul which is what Islam came to correct. This is the essence of the message of Islam. Once judged, humans are sent to their eternal existence in the hereafter commensurate with what they earned in their earthly life. In simple terms used by common folks, it is understood that individuals with good deeds outweighing bad deeds end up in Al-Jannah (paradise – الجنة) and those with the reverse end up in Jahannam (hell – الجحيم/النار/جهنم). This is the ultimate justice promised by God (الله - سبحانه و تعالى) in every message to human communities since Adam (آدم - عليه السلام) *descended* to earth. Nonetheless, this is an article of faith that can be objected to by atheists. But what could the physical material meaning of the expressions Al-Jannah (paradise – الجنة) and Jahannam (hell – الجحيم/النار/جهنم) be? Taking the Qur'anic narrative of the unfolding of the divinely enacted creation process one more step forward in terms of modern science, it is easy to find a compelling description of these expressions. The second creation or the return implies the generation of a vast universe. Modern cosmology tells of observed hellish parts where unspeakable conditions prevail and fantastic parts where life may actually exist within the familiar universe. It does not require much imagination to think of a similar construction of the universe in the hereafter. It is trivial to attach the label Al-Jannah (paradise – الجنة) to the privileged part and the label Jahannam (hell – الجحيم/النار/جهنم) to the nasty one. The vastness of that universe, extrapolating from the familiar one, accommodates the descriptions given these physical places in the Islamic tradition. It may also answer the curious question raised by unbelievers about the inefficiency of God's (الله - سبحانه و تعالى) creation of such immense observable universe to simply create life on such a negligible planet earth. It could simply be the prototype of the real universe in the hereafter if one wishes to speculate or it could be the crucible of places in waiting for the resurrected masses of humanity but this makes no difference to the validity of the previous arguments anyway.

Virgin birth of Jesus (عيسى – عليه السلام)

The virgin birth of Jesus (عيسى – عليه السلام) always existed as a fundamental article of faith in Christianity. It is also one of the thorniest issues used by atheists to denigrate the Christian religion. Obviously, Jews have no problem with scoffing at the whole premise of Jesus Christ (عيسى – عليه السلام) the person let alone his virgin birth. However, for a host of obvious political and survival reasons they ceased raising any squabbles with Christians over this issue for almost the entire past two millennia. On the other hand, atheists contend that modern science makes a mockery of this claim. Western Christians by and large never had a counter argument since the time of enlightenment and the age of reason. Those who lost confidence in Christianity found it an easy excuse to become atheists. Devotees to the Christian faith nonchalantly overlooked the matter and sometimes responded with violence. However, the well-educated among them faced an acute dilemma since they have very high regard for science while not willing to renounce their faith either. They decided to adopt both attitudes at once. The patently unreasonable combination was accepted under the allegation that the two fields are mutually exclusive and arguments of one should not by any means infringe on the other. This virgin birth matter (as well as numerous other similar ones) was considered allegorical and was taken at face value but not as a fact. This way they absolved themselves of having to present a logical scientific explanation for their faith. This is the situation at the present time as far as Western mindset is concerned. At the other end of the Christian spectrum Eastern Christians vehemently adhere to the virgin birth of Jesus (عيسى – عليه السلام), most ironically feeling solace in falling back on the beliefs of the Muslims among whose communities they live to avoid any conflicting doubts. It is most intriguing that at the present time, it is probably only Muslims that believe in the virgin birth of Jesus (عيسى – عليه السلام) as a matter of absolute fact since it is explicitly and unambiguously mentioned in the Qur'an. That does not necessarily prove the point to an atheist regardless of the argument repeatedly presented in this endeavor about the Islamic principle of accepting what is difficult to understand along with the countless statements that are understandable due the wholeness integrity of Islam. Then, how does Islam find a solution to this contradiction? As always, the solution is embedded in the consistency of Islamic statements with the findings of science. As mentioned before, proving the viability of the Qur'anic premise of resurrection is rooted in employing the evolutionary principle of the unique idea of a single origin of life. Biology, microbiology and theory of evolution and other related fields convincingly constructed a tree of life that encompassed all living things. According to this system in a crude way, what differentiates one species from another is basically the kind of mutations they experienced. That is to say that theoretically one species can revert to the other given the right combination of mutations. Thus, characteristics of one species are somewhat hidden in the DNA structure of the other. Keeping this in mind, it should be useful to recall that biologists established the fact that some species (albeit rather primitive) propagate by replication as well as non-sexual means and some do not need two parents to reproduce. While the probability of a sexual creature adopting a non-sexual method to propagate is evidently extremely low, it is by no means zero. It should be obvious by now where this argument is headed. In principle, there is no scientific impossibility that Jesus (عيسى – عليه السلام) was born in a non-sexual way while such a possibility is vanishingly small albeit non-zero. It goes without saying that there is only one single Jesus Christ (عيسى – عليه السلام) in the entire history of human kind with the unique distinction of being brought to this life in a process of non-sexual virgin birth. Therefore, the virgin birth of Jesus (عيسى – عليه السلام) is not an unscientific myth but rather a confirmation of the statistical nature of creation which is the bedrock assumption of evolution to begin with. Denying this possibility is unscientific. Now invoking the statements of the Qur'an, one finds amazing hints relating humans to Apes and pigs; the two mammals with the closest DNA structure to humans. A not too subtle an indication of evolution with all its implications. It is advisable to remind oneself that these statements were uttered in excess of fourteen centuries ago by an illiterate person of the highest character who never took personal credit for them but rather asserted his role as only a conveyor of the message from God (الله - سبحانه و تعالى).

Qur'anic Narrative and Modern Scientific Findings

EXPLANATION AND PREDICTION

It is clear from the scientific revolutions that took place at the beginning of the twentieth century and what followed that science is not about a search for the absolute truth. It is about coming to grips with it through a succession of theories replacing each other, each taking human knowledge a step towards that truth. Einstein explained this fact saying "Every theory is killed sooner or later... But if the theory has good in it, that good is embodied and continued in the next theory." Ne theories are normally proposed to find a way out of a deadlock in science. When an impasse in reached in a branch of science when the standard universally accepted approaches fail to explain certain phenomena, radical new approaches are proposed. This process usually ushers a paradigm shift in that field. The new suggested alternatives are normally subjected to rigorous examination to validate or invalidate them. As a matter of course, a new theory explains the phenomena that the standard approaches could not explain. However, it is historically known that this could be and has been frequently done by adding certain assumptions to the existing universally accepted theory. This inevitably results in a plethora of solutions with equal claims to have solved the problem(s) at hand. The most famous example in physics is the multitude of attempts over many centuries to preserve the concept of "Ether or Aether". As some fundamental physical theories were based on the assumption of its existence, physicists spent many desperate efforts over several centuries to try to validate its reality. The whole structure of

Newtonian mechanics hinged on the existence of ether as the medium through which forces of nature propagate. Many Brilliant experiments, the most famous of which is the "Michelson-Morley experiment", were designed to provide quantitative measurements of its effects. These proved to be futile attempts to measure something, as it turned out, that has no reality. Finally, in 1905 Einstein proposed the "Theory of Relativity" which did away with this concept altogether in the meantime resolving the impasse facing classical physics at the time. However, many other solutions based on additional assumptions to Newton's original ones existed and could also get around the deadlock while preserving aether's reality albeit in a most scientifically distasteful way. Additionally, the theory of relativity had its own brand new assumptions. Contrary to popular belief, when Einstein published his theory, it made no impact and was largely ignored by physicists at the time. It was considered another attempt to *explain* existing phenomena albeit in what was then considered unacceptable unorthodox way. It is interesting to note that there are a minority of scientists even at the present time who do not accept relativity's most fundamental assumption of the constancy of the speed of light and it being the upper limit to any physical speed attainable by material objects. Popularity is not a good substitute for sound science even if it is that of relativity and Einstein. Newton's fame and popularity among scientists is a perfect example. However, what gave Einstein's and before him Newton's theories their acceptance is not their explanations of certain phenomena since many other theories existed to explain these very same phenomena even in a more contorted way. The greatest successes of both Newtonian and relativistic theories rest on their *predictions* of up to that time unknown phenomena. When these phenomena were verified, universal acceptance followed suit right away. For Newton's gravitational theory its validation was the calculated return of Halley's Comet in 1758 which took place exactly as predicted. Newton's theory explained to a very good approximation the orbits of the then known planets within the solar system. However, it failed to account for the movement of the perihelion of Mercury (the point on its elliptical path which is nearest to the sun). This discrepancy was explained more than two centuries later by Einstein's "General Theory of Relativity". However, due to the extraordinary success of Newton's theories and the worship-like attitude of scientists towards him, numerous assumptions existed to help explain this anomaly. That is to say that general relativity was not the only plausible explanation. Einstein's general theory of relativity on the other hand, **predicted** the bending of light passing near massive objects such as stars due to space curvature. The expeditions to examine this phenomenon during the sun eclipse of May 29, 1919 gave credence to the theory. However, the measured parameters were exceedingly small and not all scientists (especially in Nazi Germany) accepted relativity. The ultimate vindication came with the discovery of pulsars in the 1960s. Currently the "General Theory of Relativity" is considered the cornerstone of all cosmological theories. Some noted scientists still do not accept relativity as it is only verified by extremely complicated cosmological experiments that are subject to challenge in their conclusions. The upshot of this discussion is that in science it is not enough for a theory to *explain* some existing phenomena. The ultimate proof of the validity of a scientific theory is to *predict* some unknown phenomena that can be experimentally verified later on. High Energy and Particle Physics are full of such examples. The intriguing aspect of Einstein's scientific contributions is that by the year 1919 he had completed the theory of relativity in its special and general forms. From that time till his death in 1955 he tirelessly worked on what he hoped to be a theory unifying all the fundamental forces in nature. That effort was disappointing though. Although he was also one of the very early fathers of quantum physics (he actually won his Nobel Prize for physics for the photoelectric effect), he never accepted its statistical nature leading to the concept of uncertainty and the absence of deterministic causality at the elementary particle level. Dismissing the fundamental principle of quantum physics he is known to have stated "I find the idea quite intolerable that an electron exposed to radiation should choose *of its own free will* not only its moment to jump off, but also its direction." He considered this principle a personal affront adding "In that case, I would rather be a cobbler, or even an employee in gaming house than a physicist." He is probably most famous for his comment that "God (- الله سبحانه و تعالى) does not play dice with the universe" and "God (- الله سبحانه و تعالى) is subtle but not malicious". He never wavered in his opposition to quantum physics in the face of its phenomenal success in describing and **predicting** natural phenomena. Quantum physics is the only theory known to humanity that was never proved wrong in the overwhelming number of experiments specifically designed to check its validity. Thanks to his stubbornness, Einstein after announcing general relativity became an odd man out within the community of physicists. Although he never lost his standing as a unique genius, his unified field ideas were summarily dismissed by almost everyone. The sad fact is that in reality he did not contribute anything useful to physics during the last three decades of his life. His numerous thought experiments (those which can theoretically be carried out but technically are impossible to experimentally set up) to undermine quantum physics failed one after the other. From very early age Einstein established for himself that things have to be logical and explainable in a scientific way to be accepted. As such he is quoted to have said about his youth "Through the reading of popular scientific books I soon reached the conviction that much of the stories in the Bible could not be true". Paradoxically his refusal to bow to the reality of observations validating the quantum theory went against the essence of science of which he firmly believed in during his youth and became a leading figure and an icon in his adulthood. Unlike absolute religious doctrines regardless of their acceptance, atheists base their arguments not on scientific facts as implied in their arguments but on *plausible assumptions* albeit with assumed high degree of probability according to the state of human knowledge at the corresponding instance. That approach ignores other plausible solutions to the same problems which may turn out to acquire higher probability with advancing human knowledge. In other words, atheistic arguments are by necessity temporal. Age of reason's prevailing atheistic arguments which were taken at that time as established facts are decidedly scientifically wrong when viewed from modern physics angles. Phenomena taking place within the micro-universe of atoms and elementary particles are governed by the laws of quantum mechanics and assume probabilistic nature. These laws associate with them uncertainty and indeterminate existence. Almost all such phenomena are humanly counter-intuitive. Observed phenomena in the macro-universe on the other hand cannot be described by quantum mechanics because of the incalculable number of elementary constituents involved which negate the probabilistic nature of the problems to be solved. Nonetheless, it is well established that classical approach simply represents a special case of the more general quantum one. Because humans exist in this macro-universe, their intuition is deterministic following the laws of classical/Newtonian mechanics. Phenomena in both the micro and the macro universes are well understood employing their corresponding laws. However, how nature evolves from the micro-universe to the macro-universe is totally unknown. Scientists and philosophers alike tried to tackle this problem for over a century now but failed so far with no clear methodology to be adopted in sight. Most physicists accept the unexplained idea of the collapse of the wave function derived from "Schrodinger's Equation" associated with an elementary particle on observation to give it a deterministic (probability of one) nature. Thus, giving the observer an essential role in the transition, but in the meantime creating a host of new physical and philosophical problems. Although transition from one to the other is shrouded in mystery, validity of the existence of both universes is not questioned by anyone and is taken for granted.

Arguments concerning God (الله - سبحانه و تعالى) and His attributes are used by Muslim thinkers to explain any and every occurrence in the observable universe. In the past, logic and mental exercises were deployed for that purpose covering the macro-universe experienced with sensory faculties. With the discovery of the micro-universe, humans noticed astounding number of coincidences without which life and even the universe itself would not exist. The slightest deviation in so many fundamental physical and cosmological parameters would preclude existence of life and by extension the physical universe to be observed. Different scientists and philosophers give their explanations to this situation. Atheists find it convenient to attribute everything to chance. They point out the prevalence of randomness in nature. Probabilistic nature and uncertainty form the indisputable foundation of the micro-universe. To get around the absurdity of attributing creation of the universe to pure chance, some advance the concept of "Multi-Verse" to account for the observable universe as the only one among up to an infinite number of potentially existing universes just happened to have the right parameters to allow life and hence be observed. By implications, other universes are necessarily unobservable. These are certainly all *plausible* explanations but they are decidedly *not facts*. Currently, Muslim thinkers are not puzzled in the least by these coincidences. They see God's (الله - سبحانه و تعالى) will at work. Additionally, they extrapolate such arguments from the Qur'an which every single Muslim that ever existed or will exist till the end of time consider the literal word spoken by God (الله - سبحانه و تعالى). However, they fail to explain how God's (الله - سبحانه و تعالى) will is translated into actual physical laws. This is identical to the situation faced by scientists (especially the atheistic ones) and their failure to give concrete answers to how the macro-universe evolved from the micro-universe. Both groups give *plausible* explanations with atheists patting themselves on the back as custodians of the ultimate truth while excluding all other explanations than theirs. Plausibility of atheistic explanations is by nature time dependent. It almost always changes with the progress of human knowledge. On the contrary, plausibility of Muslims' explanations falls back on eons long of deep rooted human history and humanity's consistent beliefs in an all-powerful higher divine authority outside space and time. Most importantly, they depend on the validity of the Qur'an and the Sunnah (السنة) and verifications of their **predictions** over the past fifteen centuries. From a strictly scientific point of view, there is absolutely no logical reason to give more credit to the plausibility of mere scientific theories that keep changing with the acquisition of more knowledge over that of Islamic concepts that stay the same in time while continuously subjected to verification of their **predictions**. All living things recognize the world surrounding them through their respective senses. Humans are known to have five senses they use to interact with their environment. Step-by-step patiently studying their own and other animals' anatomies, scientists discovered what they always intuitively suspected that all their senses are controlled by the brain. This fact clearly appears in common expressions in every language since immemorial times. With the tendency to specialize, the separate science of neurology was developed as a branch of biology to study all phenomena associated with this vital organ. Gradually it was discovered that every part of the brain controls some aspect of a person's bodily as well as mental activities. It was also discovered that the brain consists of billions upon billions of neurons and synapses that harmoniously cooperate and *electrochemically* interact to produce the intended result or sensation. Brain activities which can currently be recorded and studied are reduced to neurons responding to stimuli initiated by electromagnetic signals received from the world outside. Thus, the circle is closed by returning to studying the physics of these signals. In other words, neurology is applied biology which in turn is applied chemistry. Chemistry is nothing more than simple quantum physics problem as explained by the greatest American born physicist Richard P. Feynman. Thus, studying human sensory systems is reduced to issues of behavior of elementary particles which is governed by quantum physics. But quantum physics is by its very nature probabilistic and as such allowing for *free will*. Therefore, sensory perceptions are nothing more than the expression of physical interactions among elementary particles (photons, electrons, etc.) in terms of human language. For humans each sense is associated with certain words and expressions. For example, vision contributed to the development of words like color (red, green, yellow, etc.), beautiful, ugly, etc. The same is true of the other four senses. It is important to understand that these words and expressions are human conventions with absolutely no intrinsic physical values of their own. Eons of usage by humans created the illusion of their fundamental worth in uniquely describing nature. That means that reality as perceived by humans (and probably other living things) is a human invention conventionally accepted, not a physical unassailable fact, based on the surrounding environment experienced through the human sensory system. Elementary particles behave according to the laws of quantum physics with its probabilistic undergirding. However, due to the huge number of contributing elements involved, it was easy to express sensory perceptions in terms of classical deterministic physics with its Newtonian underpinnings. The unavoidable conclusion here is that while it is extremely convenient to handle daily observations and practices using conventional deterministic classical laws, to actually reach fundamental understanding of nature one has to use the probabilistic laws of quantum physics. The extreme difficulties involved in pursuing this approach are understandable but that does not preclude their basic truth. Put another way, sweeping judgments on any observed phenomenon or simply an elementary assumption based on common human sensory perceptions are necessarily false. *Convenience does not under any circumstances overrule science.*

Having this indispensable truth in mind; notwithstanding atheists' protestations one can unhesitatingly state that *all so called celebrated atheistic arguments about the nature of God* (الله - سبحانه و تعالى) *are based on human sensory perceptions that do not amount to scientific certainty.* On the other hand, Islamic arguments concerning the same issue are based on affinity to scientific speculations extrapolating ironclad scientific findings. More importantly, they are also based on explicit Qur'anic statements with the veracity of the Qur'an itself irrefutably established time and time again in recognition of past events in the many centuries elapsing since its revelation. Atheists proudly present and advance their arguments as solid conclusions to acceptable sound scientific theories. While science is unquestionably laudable, it is for fundamental reasons a very poor crutch. The universally conceded incomplete nature of any scientific pursuit disqualifies it from being an ultimate judge on crucial issues. Obviously there is nothing more crucial to pass judgment on than the existence of God (الله - سبحانه و تعالى). Science is by far the greatest tool humanity has developed to solve life's puzzles but its limitations are palpable to all. There seems to exist some natural phenomena the explanations of which lie beyond its materialistic analytical tools. While biological and chemical structures of living things from the very primitive to the very complex are well understood, what gives them conscience is totally unknown and is even considered unknowable.

To compellingly elucidate this point one has only to explore the inability of science to explain the phenomenon of sleep among living things. Sleep is as old as life itself. It is currently accepted by biologists that all living things sleep. Humans sleep, animals sleep, fish sleep and even nematode worms sleep. It is also agreed that plants sleep albeit studies are far less comprehensive in this field. Although the utmost purpose of all living things is survival, they invariably submit to periods of unconsciousness where they are exposed to danger. Whatever sleep gives to the sleeper is worth tempting death time and time again for a lifetime. Scientists attribute the need to sleep to increase in "sleep pressure" in exactly the same undefined and unproven way as cosmologists speak of "dark matter" and "dark energy" when faced with fundamental problems they cannot solve. As in the case of dark matter and dark energy, the problem of explaining the need to sleep is solved by inventing unquantifiable new expression; the "sleep pressure". It was assumed that sleep has to do with variations in brain activities. To study brain activity during sleep, researchers use electroencephalographs (EEG) to record brain waves. They describe cycles of slow activity followed by rapid eye movement (REM) intervals during which dreams take place. However, it was determined that jellyfish also sleep observing the same routine. They do not have brains but only primitive nerve net. They also belong to one of the most ancient animal groups indicating the intimate connection between life and sleep. Thus, one can clearly see that science is incapable of explaining a phenomenon as common as sleep that all members of humanity who ever lived, not to mention other living things, experience every single day during their lives. In parallel, it is fascinating to contemplate what the Qur'an had to say about the nature of sleep. In Ayah 42 of Surat Al-Zumar (سورة الزمر آية 42) for example it is stated that on death or in "*sleep*", all souls go back to God (الله - سبحانه و تعالى). However, He keeps those He ordained death for and sends the others back to life for a predetermined period of time. That is to say that there is no any physical distinction between "*sleep*" and "*death*". Nonetheless, the most illuminating expression in this Ayah is that both phenomena represent "*going back to God* (الله - سبحانه و تعالى)". The Ayah is ended by an exhortation to thinking humans to recognize the process as a sign of God's (الله - سبحانه و تعالى) omnipotence. Thus, in very few words the Qur'an (again the literal words of God (- الله سبحانه و تعالى)) gives full explanation for both sleep and death. How would one understand the repeated process of "*going back to God* (- الله سبحانه و تعالى)"? One *plausible* explanation can be found in what cosmologists currently speculate about the structure of the universe. It is intriguing that almost every person reaching advanced age objects to identifying with the old frail person in the mirror. Additionally, it should be obvious to everyone, regardless of the intellectual capacity, that a human being while growing up adds matter to the body that was not originally there. Biologists tell that the human body (so is every living thing) is continuously generating new cells to replace dead ones all the time. What that means is that there is a difference between what a person in *reality* is and what is observed as his/her three-dimensional material body. Therefore, humans (and other living things) are supposed to consist of both body and soul. Everyone understands what human body is but different cultures and different belief (or disbelief) systems vary in their definition of what a "soul" represents. Islam uses the Arabic word "Ruh" (الروح) to designate this entity. When Mohammad (محمد - عليه الصلاة و السلام) was asked about the nature of this entity, the Qur'an in Ayah 85 of Surat Al-Isra' (سورة الإسراء آية 85) found no other possible response to give him than to state the obvious that it is something within the domain of God (الله - سبحانه و تعالى)'s knowledge. In Islam an inquiry mandates an answer. The reason for the generic one given being details beyond questioners' understanding and lack of any available understandable similes as is the case with countless other examples in the Qur'an. Incomprehensible answers solicit rejection and rejection of the word of God (الله - سبحانه و تعالى) represents disbelief and solicits punishment. An underpinning of Islam is that God (الله - سبحانه و تعالى) is Merciful and Compassionate and as such, He would not punish His servants over things they could not possibly understand. The Ayah then clarifies the reason for such a generic answer asserting the limits of human knowledge. Interpreters of the Qur'an debate the meaning of this clarification. Conservatives insist that it implies the unknowable nature of "Ruh" (الروح). However, according to the Arabic language there is no absolute negation involved in the Ayah. Therefore, others more inclined to explore deep scientific facts admit the limitations of human knowledge, particularly at the time the question was posed, but point to its progress hoping to eventually reach the point of figuring out this issue. After this digression to discuss the "Ruh" (الروح), it is time to go back to cosmologists' speculations concerning the structure of the universe and keeping in mind that the "Ruh" (الروح) represents the reality of a human being not his or her physical material body. However, it is physically undetectable by human senses in this three-dimensional world. Cosmologists convincingly speculate that the present three-dimensional universe is simply part of a more general

"Multi-Dimensional" universe (this is the essence of string theory). One can legitimately extrapolate this to *plausibly* assume that while human body exists in this three-dimensional world, the "Ruh" (الروح) resides in a higher dimensional one and essentially physically controlling all bodily activities. As death and sleep states are identical according to the Qur'an and both represent states of unconsciousness, *"going back to God* (الله - سبحانه و تعالى)*"* expressed in the above mentioned Ayah can be assumed as indicating the "Ruh" (الروح) reverting to a higher dimensional universe. Death in this case means not returning to the observable three-dimensional world while sleep means it does. If true, the mechanisms involved in this process are way beyond the state-of-the-art human knowledge and that is why science has been and still is incapable of explaining the very mundane process of sleep. In this case, it is easy to explain why sleep is an essential activity to all living things including humans without resorting to ill quantified definitions of any kind. Aging and ultimately death can be in this sense described as the gradual detachment of the "Ruh" (الروح) residing in a higher dimension from the physical material body residing in the familiar three dimensional World. This process is necessary for recording deeds, enforcing accountability and passing judgment as is ascertained in Islam. *It is not claimed here that this is the ultimate explanation of this phenomena but it is emphatically claimed that Islam through the Qur'an can be interpreted in a consistent way with science to give a sounder solution than inventing unquantifiable parameters.* If the concept of the "Multi-Dimensional" universe is valid, it may explain a lot more than just sleep and death. It could *plausibly* explain Adam's (آدم - عليه السلام) and Eve's (حواء) departure from "Al-Jannah - Paradise" (الجنة) which exists in a higher dimensional world to earth which is in this observable three-dimensional world. Additionally, life after death can simply be returning to "Al-Jannah - Paradise" (الجنة) in the higher dimensional world where the dimension of time does not play a role leading to eternity. Furthermore, an intriguing possibility is to explain Islam's uncompromising affirmation of the fate of Jesus Christ (عيسى – عليه السلام) as ascending to another higher dimension rather than being crucified where he has joined and will be joined by those martyrs who were killed defending the message of God (الله - سبحانه و تعالى) enjoying privileged life before the hereafter (i.e. before the cosmologists' big crunch) as stipulated in the Qur'an in so many places. However, these are all scientifically plausible explanations rather than facts. *Whatever the true nature of cosmologists' sensible speculations turns out to be, there is no doubt that they do not contradict Qur'anic conceptions. Therefore, one can once again easily confirm that Islam can be unquestionably reconciled with findings of science.*

These were several examples of Qur'anic **predictions** and explanations for things that have already happened or are daily experiences. The Qur'an gives **predictions** (like any scientific theory to be verified) that are going to take place towards the end of human existence. For example, Ayah 82 of Surat Al-Naml (82   )

«و إذا وقع القول عليهم أخرجنا لهم **دابة من الأرض** تكلمهم أن الناس كانوا بآياتنا لا يوقنون»
سورة النمل – آية 82

"Thus, when the word of the hour of Doom shall come to pass against the disbelieving among them, We shall bring forth for them **Dabbah from Al-Ardh** that shall speak to them indicating that people no longer believe in Our signs." Surat Al-Naml (Ants) – Ayah 82

The underlined words are an example of the necessity of generalizing the words of the Qur'an beyond everyday common use. The first word "Dabbah" (دابة) is commonly used in Arabic to indicate every living being. It is derived linguistically from the verb conveying walking or moving by whatever means on earth. The meaning of the other word "Al-Ardh" has been exhaustively analyzed in the text to mean more than just earth but rather the whole material universe. The traditional interpretation of the Ayah is that a beast will be brought forth (whatever that implies) from earth to warn people in their own way of speech when Dooms Day approaches. According to the suggested new paradigm in generalizing the meaning of Arabic words, this Ayah should be interpreted to indicate a creature that is not particularly biological but is made of the substance of the material universe (matter) and is capable of communicating with human beings. Its task is to warn humans of the impending doom due to their disbelief. One can see the similarity between this description and the anticipated production in the not so distant future of robots with an advanced degree of artificial intelligence capable of understanding human logic and speech. Since these robots by definition have no free will but function according to logic and reason, they by default are incapable of disbelief. As such, they warn humans of the dire consequences of not believing in God (الله - سبحانه و تعالى). There is absolutely nothing in the Qur'anic or Arabic usage of these generalized words that would not allow such interpretation once the inevitable paradigm shift takes place. Since it is assumed throughout this endeavor that adopting the generalized meanings of the Arabic words is a matter of time through the inevitability of a paradigm shift in Islamic thought, one can recognize a Qur'anic **prediction** of building artificially intelligent non-biological robots. That was foretold more than fourteen hundred years ago which is paradoxically still a human challenging but reasonably realizable dream. For those atheists who demand **predictions** to legitimize any statement one can only offer another Qur'anic utterance of very few words among countless others. Muslims regardless of their degree of intellect or intelligence see absolutely no surprise in this situation. For them the Qur'an is the literal word spoken by the All-Knowing God (الله - سبحانه و تعالى) and therefore cannot be wrong while humans may collectively take centuries to reach that degree of comprehension. Within the context of Islam, it will take humanity till the end of time to completely understand the full meaning of the Glorious Qur'an.

Islamic Predictions, Foretelling of Events and Prophecies

Rejection of religion and faith by atheists and hence denial of God (الله - سبحانه و تعالى) is solely based on the argument that what cannot be observed (in the universal sense of being observed by whatever means) does not exist. To give atheism a veneer of respectability, modern day scientists added their criteria of trades to the abstract assumptions of their enlightenment intellectual predecessors. Thus it is currently fashionable among predominantly atheists of the scientific strand to condition the acceptance of any phenomena particularly religion and all its perceptions on the physical proof of their veracity. There are two implied basic prongs to this approach adopted from the normal practice of rigorous science. The first is that a theory has to clearly explain nature as presently observed and understood. The second is that it should make **predictions** that can be falsifiable through available experimental methods and technologies. These are the venerable criteria that contributed to the astounding progress in human knowledge during the past couple of centuries. No intellectual, thinker or scientist can in good conscience deny the paramount value of this methodology. Since religion is considered a natural phenomenon, it is futile to claim its exceptionality. Religion and in particular the concept of God (الله - سبحانه و تعالى) must be subjected to the same methodology. Westerners applied this approach to their Christian religion and in short order brought its demise as a truthfully natural phenomenon in the meantime buttressing atheism's presumed foundations. Nonetheless, there are two fatal flaws in such attitude. The first concerns Western intellect self-centeredness assuming the universality of its beliefs and consequent inferences. The second fundamental flaw automatically followed naturally. Islam as a Christian heresy in Western mindset was dismissed out of hand and was never scrutinized on its own merits. Amazingly, Islamic tradition as represented by the Qur'an, and the Sunnah (السنة) and their countless interpretations has been progressively available for fourteen centuries and most certainly during the past couple of centuries. This very rich tradition makes abundant predictions (as far as human knowledge is concerned while at the same time being absolute truths as God's (- الله سبحانه و تعالى) literal words) about historical, social and even scientific events that are permanently out there in writing open to scrutiny. Rules of the Arabic language undergird these facts and their interpretations over time. Everything involved is rigorous and well established certainly before Mohammad (محمد - عليه الصلاة و السلام) received God's (- الله سبحانه و تعالى) message of Islam. This endeavor is a call to apply to Islamic tradition these venerable criteria in a systemic process to scrutiny its concepts and perceptions. Thanks to the countless examples of such predictions it is implausible that one can give a comprehensive study. However, representative examples are herewith given with the hope that others would contribute more. It is to be emphasized here that *in no circumstances did the tradition of Islam fail to withstand such scrutiny* as can be seen from the following few examples. Islamic tradition in its most authentic representation consists simply of the Qur'an and the methodically verified sayings and acts of Mohammad (محمد - عليه الصلاة و السلام). Scholarly interpretations are temporal in Islam unlike other ideologies where some humans speak under the cloak of divinity. Therefore, checking the explicit historical **predictions** against what actually took place later and subjecting both the Qur'an and the Sunnah (السنة) to scrutiny of the most recent findings of science should serve as a potent tool to prove the veracity of Islam as a faith and a way of life. This is done in appreciation of the scrupulous demands of modern science and in the meantime in refutation of the frivolous claims of atheism about Islam.

A) THE QUR'AN

There are numerous **predictions** in the Qur'an concerning events that were taking place during the lifetime of the community of companions that experienced first-hand the process of revelation. These are well known events in the history and tradition of Islam and could be found fully mentioned in many other works. However, a very quick narration of some of them should suffice here. It is commonly known that the Qur'anic revelations during the 13-year period in Makkah concentrated on building believers' Islamic character with promises of rewards in the hereafter. The 10-year period of Al-Madinah witnessed the struggle to uphold Islam and as such was full of events that all Muslims and others participated in and became eye witnesses to. The explicit Qur'anic promise of entering Makkah to perform the rituals of "Umrah - العمرة) at Al-Hudaybyah (الحديبية) that was fulfilled the following year against the wishes and all efforts of Quraysh (قريش) is another example. These are events that atheists can still argue with their nature as political maneuvers that simply succeeded. What no one can argue with is what the Qur'an explicitly mentioned in at the beginning of Surat Al-Rum (سورة الروم) concerning the centuries long fighting between the Romans and the Persians linking the final battle to Islam's prevailing over the unbelievers of Makkah and their ultimate defeat when the Muslims under the leadership of Mohammad would successfully and victoriously reenter the city within a few years. Both linked events came to pass in exactly the way they were mentioned/**predicted**. After the long siege of the fortress of Khyber (خيبر), Mohammad's (محمد - عليه الصلاة و السلام) unexpectedly declared that the next morning he would give his banner to the individual that would breach the defenses. That person turned out to be Ali ibn Abi-Taleb (على إبن أبى طالب) that no one thought of due to his illness at the time. Lo and behold, Khyber's (خيبر) fortress walls came tumbling down exactly as described. The ultimate demise of the Persian Empire and not the Roman one was spelled out in details by Mohammad's (محمد - عليه الصلاة و السلام) when his messenger to both was rebuffed. There are countless other examples of Qur'anic **predictions** of events cotemporaneous to Mohammad's () life as well as those to take place long after his passing that are not essential to enumerate here but can be found in many other works.

B) Prophet's (محمد - عليه الصلاة و السلام) Tradition (Sunnah - السنة)

When Mohammad (محمد - عليه الصلاة و السلام) faced fierce resistance within his immediate community in Makkah to his message after ten years of trying, it became obvious that he reached a dead-end and he decided to go on the road to the closest established prosperous community of Al-Tai'f (الطائف). He was received with far worse abuse and was humiliated by its residents. This is unquestionably the lowest point in his career especially since all this happened after the loss of his wife and uncle who provided him with protection and consolation. Having reached the end of his patience, he naturally resorted to praying to God (الله - سبحانه و تعالى) asking for some relief. As soon as he finished his prayer, an angel appeared to him with the offer to mete out total physical destruction to the communities that abused him as was the norm with previous prophets and messengers of God (الله - سبحانه و تعالى). Mohammad (محمد - عليه الصلاة و السلام) vehemently turned down the offer with the emphatic *prediction* that the abusive prominent personalities egging their followers to undermine his work are the problem but their sons would be the ones to uphold Islam despite their participation in persecuting it at that time. Within one or two decades it came to pass that Khlid ibn Al-Walid (خالد إبن الوليد), A'mr ibn Al A'as (عمرو إبن العاص) and Ikremah ibn Abi-Jahl (عكرمة إبن أبى جهل) to only mention a few of the great Muslim military leaders that vanquished the Persian Empire and brought to heel the Roman Empire in Al-Sham (الشام). Absolutely nothing at the time Mohammad (محمد - عليه الصلاة و السلام) had **predicted** these changes of heart could have driven him to hope for such change bearing in mind that they and their fathers were the worst nemeses of Islam and Mohammad (محمد - عليه الصلاة و السلام) personally. One can dismiss the appearance of the angel as mythology but one cannot escape the reality of history no matter what. This is a most authenticated narrative in Islamic tradition as it is also intimately associated with the event of Isara' and Me'raj (الإسراء و المعراج) which is among the most prominent incidents in Islam. History tells the accomplishments of these military leaders in full details as well as their initial antagonism to Islam. When the time has come for Mohammad (محمد - عليه الصلاة و السلام) to immigrate to Yathrib (يثرب) at the command of God (الله - سبحانه و تعالى), he was chased by many horsemen hoping to collect the reward for his capture. One such individual by the name of Soraqah (سراقة) caught up with Mohammad (محمد - عليه الصلاة و السلام) and his companion. Only few feet from arresting the fugitives, Soraqah's (سراقة) horse's legs kept sinking in the sand. As hard as he tried, he could not reach Mohammad (محمد - عليه الصلاة و السلام) to capture him and collect the promised lucrative bounty. Seeing how much Soraqh (سراقة) is disappointed, Mohammad (محمد - عليه الصلاة و السلام) advised him to turn around with the promise that he would one day instead get the crown jewels of the Persian Emperor. It is utterly inconceivable that a fugitive running away for his life could make such a promise on his own to a man in hot pursuit within a mere few feet from his prey and is determined to win a reward. More than three decades later after Mohammad's (محمد - عليه الصلاة و السلام) death and during the Caliphate of Omar ibn Al-Khttab (عمر إبن الخطاب), the Persian Empire was vanquished and the crown jewels of the emperor among other things were brought to Omar (عمر إبن الخطاب) who promptly delivered them to

Soraqh (سراقة) as promised. It is not useful to debate anyone who dismisses what went on during the chase as mythology and even the promises made. However, Omar's (عمر إبن الخطاب) delivery of the Persian emperor's crown jewels to Soraqah (سراقة) is an authentic event taking place in front of a multitude of individuals with claims on the booty. Soraqah (سراقة) who is now a Muslim but with no record of any unusual deeds deserving such a magnificent gift had Mohammad's (محمد - عليه الصلاة و السلام) promise/**prediction** though. This is an integral part of the Hijrah trip meticulously recorded in the Islamic tradition. When a coalition of anti-Islam tribes in the fifth year of Hijrah decided to attack Al-Madinah (المدينة المنورة) to rid themselves of the Muslims once and for all, it was decided to dig a trench around the town at the suggestion of Salman the Persian (سلمان الفارسى). During the work a boulder fell on Ammar ibn Yasser (عمار إبن ياسر) and other companions took him for dead. However, Mohammad (محمد - عليه الصلاة و السلام) dismissed their concerns and assured them that he is to stay alive till killed by the transgressing group. Ammar Ibn Yasser (عمار إبن ياسر) lived till his nineties and joined Ali ibn Abi Taleb (على إبن أبى طالب) in fighting Mo'a weyah Ibn Abi-Sofian (معاوية إبن أبى سفيان) and was killed more than half a century later. During the digging of the trench Mohammad (محمد - عليه الصلاة و السلام) announced the demise of the Persian and the Roman Empires with certain successive blows to the rocks. Muslims under siege were in a very desperate situation that drove some weak individuals to suspect that Mohammad's (محمد - عليه الصلاة و السلام) promises of conquering the mighty empires were mere flights of fancy under the circumstances. History testifies to the veracity of his **predictions** after scores of years have passed concerning the Persian Empire. It is interesting to know that this pronouncement gave the great Ottoman ruler Mohammad the conqueror (محمد الفاتح) the impetus to acquire the honor of fulfilling the Prophet's (محمد - عليه الصلاة و السلام) **prediction** by attacking Constantinople and finishing off the last vestiges of the Roman Empire in 1453 close to seven centuries after that solemn **prediction**. Recorded history while mentioning numerous **predictions** by great military personalities such as Alexander the Great and Adolph Hitler of empires to last for thousand years, they actually collapsed in short order. No other **prediction** that came to pass seven centuries later is known in all the recorded human history. After the collapse of the siege of Al-Madinah (المدينة المنورة) in the fifth year of Hijrah, Mohammad announced that this battle was the last time ever that his Muslim community would be invaded. He and his companions took the fight to the lands of the unbelievers and his **prediction** stood the test of time. Eight years later he passed away without the slightest deviation in his **prediction** regardless of the tenuous situations the Muslims found themselves in during that period. Mohammad's (- محمد عليه الصلاة و السلام) emphatic pronouncement after the collapse of the siege of Al-Madinah that the Muslim community of his time will never be attacked again in their town regardless of any efforts by the unbelievers and their collaborators within Al-Madinah is an example of such contemporary **prediction**.

The Final Verdict

This analytical and commonsensical study should have by now arrived at the intended destination after a long trip of preparing the ground for exploring the claim that nothing found out by modern science in any field of human intellectual and/or scientific endeavors is in clear contradiction with the Islamic sacred tradition unlike the state of affairs with the Jewish and Christian ones. The most fundamental underpinning of this claim rests on the regularly repeated principle in the Qur'an itself that the Islamic sacred tradition can *only* be interpreted according to the rules of the ***Arabic language*** which is the ***original medium*** this tradition appeared in. That principle in turn requires that words and phrases of that tradition must have global meanings which are progressively understandable at all times regardless of the knowledge level of the associated folks at any point in time. As currently experienced, ignoring this maxim leads to intellectual stagnation prompting an urgent call for a "paradigm shift" in Islamic thought especially in the lexicon. In the following part of this study, an attempt is made to subject certain aspects of the Islamic tradition to the scrutiny of the unanimously accepted findings of modern science. It should be obvious that such an attempt cannot be exhaustive covering all aspects of that tradition. That is because these aspects are vastly more than can be explored by a single study as well as the fact that these explorations are time dependent. The Islamic tradition will progressively open itself to new interpretations in conjunction with the advance of human knowledge till the end of time or the demise of humanity. It is essential to remind oneself that in no way does this paradigm shift affect Islamic rules and regulations governing the daily lives of people or simply the "Shari'a" (الشريعة) as explained before. What follows is a very modest effort to show the absolute validity of the compatibility claim between Islam and science. It is sincerely hoped that this effort will stimulate others to do the same; in the meantime, enriching human intellectual pursuits and humans striving to get to know God (الله - سبحانه و تعالى) their Creator and Sustainer in a meaningful way.

INTRODUCTION

Every single word contained in the Qur'an in the Islamic tradition is believed to be the "literal word spoken by God (الله - سبحانه و تعالى)". It was transmitted to the Archangel Gabriel by God (الله - سبحانه و تعالى) Himself with orders to convey it to Mohammad (محمد - عليه الصلاة و السلام). This is simply the process of "revelation" as understood by Muslims. Mohammed (محمد - عليه الصلاة و السلام) is not by any means unique. All Prophets and Messengers of God (الله - سبحانه و تعالى) to humanity received revelation in the very same manner. Obviously some of the various steps of the revelation process are beyond scientific verification due to the limitation of human knowledge and technology. These steps are subject only to belief and from an Islamic perspective it is pointless to argue over them. But Mohammed's (محمد - عليه الصلاة و السلام) reception of the revelation from the Archangel Gabriel is witnessed by a multitude of Muslims and non-Muslims alike and is recorded in its minute details. Therefore, it can be interpreted in a rational methodical way. In Islam God (الله - سبحانه و تعالى) is All-Knowing hence, every word in the Qur'an is the truth, the whole truth and nothing but the truth. The only permissible human contribution to the Qur'an is to interpret it. That process started during Mohammed's (محمد - عليه الصلاة و السلام) life and never stopped and will continue till the end of time. When his companions volunteered their interpretations, it is very well established that Mohammed (محمد - عليه الصلاة و السلام) himself rarely participated in this practice. That is not surprising as the Qur'an itself in Surat Al-Najm (Every Star) Ayahs 3 and 4 (*و ما ينطق عن الهوى*إن هو إلا وحى يوحى» سورة النجم آية 3-4) described Mohammed's (محمد - عليه الصلاة و السلام) utterances as not of his own whims but they are revelations from the Almighty (الله - سبحانه و تعالى). As such whatever interpretation he gives should be considered the final word with no room for further elaboration by anyone. That is not what the Qur'an is about. It is understood that as humanity acquires more knowledge over time, better understanding of certain statements contained in the Qur'an would be gained and more accurate interpretations would be obtained. It is also understood that the Qur'an *is and only is* the Arabic text. Therefore, any interpretation is strictly bound by the rules governing the meaning of words of that language. These were derived and well established in ancient times long before Islam. This is the only admissible approach to the Qur'an acceptable to Islam. This approach has been unequivocally established by the Muslims long before they encountered non-Arabic speaking peoples including the Muslims among them. Unqualified pseudo-scholars (some with very fancy titles particularly in the West) over the centuries till the present time attempted to undercut Islam by giving their twisted interpretations in stark ignorance of the rules; tossing real scholarship to the wind. Nonetheless, nothing as much as dented the fundamental concepts of Islam due to these feeble pathetic efforts over the past fourteen centuries. The Qur'an is not a book of science but rather a book of absolute truths as far as Muslims are concerned. It nonetheless includes a whole host of extensive details pertaining to what nowadays would most certainly be considered scientific statements. From an Islamic perspective these are not exactly *predictions* but are rather facts as divine words by God (الله - سبحانه و تعالى) who is not bound by time. These facts *explain* all physical phenomena that humans (who are obviously bound by time) still strive to understand. However, human knowledge has not yet attained the level of complete understanding of any of these phenomena. Therefore, from a secular non-Islamic stand point they could eventually be considered scientific *predictions*. It is essential to bear in mind that these explanations/predictions were stated more than fourteen centuries ago when humanity did not know much of true scientific value about anything. These statements survived (as did the entire text of the Qur'an) intact over this very long period and according to Islamic belief will be valid until the end of time. Subjecting the Qur'an (and hence Islam itself) to the scrutiny of science boils down to essentially validating or invalidating these numerous Qur'anic *statements/predictions*. This is the goal of this undertaking. The methodology is simple and can be described in the following procedure. Those Ayahs of the Qur'an that hint at a physical phenomenon are found regardless of the context of their occurrence within the text. Observing the most general meanings of the included words in the Arabic language, a plausible interpretation of these statements is given. ***However, definite presumably final interpretations are strictly not claimed.*** The consistency/inconsistency of these plausible interpretations are then checked against the established scientific facts bearing in mind that even universally accepted theories do not necessarily represent facts unless and until they are verified by unambiguous experimental results. That should help determine the validity or the speciousness of the Qur'anic text which is the irreducible foundation of Islam itself. Hence, one can come to the conclusion whether or not Islam can survive the rigorous scrutiny of science. That is the crux of this work. In carrying out this process, atheistic arguments are implicitly or explicitly discussed as a matter of course. It goes without saying that these new plausible interpretations would significantly differ from those adopted by Islamic scholars for centuries. They may even be at odds with them and doubtless will raise the ire of some intellectually stagnant members of the religious establishment. That would hopefully stimulate enlightened discussions to support or to refute these findings. If nothing else, this outcome makes the effort worth pursuing. Whatever new interpretations result, new era in Islamic thought would have been launched. Nonetheless, the glory of the great works and scholarship of the past would be strictly preserved. The simple fact that great scholars of past did not have access to the knowledge currently available naturally limited the scope of their conclusions. As in physical sciences, new conclusions should reduce to the old ones when subjected to the limited conditions the latter were developed under. Clearly one should expect that by the passage of time, these new interpretations would be superseded by even newer interpretations. This is the only shield against stagnation and becoming irrelevant. It goes without saying that human knowledge at any point in time is never final.

Since the beginning of humanity's existence on earth, it has been contemplating the purpose of this existence and that of its environment as well as the universe as a whole. For millennia this effort took the shape of mental and pure thought exercises. This in turn resulted in ideas and philosophies that espoused assumptions and their counterparts at the same time. There were no legitimate reasons why one idea should be more justified than the other. The situation has drastically changed in the past century and half when laws of nature and the physical universe became more and more understandable. What is warranted by nature became irrefutable. Quantifiable natural law replaced ideas as uniquely accurate. While individuals argued over ideas forever, they accept these laws once they are shown to work and correctly explain physical phenomena. Unlike ideas and philosophies, natural laws are accepted everywhere and in any culture by any language. Unexpectedly most of humanity's primitive notions were shattered and new realities took hold. However, a paradox currently exists since most newly discovered natural laws run counter intuitively to human long and dearly held notions. This manifests itself in the fact that while the practical applications of the new knowledge touch the very fabric of the common individual's everyday life, belief in and understanding of the new concepts is still confined to the select few who constitute the scientific elite due to the complexity of the issues involved. These issues go far beyond humanity's immediate observations and mundane daily experiences. Within this period of a century and half so far, humankind learned so much about the basic building blocks of matter, the structure of the universe, humans own intricate biology and genetics and gained better handle on controlling the forces of nature and its environment at large. Ancient superstitions about forces of nature are readily becoming relics of the past to a very large extent even among primitive populations and cultures. It is determined that all matter, animate or inanimate regardless of how complicated its structure, can be reduced to a limited number of fundamental elementary particles. These particles in turn are nothing more than materialized energy. It is common knowledge in Nuclear Physics that when an electron and a positron combine, they are annihilated producing an energy burst of 1024 eV. Alternatively, having that much energy available, an electron and a positron are created with each having 512 eV. What is baffling, is that out of nowhere an electron and a positron (its anti-body) can materialize since their combined existence adds up to zero preserving the physical status-quo-ante. The same is true although less probable due to the amount of energy involved for all elementary particles. That convinced physicists and cosmologists that the vacuum is a huge store of energy rather than a void as was always believed. Annihilating and creating matter is a common practice in High Energy Physics. This is an absolutely scientific fact although the terminology may be confused with religious discourse. It is remarkable to understand that each category of elemental particles consists of identical members. Every electron, proton or neutron for example is identical to every other electron, proton or neutron in existence. However, combined in a specific way they form the countless varieties of physical objects and creatures that constitute the observable physical universe. Dimensions, age and shape of the universe according to modern day Astrophysics are arrived at and measured in an approximate manner with room only for more accurate descriptions pending more accurate technological means. At the present time, the origin of the universe according to the "Big Bang" theory is a widely recognized concept. On the other hand, it is determined beyond any doubt that since its point of origin, the universe has been expanding and it will reach an ultimate size where it starts contracting in a "Big Crunch" according to the most probable cosmological scenarios. Time according to modern understanding is associated with the beginning of the universe and expressions of time before the Big Bang are meaningless nonsensical statements. That is to say that time is merely an effect of the big bang event and has no intrinsic meaning in itself. Since the expansion of the familiar three dimensional Universe is a determined fact it can be reasonably assumed that there is an on-going process in motion that will eventually end with the Big Crunch. Therefore, all natural events taking place in the universe are part of this on-going process and are not particularly unique. In other words, natural disasters (such as earthquakes, floods, hurricanes, etc.) are simply part of the on-going process of evolution of the three dimensional universe and are not meant for anything else. However, strictly speaking that does not necessarily negate the plausibility of a coordinated cosmological plan binding together the demise of certain communities with some physical calamity. This is a sensible way to answer the age old questions of why bad things happen to good people and why young children acquire incurable diseases and other numerous human protestations against fate and the fairness of life. There is no ill will intended in these processes but rather an evolving process that humans are but an extremely insignificant part of, notwithstanding their illusions of grandeur as the purpose and cause of all that goes on in the universe. The fault is not in nature but in humans' idea about themselves and their position in the universe. A simple on-going process in the three dimensional universe such as the growth of a plant involves the combining and breaking up of countless numbers of molecules. It is legitimate to assume that each stable molecule forms a whole world unto itself. The process of combining it with others or even breaking it up does not involve in any rational way malice of any kind towards that molecule on part of the grower albeit involving the absolute alteration of its existence. To take the issue another step farther, one can consider modern day nuclear medicine where atoms are smashed to produce elemental particles such as neutrons to treat cancer or for any other highly regarded medical applications. Nobody ever considered this is done with malice towards these billions upon billions of atoms in spite of the fact that they are deliberately pushed out of existence. If human beings assume that they are entitled to the privilege of existing in the best way that suits them (e.g. no disease or pain of any kind), it should be the privilege of atoms and molecules (which form integrated whole worlds in themselves) to exist in the same manner. The same argument should be valid when the four dimensional universe (three space and one time) is considered. What goes on in the constituent three-dimensional universe is but a process that carries no malice even if it involves pain and suffering to humans. Humans have no right to think that it is unfair for natural disasters to take place or that children should have incurable illnesses or any similar self-centered protestations. The whole matter should be treated as part of a process devoid of malice towards humanity or an argument against a Just and Fair Supreme Being.

The Familiar Universe and Beyond

Classical physics and mathematics require that forces in the three dimensional world obey an inverse square law since the lines of force are spread out over an area proportional to the square of the distance. For a four dimensional world, the concentration of the lines of force are spread over the area of a sphere and hence will decrease with the cube of the distance. Therefore, in a four dimensional world forces will obey an inverse cube law. Thus, forces' strength in a higher dimensional world will fall extremely rapidly with the distance. On the other hand, lower dimensional worlds will not allow the existence of complex structures. For example, in a two dimensional world (any surface or plane) it is impossible to have a complicated network without the wires crossing and an object cannot have a channel through without dividing into two. These are the obvious qualitative reasons why humans and other living beings can only inhabit the familiar three dimensional Universe. However, there are compelling mathematical reasons as well. The clear consequence of this argument is that humans and other common living beings in their current structure cannot inhabit higher or lower dimensional universes. Gravity and electrical repulsion are the most familiar forces that obey such inverse square law in the three dimensional world. Scientists realized this criterion and both Newton and Faraday primarily based their theories of planetary movements and electrical repulsion on that principle. However, they continued applying two dimensional analyses (especially plane geometry) as excellent approximations to solve three dimensional problems. In other words, life can only exist under the observable conditions and the corresponding governing laws clearly show these facts. It is fascinating to notice that there is a long list of physical laws and conditions that, varied slightly, would have resulted in a very different universe or no universe at all. One can easily make the connection between the cosmological multi-dimensional universe or the many worlds proposals and the Islamic concept of Al-Sama' (السماء) in either the singular or the plural form. Consequently, in this suggestion birth represents the transition to Al-Ardh (الأرض) from some temporary existence while death represents the transition in the other direction not necessarily to the same existence.

Resurrection is therefore the summoning act to face God (سبحانه و تعالى) and account for individual deeds. That approach, immediately explicate the Qur'anic description of the life of martyrs after death as well as what has happened to Jesus Christ (المسيح عيسى إبن مريم – عليه السلام) as opposed to crucifixion since that act is explicitly dismissed in the Qur'an. Whatever the situation may be, it is clear that a more global meaning for the word Al-Sama' (السماء) than the mundane everyday usage is a must not just to reconcile the Qur'an with science but also to elucidate any interpretation of the Qur'an and the Sunnah (السنة). Some aspects of this theory and the concept of multi-verse can easily provide confirmation (or at least the plausibility) of what is known in the Islamic tradition and explicitly described in the Qur'an as the "Night Journey" Al-Isra' wa Al-Me'raj (الإسراء و المعراج) embarked on by Mohammad (محمد – عليه الصلاة و السلام). That journey took him to Jerusalem and then through the various levels of Heaven to visualize Al-Jannah (الجنة) or Paradise and Jahannam (جهنم) or Hell. Details of that narrative were decidedly out of reach of the comprehension of all who heard it firsthand. Only unshakeable trust in Mohammad (محمد – عليه الصلاة و السلام) and belief in the Qur'anic revelation that followed provided a litmus test to differentiate between genuine and spurious belief in Islam among his fellow Arabs. It also qualified those who accepted his story for the struggle against the forces of unbelief that ultimately upheld the message of God (الله - سبحانه و تعالى). With the invention of high speed transportation means in the twentieth century, the plausibility of overnight trip to Jerusalem is obvious to everyone. Travelling through levels of Heaven may be likewise obvious if one accepts the concept of teleportation or leaping into one higher dimension after another (as stipulated by modern physics) and back again. Mohammad (محمد – عليه الصلاة و السلام) never claimed having the power to accomplish such a feat. He attributed that capability to his trip companion; the Archangel Gabriel Who is presumably in this context supposed to exist in a higher dimension. According to well established mathematical theories, higher dimensional beings can easily visualize lower dimensional beings but not vice versa. What humans who are three dimensional beings can see are only three dimensional sections of a higher dimensional being. That may explain the fact that at certain times some of Mohammad's (محمد

(عليه الصلاة و السلام –) companions could clearly see the Archangel Gabriel with their naked eyes as established in the tradition. This is another example of subjecting Qur'anic and in general Islamic perceptions to the scrutiny of science and their irrefutable confirmation time and again. Nonetheless, it is essential to understand that only plausibility rather than actuality is discussed here. These purely scientific speculations and ideas of higher dimensional spaces and teleportation, etc. have been generally accepted by the public and have currently permeated popular imagination to give birth to numerous science-fiction contributions. As always one reaches the conclusion that nothing modern science accepts as most probable contradicts in any meaningful way with what Islam has unequivocally asserted more than fourteen centuries ago. Looking at the three dimensional universe from the four dimensional universe perspective, one can immediately realize that interaction and influence occurs in a direct manner and in zero time. One dos not have to tour the entire three dimensional universe or to travel immense distances in it to get to the four dimensional universe. It is, if the expression means anything, right there. It is clear that a creature in the four dimensional universe is constantly aware of its constituent of the three dimensional universe and it incessantly interacts with it. In this way, the Islamic belief in revelation and the incident known as the Al-Isra' wa Al-Me'raj (الإسراء و المعراج) or the "Night Journey" can be easily interpreted as narrated in the Qur'an in Surat Al-Isra' (سورة الإسراء) and Surat Al-Najm (سورة النجم). Interaction between a higher dimensional universe and the three dimensional universe constituent is an obvious possibility. Modern science that led to all these advances in human knowledge came about with the fundamental breakthroughs of the as yet competing Quantum and Relativity theories. While the quantum approach is statistical in essence with the "Principle of Uncertainty" at its heart, the relativistic approach is deterministic. Since both theories unambiguously explain a wide range of physical phenomena, a relativistic quantum approach is currently used by researchers to account for phenomena at the elementary particle and sub-atomic levels. Quantum theory established the identical nature of all elementary particles belonging to the same category. Special Theory of Relativity established the equivalence of energy and matter described in the most famous and popular equation "$E = MC^2$" that caught the public imagination. General Theory of Relativity explained the geometry of space and related the fundamental force of gravity to curvature of space-time. Both approaches have fundamental philosophical implications and resulted in a few paradoxes as well. A unified theory is still out of reach however. All modern approaches to the study of the three dimensional universe put an upper limit to the speed any matter (elementary particle) can attain. This is the speed of light. That fact is repeatedly ascertained both theoretically and experimentally. The most profound result of the relativity theory, and yet the least spoken of, is the demolition of the concept of absolute time and the establishment of the fact that the universe is "**Multi-dimensional**". The four dimensional space-time became part of the lexicon of popular culture. Numerous theories attempting to explain the structure of the universe sprang up with multi-dimensionality of the universe as a common thread. It is clear that the number of dimensions for the structure of the universe in these theories varies but the concept of multi-dimensionality is irretrievably firmly established. Two dimensional phenomena are well known in physics. An example is what is known as the "Quantum Hall Effect". It describes a two dimensional quantum phenomenon taking place within ordinary three dimensional physics. Here there is a large energy barrier constraining the relevant system to a two dimensional surface, and the quantum physics of this lower-dimensional can be apparently oblivious to the extra third space dimension because its contents do not possess the energy sufficient to surmount this energy barrier. Ancient mathematicians described zero dimensional space (a point), one dimensional space (a line) and a two dimensional space (a plane). Zero and one dimensional spaces are obvious to even the least educated individual. Line spaces (one dimensional) are universally used to define distances. Two dimensional spaces were studied during the classical periods and resulted in Euclidean or plane geometry. Extensive hypotheses governed the study of surfaces of all shapes. The concept of "area" is associated with this type of space. It is rather an interesting fact that while every school child everywhere in the world going back in time for countless centuries studied these concepts and took them for granted, in reality these spaces do not actually exist in nature. Humanity's mindset deals exclusively with and is intimately accustomed to three dimensional phenomena in its daily experiences. Humans invented these other universes to help them understand their immediate world. Eventually, relations among complex three dimensional objects are developed and the concept of volume evolved. Dimensionality of a universe describes the number of coordinates one needs to know to locate an object in its corresponding space. It is customary among mathematicians and physicists to express a position of an object in three dimensional spaces in Cartesian coordinates (x, y, z) or polar coordinates (r, θ, Φ). In this way an object would be entirely defined in this three dimensional space. Although zero, one and two dimensional universes were familiar, it was common knowledge that humanity lives in a three dimensional universe and that was all there is. Modern Physics brought about a fourth dimension (time) indisputably ascertaining that the universe is four dimensional. Astoundingly, showing that as the three dimensions are exchangeable (i.e. anyone can be projected or transferred into anyone else) in the three dimensional universe, time can be projected or transferred into any of the space dimensions in the four dimensional space-time. *There is no reason not to think that the all-encompassing universe consists of more than four dimensions. Humans are only aware of the three dimensions they deal with in their immediate experiences due to their lack of knowledge assuming (as they have done since their early moments on this earth) that time is absolute.*

A one dimensional universe (line) is formed by combining an infinite number of zero dimensional ones (imagining infinite number of points put next to each other). Similarly, a two dimensional universe (plane) is made of an infinite number of one dimensional universes as well (imagining infinite number of lines put next to each other). The familiar three dimensional Universe is understood to consist of an infinite number of these two dimensional universes put on top of each other. Although not strictly aware of it, modern humans gain knowledge about their three dimensional universe through the study of two dimensional ones conceiving of two dimensional creatures living within their two dimensional universe. However, they realize that these creatures do not exist and that their three dimensional universe is the "**lowest**" possible form of existence. The Qur'an stipulates human habitation after exiting Al-Jannah (الجنة) to be located at such lowest level in numerous Ayahs. However, Surat Al-Tin (سورة التين) is unequivocal about that fact. One can again reasonably use the current scientific speculations about the multi-dimensionality of the universe to interpret Surat Al-Tin (سورة التين). For the sake of clarity, the hypothetical two dimensional creatures are not aware of the third dimension and whatever is included therein even when they themselves are building blocks of that three dimensional universe. No matter how extended their universe is, they are confined to the available two dimensions and cannot escape into the third dimension on their own without some radical transformations. While they are not separated from the third dimension they cannot conceive of its existence unless they are endowed with thinking powers and a healthy imagination to indirectly develop the necessary proofs of its reality. By the same token, the same should hold true of the three dimensional creature (the human being). *A human being is nothing more than a tiny building block of the four dimensional universe.* Fortunately, though, human beings are endowed with insatiable curiosity and learning powers and keep on discovering the physical laws governing their three dimensional universe and how they, so far, relate to the contiguous four dimensional Universe they are part of. They now understand that they, **in their current physical existence**, are constrained only by the speed of light and that time is the barrier between them and the four dimensional universe. Plainly speaking, humans know at the present time that to proceed along the fourth dimension, they have to overcome the speed of light constraint and accomplishing this, they enter into a universe where time has no unique meaning and can extend indefinitely in exactly the same way the other three space dimensions extend. The second part of this understanding is familiar to almost every human being as the notion of immortality. The first part can only be pondered and comprehended by a select group who deal extensively with the study of the laws of Physics. That is if one limits oneself to dealing only with facts. On the other hand, the very same issue has been dealt with since ancient times in contemplations and pure thoughts leading to Philosophy and Sufism. To physically break the speed of light barrier, science mandates that a *material object* requires the acquisition of an infinite amount of energy which is impossible. Extrapolating from this fact one can theorize that a ***non-material object*** (made of light photons for example) can accomplish this feat with ease.

To summarize, in order for humans to proceed from their current residence in the three dimensional universe **(as living human beings)** into the higher status four dimensional universe **(as immortal beings)**, they have to separate from their material object self and become a non-material object. This is not a new issue. This is the age old question of the nature of sleep and death. Humans presumed to solve this dilemma by giving the material body of a human being a "soul" and religions confirmed the concept for them. Even those who profess to be atheists believe that a human is certainly more than what this conglomerate of molecules known as the human body represents. They speak of creative self beyond the physique which can be totally decrypt while the person can be amazingly thoughtfully productive. Almost everybody at one time experienced the phenomenon known as déjà-vu and/or visions (or dreams) that unfolded in reality some later time. Since the time dimension is a fact, one can explain these phenomena as experiences of the soul (or the creative self) in the four dimensional universe that are projected into the three dimensional everyday universe of human beings. The direct implication here is that sleep and death are exactly the same event where humans return from the four dimensional universe in one (sleep) to the familiar three dimensional Universe and not from the other (death). The Qur'an dealt with exactly this issue in Surat Qaf (سورة ق). A four dimensional universe consists of the familiar three dimensional Universe with an added time dimension. It is to be noticed that dimensions in the one, two and three dimensional universes are identical in nature and measure the same quantity albeit in different directions. The fourth dimension however has a decidedly different nature. While humans in this three dimensional universe do not actually experience the one and two dimensional universes, they do indeed experience the fourth dimension or time. Although one must submit to this logic, there are profound Implications to accepting this hypothesis. As with lower dimensional universes where any point can be treated as a point of contact with the immediate higher dimensional universe, any point in the three dimensional universe can be considered a contact point with the four dimensional universe. It is essential to keep in mind that the fourth dimension is time and that it can be transferred into any of the other three space dimensions with the speed of light as the linking factor. This is hinted at in Surat Al-Noor or "Light" (سورة النور). Light holds the key to the puzzle of the formation of the three dimensional universe and its continuing existence. No current theory determines with certainty the nature of light whether it consists of quanta belonging to the category of matter or wave being non-matter. Both natures are exhibited without a shadow of a doubt. Duality of light is a given at the present time since what phenomena can be easily explained using the wave nature cannot be explained using the quantum nature and vice versa. Consequently, one can presume that existence in the four dimensional universe, and maybe higher, consists of objects (creatures or inhabitants) made of light which materialize into matter on descending to the three dimensional universe. That would be consistent with the scientific physical findings so far but bridges the gap between the two universes. Pure light is independent of time and such transition therefore, is made in zero time without breaking any of the known physical laws. It is also essential to understand that physical laws in the four dimensional universe do not follow three dimensional rules but rather are the more universal form of physical laws where the three dimensional ones that govern the familiar universe are their mere approximations. If one subscribes to this hypothesis, then the concept of prophets and messengers receiving revelation through the agency of the Archangel Gabriel makes perfect physical sense and no mythological or magical powers are needed or are at play during this phenomenon. The Big Bang theory of the origin of the universe divides the process of forming the universe into several phases. Among the first of which the change of energy into matter in the form of elementary particles took place. This process has been repeatedly simulated at High Energy Physics laboratories such as CERN in Geneva, Switzerland. The time taken to accomplish this phase of the big bang theory interpretation of universe's creation was extremely short by everyday standards in the order of factions of the billionth of a second. Other phases followed suit with extended periods of time. These phases exhibited the formation of atoms, molecules, substances, compounds leading to the formation of

Galaxies, stars and planets. On planet earth, inanimate objects took shape. The current phase started with the beginning of life and the formation of animate objects and living creatures. Eventually the first human appeared as the latest known object with conscious. This final phase is approximately five billion years old at the present time. Formation of the familiar three dimensional Universe can be interpreted as an on-going process (conceivably out of many) in the four dimensional universe. Time as known in this universe started with the Big Bang. The current process of evolving the three dimensional universe involves countless numbers of sub-processes none of them with malice towards humans regardless of their own point of view.

THE PROCESS OF CREATION AND ITS STEPS

Atheists consider the self-generation and evolution of the universe without the need for any external supreme power initiating or maintaining the process. This is the materialist approach to the existence of the universe. They can relate to the latest scientific discoveries and find them consistent with their beliefs. On the other hand, the three major religions contend that a Supreme Being does exist and it was He who created this universe. They mention the process of creation in their associated sacred books. The Old Testament of the Christian Bible (supposedly the Jewish bible) essentially narrates the story of creation in the Book of Genesis. The Qur'an has numerous references to the creation process. While the details radically differ, there is a consensus among the sacred books that the universe was created in *seven days*. The Book of Genesis relates these days to human experiences as in the seven days of the week. It gives details as to what happened on each of these days ending with God (الله - سبحانه و تعالى) resting on the seventh day. Consequently, Jews have Saturday as their Sabbath while Christians choose Sunday. Muslims have no particular holy day as the day of rest since they dismiss out of hand without a second thought the concept that God (الله - سبحانه و تعالى) needed to rest after the hard work of creating the universe. The Qur'an uses the word (يوم) which indicates a day in common usage in its various forms approximately 461 times. However, it does not limit its usage exclusively to that end. Neither does Arabic literature, ancient or modern. Therefore, the Arabic word (يوم) used in the Qur'an in association with the process of creation cannot be interpreted as indicating a normal human day. It rather means *"phase"* or *"stage"*. The proposition then becomes that the universe is created in seven phases or stages. It gives details of which phases involved the creation of what. This would be consistent with the current understanding that an Earthly day is different from a Martian one which is certainly at odds with a galactic revolution or a day when it comes to measurement of time. That is to say that the Qur'an speaks of an initial phase (يوم) that lasted fractions of a billionth of a second and also of the last phase (يوم) that lasts several billion years. Materialist denunciations of religious interpretations of the process of creation may find some ground to uphold them as far as the Book of Genesis is concerned. They nonetheless have no basis whatsoever when it comes to the Qur'an. That is due to the rules of the Arabic language and its employment in the Qur'anic narrative. In Biology, it is determined that what differentiates one species from another is its genetic structure and what differentiates one member of the species from another is its DNA. One has to bear in mind that all these very complex structures are in fact made of the same identical elementary particles nonetheless. One can explore the Islamic concept of Adam's (آدم - عليه السلام) creation by a unique act of God (الله - سبحانه و تعالى) as part of the issue of creation in general in light of what is scientifically known or universally accepted by scientists as inevitable even if not proven yet (due to lack of adequate technical tools) to see if there exist any contradictions. This is carried out according to the latest findings of modern physics and cosmology without getting entangled in technical details. In this regard, one has to keep in mind that space and particularly time are human concepts wholly subjected to conditions of the known three dimensional Universe in which humanity resides. For example, according to the special theory of relativity, speed of light is a universal unchanging constant. Thus, one has to understand that what takes billions of years measured by humans happens in zero time as far as objects (like photons) move with the speed of light. Therefore, no real physical laws are actually trampled upon when one speaks of God (الله - سبحانه و تعالى) creating anything even the whole universe in zero time if He wills it. Another very important point to keep in mind is that a constant speed of light is only a fundamental parameter of the familiar three-dimension universe. However, currently most cosmologists speculate that, the familiar three-dimension universe is nothing more than a part of a multi-dimensional universe of higher dimensions. If that is true, then there is no reason to mandate a constant speed of light within the higher dimension universes. The well-known Islamic concept of God (الله - سبحانه و تعالى) creating the universe(s) in seven days hinges on the definition of what a "day" signifies. From a human perspective, there is no universal meaning to that word since there is no universally absolute meaning to time. Conventionally, humans measure time according to their earthly experiences (such as the repetitive rising of the sun or the moon) and extend that convention to measure all other events. This is for example what it means to say that the age of the universe is approximately 13.7 billion years. Therefore, the Qur'anic use of the word "**DAY**/(يوم)" in describing the process of creation does not imply measurement of time as humanly experienced since it is associated with God (الله - سبحانه و تعالى) not humans. It should rather mean a well-defined change in the character of the universe during its evolution or simply its transition from one "**PHASE**" to the next. In other words, the on-going creation process of the universe according to the Qur'an progresses in seven phases. Each phase has its own time (humanly measured i.e. calculated from cosmological equations) ranging from the extremely small (Planck time $\sim 10^{-43}$ seconds) to the extremely large (Proton decay time $\sim 10^{31}$ years). Therefore, different Qur'anic *days* have different time durations as measured by humans. Cosmological theories stipulate that after the "Big Bang", the universe starting from a point with tremendous energy was in a state of flux, was expanding, was opaque and was made exclusively of radiation. At a specific stage of expansion, photons (light) could escape and the universe became transparent. At another stage of expansion, forces acting on this radiation changed dramatically to allow the formation of elementary particles that eventually produced Hydrogen and then Helium atoms. Construction of galaxies and planetary systems ensued. Forces acting on materials of the structural constituents of the early universe produced all currently known elements of the periodic table. The universe will keep expanding till it either reaches thermal equilibrium where it becomes dead and absolutely nothing happens or it reverses its direction contracting to end in a "Big Crunch".

It is quite fascinating that whether in the Qur'an or in cosmology one time and again comes face-to-face with the number *seven* (or six and then the present state) when dealing with the process of creation. The Qur'an is explicit about that whereas cosmology adopting varied approaches speaks of phases in the creation process without paying attention to their numbers or attaching any significance to it. However, it is of the utmost importance to understand that these are not arbitrary phases chosen for convenience. They are rather distinct states separate from each other by clear transitions that radically change the character of the universe and its constituents. The last of these transitions leads to the present state of the universe lasting without significant change until its doom in a "Big Crunch" as most cosmologists speculate/believe depending on the value of the cosmic density. Although all three monotheistic religions make the assertion that the creation process one way or the other lasted for *"seven days"* in their sacred books, it is curious that no one has ever made that observation before. Although there is practically general agreement among cosmologists that the creation process started with a "Big Bang", their approaches diverge in describing the reasons for the transitions from one phase to the next. There is also fundamental disagreement on the big bang singularity and even its existence. Nonetheless, once the big bang arguments are out of the way the *seven phases* are universally agreed upon. Different cosmologists describe these phases in different terms depending on their interest but they all agree on the fact that the universe passes through **"seven phases"** to reach the present stage and will continue to expand without any more phase transitions till the end (whatever that means) because of its thermodynamic development.

Creation of the universe as a unique process is qualitatively and quantitatively described by Steven Weinberg who is a highly regarded physicist, a Nobel laureate that has made fundamental contributions to modern Physics and a well-known skeptic of the concept and the very idea of religion and faith in general. He finds no purpose in the creation process or any higher authority behind it. He concludes his landmark book describing the creation of the universe from cosmology point of view *The First Three Minutes* by commenting that he believes that the more the universe is comprehensible, the more it also seems pointless. However, for him the effort to understand the universe is one of the very few things that lifts human life a little above the level of farce, and gives it some of the grace of tragedy. Interestingly, in the same book he gives a very succinct description of these *seven phases*. Transition from one phase to the next is based on the temperature of the universe which changes (cools) due to its expansion. The big bang event where temperature of the universe was infinite is left alone as there are no scientific tools in terms of physical laws in existence to deal with it. Qualitatively, the highlight of his analysis describing these phases taking the big bang as zero time is:

1) One-hundredth of a second after the big bang, the temperature of the universe cooled to **10^{11} degrees Kelvin** and the universe is in a state of nearly perfect thermal equilibrium. The only particles in existence are electrons, positrons and the massless photons, neutrinos and anti-neutrinos since their threshold temperature (temperature above which a particle can be freely created out of thermal radiation) is significantly below 10^{11}. Additionally, very small numbers of protons and neutrons exist due particles' interactions (one neutron or proton for every billion photons or electron or neutrino). Although the neutron is slightly heavier than the proton, under such conditions rates of protons converting into neutrons and vice versa are exactly the same. Simple calculations give the universe a density of 3.8 billion kilograms per liter (for comparison, water density under normal terrestrial conditions is 1 kilogram per liter).

2) At 0.11 seconds, the temperature of the universe has dropped to **30 billion degrees Kelvin** (due to expansion) which makes it significantly easier for the heavier neutrons to convert into the lighter protons than vice versa. Thus, the nuclear particle balance at that point has greatly shifted in favor of the protons.

3) At 1.09 seconds, the temperature has dropped to 10 billion **degrees Kelvin** where the decreasing density and temperature

allow neutrinos and anti-neutrinos to be free particles no longer in thermal equilibrium with electrons, positrons or photons. This designates the first *decoupling* (separation of the fundamental forces of nature). At this instant, mass density falls as the fourth power of temperature ratio to become 380,000 times that of water. Additionally, the temperature is just above the threshold temperature of electrons and positrons so they are beginning to annihilate more rapidly than they can be recreated from radiation.

4) At 13.82 seconds, the temperature of the universe has dropped to **3 billion degrees Kelvin** which is below the threshold temperature of electrons and positrons so that they are beginning rapidly to disappear as major constituents of the universe. The consequence of this process is that the temperature of the universe from that point on is that of the photons. However, deuterium (H^2) (hydrogen isotope consisting of one proton and one neutron) nucleus is loosely bound and is blasted apart as soon as it is formed preventing heavier elements from forming. This is technically known as "deuterium bottleneck". The number of protons at that stage is more than four times that of neutrons.

5) 3 minutes and two seconds have elapsed since the big bang and the temperature of the universe has cooled to **one billion degrees Kelvin** (only about 70 times hotter than the center of the sun). Electrons and positrons have practically disappeared at that temperature. The universe is chiefly consisting of photons, neutrinos and anti-neutrinos. At that point, the temperature is cool enough for nuclei like hydrogen isotope tritium (H^3) (one proton and two neutrons) as well as helium isotope (He^3) (two protons and one neutron) to form. However, *there are no stable nuclei with five or eight nuclear particles* to help form heavier elements. Collisions of neutrons and protons with electrons, neutrinos and their anti-particles have practically ceased.

6) At three minutes forty-six seconds, temperature of the universe has dropped another order of magnitude to **900 million degrees Kelvin**. As soon as the temperature reaches the point where tritium can form, almost all of the remaining neutrons are

immediately cooked into helium nuclei. When the deuterium bottleneck is passed, heavier nuclei can be built up very rapidly by the chain of two-particle reactions. Assuming the existence of a billion photons for each nuclear particle, *"nucleosynthesis"* will commence. Neutron decay would have shifted the neutron-proton balance just before nucleosynthesis began to 13 percent neutrons and 87 percent protons. After nucleosynthesis, the fraction by weight of helium is twice the fraction of neutrons among nuclear particles or about 26 per cent. At this stage in the creation process, the fundamental building blocks of the current universe (hydrogen and helium in addition to other radiation constituents such as photons, neutrinos and anti-neutrinos) are present. The percentage of helium in the universe should persist at that level since it is an inert element and its distribution across the universe should be homogeneous. That is what the latest measurements confirm and this is taken as strong evidence in support of the current big-bang based standard model of the evolution of the universe. One has to realize that only helium is produced during the early universe stage after the big bag due to deuterium bottleneck while the heavy elements are produced within the stars much later.

7) At 34 minutes and forty seconds, the universe has cooled to **300 million degrees Kelvin** where all electrons and positrons are annihilated except for the excess electrons (one part in a billion) needed to balance the charge of the protons (hydrogen nucleus). Energy released in the annihilation process is taken up mostly by photons to give them a temperature 40.1 per cent higher than the temperature of the neutrinos and the anti-neutrinos. The universe mass density has dropped to only 9.9 per cent of that of water. While the universe will keep expanding *no more phase transitions will take place.* What this means is that the basic constituents of the universe will remain the same till it either collapses in a big-crunch or reaches absolute thermal equilibrium where unchanging "dead universe" can keep expanding forever. This is the last *phase* in the creation process since from this moment on there is only interaction

among constituents of the universe and nothing else. Thus, after 700,000 years the temperature will drop to a point where electrons and nuclei can form stable atoms. Additionally, lack of free electrons will make the universe transparent to light (photons). Eventually *decoupling* of matter and radiation allows matter through violent collisions and gravity to form galaxies and stars. Internal forces within stars cause temperatures and pressures to reach values leading to the formation of elements. Due to the very unique characteristics of the "Carbon" atom and the passing of another 10 billion years, life begins. The universe kept expanding and cooling to reach the present (approximately 13-15 billion years) where the temperature of the universe is just a few degrees (approximately **3 degrees Kelvin**) above absolute zero.

Another description of the same creation process from a substantially different point of view is given in the following discussion. Assuming (consistent with monotheistic claims) that the purpose of creation is to produce conscious human beings (The Anthropic Principle), authors John D. Barrow and Frank J. Tipler in their important and controversial book *The Anthropic Cosmological Principle*, assert in section 6.4 that the hot Big Bang cosmological model contains <u>*seven times*</u> whose relative sizes determine whether life can develop and continue. According to their analysis in summery they are

1) the minimum time necessary for life to evolve by random mutations and natural selection (cannot be calculated from first principles yet)
2) the main-sequence stellar lifetime, necessary to evolve stable, long-lived, hydrogen-burning star like the Sun (~ 10^{10} years)
3) the time before which the expansion dynamics of the expanding universe are determined by radiation, rather than the matter content of the universe (~10^{12} seconds)
4) the time after which the expanding universe is cool enough for atoms and molecules to form (~10^{12} seconds)
5) the time for protons to decay (~10^{31} years)

6) the Planck time (10^{-43} seconds)
7) the present age of the universe (($\sim 15 \pm 3$) $\times 10^9$ years)

All these fundamental times except 1) and 7) are expressed in terms of constants of nature. The authors' basic purpose for these calculations is to ascertain the fact that a small change in the constants of nature would preclude creation of conscious human beings to observe the universe representing reality and is the essence of their principle. **It must be emphasized here that these seven times are not arbitrarily chosen and can be reduced or expanded in numbers but they have to essentially remain seven; no more and no less to present a unique situation for the universe to produce conscious human beings.** These times are closely and firmly associated with the fundamental constants of nature in a way fixing their numbers to only *seven*. Additionally, and most importantly, these constants of nature determine the "Standard Model" universally accepted by physicists to describe the creation process and the basic fundamental constituents of the universe. Additionally, these times are closely linked to energy and/or temperature which are used interchangeably by physicists to describe physical phenomena. It is thus a different way to look at the evolution of the universe with the specified purpose of producing conscious human beings validating The Anthropic Principle. Therefore, it is not a huge unscientific leap of faith to go from this approach to defining the Qur'anic seven days of creation.

A third approach is adopted by the British Astronomer Royal Martin Rees who is universally considered among the top leaders of modern day cosmology. In his very valuable book *Just Six Numbers* he also enumerates "*seven phases*" that followed the big bang. They start at 10^{-44} seconds eventually resulting in the universe that humanity currently inhabits after the elapsing of 13-15 billion years. He is mostly interested in the concrete physical transitions from one phase to the next that gave the universe new properties. His accounting can be summerized as follows:

1) Planck era where the universe can only be analyzed using quantum physics to establish the various fields affecting the

enormous density and extremely high energy and temperature involved.

2) Inflation period to account for the very rapid expansion of the universe resulting in the currently observed isotropy and homogeneity of the universe.

3) The period when the universe was dominated by photons. All photons and other rare elementary particles experienced violent collisions not being able to form any stable matter and the universe was opaque.

4) The period that elementary particles could form which he designated as "exotic matter".

5) The period where helium could finally be formed by fusion out of hydrogen (protons) and electrons. Since helium is a stable inert atom, its density would not appreciably change with the passage of time.

6) The "fireball" period where radiation-matter decoupling takes place. The energy and densities of the universe are such that it becomes transparent.

7) The current stage when galaxies and stars begin to form and heavy elements are created within these stars. Stars explode spreading heavy elements inconsistently across the universe and new ones are formed. Life starts and modern humans appear. It is the phase where the fundamental properties of the universe from a physical point of view (i.e. its material and radiation constituents) remain the same while it undergoes drastic changes in its shape.

These are qualitative phases describing the creation process where each phase is distinct from the next. Going from one phase to the next designates a fundamental "phase transition". Physics (i.e. the effects of the dominant one of the fundamental forces of nature) during each phase is the same until it becomes untenable and the universe has to undergo a phase transition to a new phase and so on. Length of these phases varies significantly according to the physics involved. **Because each "*phase*" is fundamentally distinct from the others, there can be no more phases to count and because of the thermodynamic of the**

universe there can be no more phase transitions in the future. There are only seven phases in the creation process.

These three approaches describe the same creation process from vastly different angles according to the basic interest of their advocates be it physics, cosmology or astronomy. However, they are nonetheless very tightly connected. They are all based on the fundamental constants of the universe which quantitatively determine the goings-on within each phase regardless of any qualitative interest. In other words, for example matter-matter interactions to form galaxies, stars and to start life are exactly the same even if that takes billions of years. Contrast that with radiation-radiation interactions during the very early universe that take an extremely short time but give rise to new phenomena that require a phase transition. One way or the other, *there are only seven phases to the creation process*.

It is to be emphasized at this juncture that most people throughout history considered the creation process as a one-off act to bring about humans. On the other hand, scientists and cosmologists in particular at the present time believe it to be an on-going process that may lead through evolution to post-human creatures. However, there is no escaping an end to that process. Here there are as many opinions as to what happens as there are scientists. What is going to happen is a matter of pure speculation. Islam gives a very definite answer to that question. There will be a doomsday (يوم القيامة و الحساب) designating the end of the process of creation. It is irrelevant whether there will be post-human creatures inhabiting the universe or not. As a matter of fact, Islam asserts in no uncertain terms the existence of such creatures (angels, jinn, etc.) here and now regardless of human conceptions of space (here) and time (now). After doomsday (يوم القيامة و الحساب), new creation process commences leading this time to eternal existence. In the language of science this new creation process will proceed under different physical laws that exclude the concept/dimension of time. Judgment on human deeds would be meted out such that its absolute fairness precisely determines the eternal residence/existence of every single human being from that moment on. Clearly, there are no contradictions between scientific findings describing the creation process and Islamic assertions. That what comes after the end of the creation process is supposedly a matter of scientific speculation rather than established facts. Muslims rationally and wisely believe in Qur'anic declarations in this regard (Jannah and Hell) Al-Jannah (الجنة) or Paradise and Jahannam (جهنم) or Hell based on past and present validations of its contentions. Additionally, it is not claimed in this work that this is an unchallenged interpretation of the Qur'an but rather that whatever compelling results science finds can always be reconciled with the Qur'an as considered by Muslims to be the literal word of God (- الله سبحانه و تعالى).

HUMAN CREATION AS A SUBTEXT OF THE CREATION PROCESS

Thanks to the unusually unique properties of the Carbon atom, the probabilistic nature of collision among elements of the universe over billions of years created precursors of life (amino acids) that eventually formed single living cells. Evolution took over to result in Homo-Sapiens many millions of years ago. It is very important at this time to point out the fact that because of the age and nature of the forces acting on planet earth, life could not have been possibly initiated on earth. Rather, it was initiated somewhere else in the universe and by sheer chance landed on earth taking advantage of its conditions to evolve. That obviously does not preclude the possibility of life (either Carbon based or different) evolving in other places in the universe under similar or other conditions. That is basically the story of creation according to science with the process up to formation of galaxies cast in concrete physical proof. The rest of the story is logical and most plausible but not proven yet. Currently most scientists find it unfeasible to get around the idea of a multi-verse to explain some fundamental aspects of cosmology. Each universe among the infinite number included in the multi-verse has its own "Big Bang" and life cycle. Only those with similar but not identical histories to the known universe could survive to produce life while others are doomed. Since the most fundamental parameters sustaining the universe(s) necessarily have to be slightly different for each universe, that fact explains why other universes cannot mutually be ever detected.

Other scientists believe that higher dimensions must be incorporated in equations describing the universe to avoid certain anomalies or singularities. Concentration of coincidences required for human existence in this universe has been termed the "Anthropic Principle". Had things been different, humans could not have existed. It may be that many different universes are possible and many may exist in parallel with this one. On the other hand, because things become simpler approaching the moment of creation, there might have been only a limited range of possibilities or even only one with everything so perfect that it could have been no other way. Science would keep contemplating the question: why these conditions and not others?

At this junction, only assumptions and assertions pertaining to the creation of "Adam" (آدم - عليه السلام) are of interest. *The most fundamental quest here is to see whether there is anything included in the Islamic concept of Adam's (آدم - عليه السلام) creation that cannot be reconciled with these findings.*

CREATION OF ADAM AND EVE AND THE "ORIGINAL SIN"

Islamic narrative of human creation is explicitly mentioned in the Qur'an as beginning with the creation of the first human "Adam" (- آدم عليه السلام) who is given a very elevated status over other creations. At the incident of his creation, he is given residence in "Al-Jannah - Paradise" (الجنة). The name "Adam" (آدم - عليه السلام) appears in the Qur'an 25 times encompassing many aspects of who and what he is as well as how he and his progeny are created and what is required of his progeny (بنو آدم) till doomsday. A single member of Adam's progeny (بنو آدم) or simply a human being is referred to in the Qur'an generically as "Al-Insan" (الإنسان). This word and its plural appear in the Qur'an in various forms 90 times. For starters the *phased* process of creating "Adam" (آدم - عليه السلام), as well as the very long period it took to complete it are averred in the Qur'an as mentioned before. It is also understood that the Angels are beings made of light and were already in existence at Adam's (آدم - عليه السلام) creation. Thus, Angels could have existed when the universe consisted of pure radiation but appeared only after the universe (whether unique or a member of the multi-verse) became transparent and are made of photons. In his first form, "Adam" (آدم - عليه السلام) resided in "Al-Jannah - Paradise" (الجنة) which could plausibly imply him having different physical nature than what he acquired later after having to exit "Al-Jannah - Paradise" (الجنة). Therefore, his creation could have commenced early on i.e. immediately after the Angels and he was made of their substance. He could have been in the known universe (at some stage in its evolution) or in a higher status (dimension) universe. Or he evolved after exiting "Al-Jannah - Paradise" (الجنة) to be made of atoms and molecules after the formation of elements of the periodic table in the early universe. Or even billions of years later when the precursors of life materialized. Any one of these assumptions is plausible and consistent with requirements of scientific findings. What is more, one can easily explain his exiting "Al-Jannah - Paradise" (الجنة) in terms of getting trapped (due to eating from the forbidden tree which is not necessarily a material thing as known to humans depending on the definition of the word "eating") in the process of transferring radiation into atoms and then following the standard *assumptions* of science. In case one adopts the multi-verse approach, exiting "Al-Jannah - Paradise" (الجنة) can be explained in terms of descending from a higher status universe to the known three dimensional Universe. Or adopting the multi-dimensional universe picture it can be explained in terms of descending from a higher dimension (higher status) to a lower dimension (lower status) part of the universe. Since there are no one or two dimensional universes, the known three dimensional one is the lowest of the low existence referred to in Surat Al-Teen (سورة التين). Additionally, that transition can also be explained in terms of life commencing somewhere else in the universe and moving to earth later. There is also nothing in the Qur'an that negates the idea that humanity followed the process of evolution in conjunction with other organisms and species. Hence, exiting "Al-Jannah - Paradise" (الجنة) can be the explanation to descending to the lowest of the low and starting from a single cell on earth where evolution used its mechanisms to eventually produce humans. In the meantime, evolution produced all sorts of creatures as by-products, if one wills, mostly to cause the appearance of humans in the best possible way to accommodate life on earth. The conclusion, as always, is that there is nothing that science can find most plausible which is in contravention of the Qur'anic narrative.

CREATION AND PROCREATION

Unlike the prevailing belief in other religions, the Qur'an states that Adam's (آدم - عليه السلام) creation went through many ***phases*** and took quite a long time as mentioned in no uncertain terms in Ayah 7 of Surat Al-Sajdah (سورة السجدة آية 7), at the very beginning of Surat Al-Insan (سورة الإنسان آية 1) and Ayah 4,5 of Surat Al-Teen 4 و 5 آية – سورة التين). The origin of human beings is one of the most contentious issues pitting monotheistic religions against atheism. The three monotheistic religions explicitly assert descent of the entire humanity from a single person created by God (الله - سبحانه و تعالى) in a unique act of creation. Atheists claim that according to the *probabilistic* (as opposed to a *deterministic scientific fact*) nature of the act of evolutionary creation, human beings can never trace their ancestry to a single being. Obviously, both claims are not based on any physically proven facts at the present time and clearly there is no reconciling the two arguments. Unfortunately, religious stand on this matter crystallized over the past few centuries to basically represent the Christian viewpoint as indicated in the Bible. That was not accidental by any means since the fight was confined to Western scientists and Western Christians. What Islam said was of no interest to the antagonists as they always ignorantly misconceived Islam as a Christian heresy. Adding to the intellectual mess, scientifically unqualified Muslim scholars and commoners alike felt offended by theories of evolution that described humans as sharing common ancestry with primates and other lower species. These Muslims could not abide the idea that human beings would be linked with lowly creations. They pointed out many Ayahs in the Qur'an that elevate humans over even angels and the assumption that other earthly creations were there for all practical purposes to serve humans which is a fact borne by nature. It is rather ironic that on the issue of the origin of humanity some Muslims start by absolutely dismissing biblical assumptions on the matter but eventually accepting their anti-scientific conclusions anyway. This contradiction can easily be traced back to the reverence they show to writings of the great ancient Muslim scholars who, for lack of any other sources, incorporated biblical arguments into their work for the sake of completion. However, currently the situation is starkly different as there is no lack of information on these subjects in modern times. Ignoring this wealth of currently available scientific information, Muslims are also closing their eyes to explicit and subtle mentions of this creation process in the Qur'an as in the Ayahs just mentioned. While there is no doubt that according to the Qur'an, humans enjoyed a prominent status above most creations at the time of creation, theirs is relegated to the lowest of the lows excepting those who believe according to Surat Al-Teen (سورة التين). The fundamental question now is how and when did this loss of status happened. One can assume without being mistaken that departing "Al-Jannah - Paradise" (الجنة) implies this degradation; a situation that would be remedied for the believers in the hereafter. That may to some extent appeal to those believing in the departure being a punishment. Equally valid is the argument that this disgrace will only take place after doomsday. Having to depart "Al-Jannah - Paradise" (الجنة) in this context is simply a naturally mandated process due to the physical change in human nature that excluded "Adam" (آدم - عليه السلام) and Eve (حواء) from residing in this environment. As long as one is pursuing *mental exercises and logic without any concrete measurable physical facts*, one can advance many other interpretations including the atheistic one that this process never actually took place. The Qur'an is abundantly clear that humans go through three distinct qualitative existences. These are: a) before birth where they testify to God's (الله - سبحانه و تعالى) glory and take an oath proclaiming their absolute belief in Him as their creator which is indicated in Ayah 172 of Surat Al-Aa'raf (سورة الأعراف آية 172), b) from birth to death or simply during their earthly (whatever that means) life where their covenant with God (الله - سبحانه و تعالى) is put to the test and c) in the hereafter where deeds are evaluated and they are judged and consequently placed accordingly in a permanent eternal existence. The first phase is a little obscure and less understood even by Muslims. The second is obvious to everyone and the third is disputed by non-believers of all strands but is well defined nonetheless. The first and third stages are organically linked according to the Qur'an as explicitly stipulated in Ayah 172 of Surat Al-Aa'raf (آية – سورة الأعراف 172). The oath taken in the first stage is the irrefutable evidence against unbelievers' claims of ignorance on Judgment Day at the beginning of the third stage. The next Ayah (173) confirms the individual accountability stemming from this oath regardless of any protestations (on Judgment Day) from unbelievers that they were simply following what their ancestors practiced and should not be held accountable. Exercising free will in the second stage leads unambiguously to the status in the third. Continuity of human existence albeit in different forms is obvious in this Qur'anic narrative. "Adam" (آدم - عليه السلام) in his first appearance took on a special form that suited his environment in "Al-Jannah - Paradise" (الجنة). The Qur'an gives an extremely strong hint that his form then is wholly dissimilar to the physical three dimensional shape he took on later on exiting "Al-Jannah - Paradise" (الجنة). The same change in nature goes for his companion/wife (حواء). It is interesting to notice that the Qur'an mentions existence of the rest of human beings only as from a male and a female not as a similar recurring unique act; this is unequivocal in Ayah 1 of Surat Al-Nisa' (1 آية سورة النساء) and the expression used for the introduction of the rest of human beings after the unique act of creation of "Adam" (آدم - عليه السلام) and his companion/wife (حواء) is "*spreading*" (بث) rather than repeated act of creation to emphasize the routine nature of this different recurring process of creation that is characteristic of most if not all living things on earth. In "Al-Jannah - Paradise" (الجنة) only "Adam" (آدم - عليه السلام) and his companion/wife (حواء) existed and a progeny is mentioned only after exiting. The inescapable implication here is that in "Al-Jannah - Paradise" (الجنة) there was no "*procreation*", as currently understood by science, which is the single fundamental process for the production of the rest of humanity on earth. One can easily see that existence in "Al-Jannah - Paradise" (الجنة) is controlled by different physical laws from those familiar to humans in their three dimensional world. That is the reason for the inevitable change in nature taking place in transiting each stage of existence discussed above or the two acts of exiting "Al-Jannah - Paradise" (الجنة) and death. On the other hand, while the Qur'an distinguishes between the original act of creating Adam (آدم - عليه السلام) and his companion/wife (حواء) and their progeny, it lumps all humans in the original act of creation of Adam (آدم - عليه السلام). That means that somehow every single human being's eventual existence was determined and took place at the unique act of creating Adam (آدم - عليه السلام). That is why the Qur'an speaks alternatively of creating Adam (آدم - عليه السلام) and any other human being from the substance of the universe while specifies the process of procreation only for the rest of humanity excluding Adam (آدم - عليه السلام) and his companion/wife (حواء). The substance of the universe is described in the Qur'an as wet dust/mud which is also used in other places to describe everyday materials familiar to humans. This dual use is an affirmation of the thesis in this work that Qur'anic Arabic words and expressions have global in addition to common usage. One can detect affinity and similarity in this expression with biologists and evolutionists referring to the origin of life as starting in the "primordial ooze". As for the process of procreation, the Qur'an explicitly only alludes to eggs and sperms (as opposed to the initial process of creation) in the description of the "fertilized human egg" as (نطفة) and the development of the fetus in several Ayahs at least 12 times within the texts of Surat al-Nahl (النحل), Al-Kahf (الكهف), Al-Hajj (الحج), Al-Mo'menoun (المؤمنون), Fater (فاطر), Yaseen (يس), Ghafer (غافر), Al-Najm (النجم), Al-Qyamah (القيامة), Al-Insan (الإنسان), Abasa (عبس). The Qur'anic narrative also explicitly mentions Eve's (حواء) creation *from* Adam's (آدم - عليه السلام) original form. The clear implication here is that both males and females are identical in form albeit different in physiology. There isn't the slightest hint in the Qur'an that in their original form either one had a privileged status above the other. After their departure from "Al-Jannah - Paradise" (الجنة), their survival in the new environment demanded specialization and division of labor. That in a nutshell simply explains any observable physiological differences between men and women while affirming in no uncertain terms their equality and single nature. In the Qur'an, the cause for departing the privileged life in "Al-Jannah - Paradise" (الجنة) and enduring the harsh earthly (whatever that means) existence encompassing pains of birth, growing up and death as well as everything experienced in between is eating from the forbidden tree. Judaism and Christianity imply similar reasons with some twist and condemn Eve (حواء) for originating this "misdeed" with Christianity giving it the denigrating term the "**Original Sin**". In both belief systems, women are looked down upon unkindly and they explain the female monthly pain as the deserved punishment for that sin. While modern rational Jews and Christians reject that attitude, they nonetheless avoid discussing the explicit anti-female texts and attitudes of both belief systems to keep their cherished Jewish and Christian affiliation. On the other hand, Islam explicitly associates the *mistake* equally to both "Adam" (آدم - عليه السلام) and Eve (حواء). **There is absolutely no "Original Sin" in Islam**. Consequently, there is no punishment whether to both genders or only to females either. Female monthly period is a direct consequence of the division of labor imposed on humans for survival. Additionally, losing the privileged existence in "Al-Jannah - Paradise" (الجنة) on exiting is described in Islam as a natural after effect to eating from the forbidden tree which made their nature incompatible with residing in "Al-Jannah - Paradise" (الجنة). They had to bear the consequences of their actions. God (الله - سبحانه و تعالى) advised them not to eat from that tree; they ignored that advice and got themselves in trouble; but God (الله - سبحانه و تعالى) gave them a way out after their repentance (in the sense of descending to eventually live on earth) exacting no punishment. This attitude is consistent with Islam's rational and compassionate approach to life in submission to the entreaties of the Merciful God (الله - سبحانه و تعالى). Before one misinterprets the phrase "eating from the forbidden tree", it is reasonable to assume that the words "eating" and "tree" are not necessarily of the same nature as what is commonly understood by every human being in this life simply because nature of "Al-Jannah - Paradise" (الجنة) where this act took place is decidedly different. These are the indisputable and irrefutable facts as they appear in the sacred books of the three monotheistic religions. However, ignoring their own appalling shortcomings some people, especially Westerners, are fond of attacking Islam and Muslims claiming they oppress women. There is no denying that practices of *some* Muslim individuals and communities are anti-woman. No one ever said that *all* Muslims behave in this way. The exact same situation can be found in the West and everywhere else in the world. Thus, these attacks are clearly and purposefully selective and are advanced by bigots. Some Islamic mandates and rules such as inheritance laws for example are used by Westerners to claim that Islam discriminates against women. In this case as in all other cases where males and females are not subjected to the same rules and regulations it has already been established that these approaches result from division of labor for survival. Men are wholly without any exceptions responsible for materially supporting their dependents particularly females while females have absolutely no such obligation. An impartial look at Islamic inheritance laws would actually find women more privileged than men. The supreme irony is that Muslim women in general don't complain of discrimination and those who are obviously repressed do complain about men not following the rules of Islam not the other way around. Westerners should strive to put their morally (or better immorally) messed up own house in order rather than claiming to be the guardians of Muslims who never asked them to interfere in their behalf anyway.

SULEIMAN/SOLOMON (سليمان – عليه السلام) OF THE QUR'AN

Suleiman/Solomon (سليمان – عليه السلام) is considered by biblical scholars as the first real historical person among the patriarchs mentioned in the Bible and as such they use his reign as the real beginning of biblical history. Among Jews and Christians, he is referred to as "King Solomon" while in Islam he is a prophet conveying the message of God (الله - سبحانه و تعالى) in exactly the same manner as any other prophet from Adam ((آدم - عليه السلام)) to Mohammed (محمد - عليه الصلاة و السلام). The story of Suleiman/Solomon (سليمان – عليه السلام) as told by the Qur'an attributes unusual unique powers to the person of Suleiman (سليمان – عليه السلام) enabling him to enact extraordinary phenomena. The bible describes him as an unusual person as well. Nonetheless, the Qur'anic version is probably more detailed and more elaborate than what is given by the bible. These attributed phenomena represent fertile ground for atheists to attack these narratives as supernatural actions clearly not permitted by the laws of nature. It is noticeable that every prophet and messenger of God (الله - سبحانه و تعالى) mentioned in the Qur'an is one way or the other associated with some out of the ordinary actions. That is why Muslims for example do not see any abnormal matters concerning Jesus' (عيسى – عليه السلام) actions to mandate his divinity. Suleiman (سليمان – عليه السلام) is mentioned in the Qur'an 17 times in 7 different Surahs frequently in association with his father David (داود – عليه السلام) who is also a prophet in the Islamic tradition. According to Ayah 15 of Surat Al-Naml/Ants (سورة النمل) and Ayah 79 of Surat Al-Anbia' (سورة الأنبياء), they were both granted sovereignty and exceptionally vast knowledge by God (الله - سبحانه و تعالى). The most extraordinary phenomena linked to Suleiman's (سليمان – عليه السلام) powers in the Qur'anic narrative are two in particular. That is his ability to communicate with ants and birds and his fulfilled order to Jinn in his service to instantaneously transport Balqees's (Queen of Sheba – بلقيس ملكة سبأ) throne (which is located thousands of miles away) to his court before her arrival. Ayahs 16-28 of Surat Al-Naml/Ants (سورة النمل) explicitly express the fact of his ability to communicate with ants and birds. The instantaneous transportation of Balqees's (Queen of Sheba – بلقيس ملكة سبأ) throne over such immense distance is unequivocally expressed in Ayahs 38-40 of Surat Al-Naml/Ants (سورة النمل). While such narrative and even remotely similar stories given in the Bible would now be described as *"parables"* and allegories by the faithful, Muslims would consider them as absolute facts that most certainly took place. Almost all Muslims, scholars or otherwise, base their judgment on their acceptance of the collective integrity of the Islamic tradition and its wholesomeness that was proven beyond the shadow of a doubt in so many other situations even if they themselves cannot give an outright explanation to these incidents. The question now is whether or not there is any objections that might be raised by modern science to the plausibility of such occurrences conceding beforehand their exceedingly low probabilities? As far as communications are concerned, biophysicists, neurologists and in general scientists involved in studying human senses currently subscribe to the notion that all brain activities are in essence nothing more than electrochemical processes. Hearing and speech are two capabilities most living things endowed with that are carried out by such activities. Once human ear receives a sound signal, human brain responds in a complicated sequence of procedures resulting in the understanding of what was communicated. The same is true of other living things according to their evolved physiology. Obviously each species has its own interpretation of what the received signal mean in accordance with its brain structure and its processing capability. It is not hard for anyone to accept the fact that pets such as cats and dogs can to some degree understand and respond to human speech and vice versa. The difficulty arises when one extrapolates that process to ants because of their diminutive sizes. However, the mechanism should be the same albeit with accommodating modifications conforming with the electrochemical nature of the process. Therefore, it should be plausible to accept the assertion that Suleiman (سليمان – عليه السلام) as a unique human being, since he is designated a prophet by God (الله - سبحانه و تعالى), was endowed with a special electrochemical and physical capabilities to receive, process and understand communicating signals from ants. There is no denying the extreme low probability of such endowment except when one realizes that the Qur'an tells of no other human being ever having such endowment. Evolution and natural selection are based on random mutations and there is no legitimate reason, other than the probabilistic nature of the process, to exclude such possibility from taking place within the physiology of Suleiman (سليمان – عليه السلام). It should also be emphasized here that the Qur'an never mentions a conversation between the ants and Suleiman (سليمان – عليه السلام) but only that he understood their cross discussion. On the other hand, what is really intriguing here is the fact that the ants were aware of who he was which may lead one down the road of presuming that other living things (probably up to a point) recognize the divine privileges bestowed upon humankind no matter how speculative this assumption might be. Therefore, there is no mystery in the scientific plausibility of Suleiman (سليمان – عليه السلام) being able to understand ants' communications in principle. However, the details of how this is done is beyond present day comprehension bearing in mind that for the sake of this study only the plausibility not the proof is needed to refute any atheistic arguments against such phenomenon. The second issue is far more fascinating as it exposes how little humanity knows of its universe after all these scientific breakthroughs; a fact any insightful inspection of the Qur'an lucidly and demonstrably shows. It also unmistakably reveals the accuracy of every single word in the Qur'an as the word of God (الله - سبحانه و تعالى) that cannot be interchanged or replaced by any other. In this instance, it took only 3 Ayahs to make that point as the instantaneous transportation of Balqees's (Queen of Sheba – بلقيس ملكة سبأ) throne over such immense distance is unequivocally expressed in Ayahs 38-40 of Surat Al-Naml/Ants (سورة النمل). The first Ayah (38) tells of Suleiman's (سليمان – عليه السلام) expressing his desire of someone from among his Jinn servants to bring that throne to his court. The second Ayah (39) gives the answer of one of these Jinn volunteering to accomplish that feat before Suleiman (سليمان – عليه السلام) ends his meeting with the Jinn and stands up to leave as he commands great powers. However, the third Ayah (40) informs that such achievement did not sound fast enough to another member of the Jinn who wanted to boast about his capabilities in the competition to serve, please and impress Suleiman (سليمان – عليه السلام) by declaring that he could perform that act before Suleiman (سليمان – عليه السلام) blinks his eye. What distinguished the second member of the Jinn from the first volunteer as stated matter-of-factly in this Ayah is his ***acquisition of more divine knowledge*** from God (الله - سبحانه و تعالى). The Ayah ends with Suleiman (سليمان – عليه السلام) looking at that throne before him and being immensely thankful to God (الله - سبحانه و تعالى) for the blessings he received as an answer to his prayers. As far as atheists are concerned, this is a farfetched mythical tale that blatantly exceeded any reason this time. For most of the faithful of the other two monotheistic religions with a fainting sliver of respect for science, this is simply an allegorical parable. On the other hand, for Muslims it is an absolute fact that took place since it is mentioned in the Qur'an which is the literal word of God (الله - سبحانه و تعالى) that is guaranteed by the Almighty (الله – سبحانه و تعالى) Himself not to be corrupted by any means till the end of time. Now, what one should accept of these three incompatible points of view? And if one insists on the Islamic stand, how can one reconcile this Islamic unshaken belief in the veracity of this narrative with findings of modern science? Present day physicists are at a loss trying to explain "quantum reality" and those who vehemently opposed quantum theory's probabilistic nature including luminaries such as Albert Einstein and even one of its own founding fathers; Erwin Schrodinger raised (and still to this day raise) legitimate questions about its validity regardless of its unbroken success in explaining every problem it faced. It was thought that the ultimate failure of the quantum theory would be in its unavoidable allowance of what became known as "quantum entanglement". In this phenomenon two objects at a distance that can even be measured in cosmological terms are closely linked as a change in one necessarily affects an instantaneous change in the other without any possibility of information transmission which is limited by the speed of light anyway. Against all odds, "quantum entanglement" has been proven as a permissible phenomenon by the laws of nature. Recent experiments to test this entanglement resulted in its confirmation. Moreover, physicists succeeded in instantaneously *transporting* particles over some distance employing this phenomenon. Transportation here means placing the same object in a completely different spatial location without actually travelling taking advantage of the scientific fact that all particles are identical. This phenomenon unleashed the talents of science fiction writers who coined the term "Teleportation". However, so far scientists succeeded only in transporting elementary particle over distances measured in centimeters. In principle, there is no physical reason not to expect more successful results achieving complex structures' transportation over much longer distances. Therefore, there is absolutely no scientific legitimacy in dismissing the Qur'anic statement of the Jinn transporting Balqees's (Queen of Sheba – بلقيس ملكة سبأ) throne over such an immense distance. The Qur'an is extremely precise in attributing this exploit to *the acquisition of exceptional kind of knowledge* not available to all Jinn. Again, this Qur'anic statement was conveyed to humanity by an illiterate person of the highest moral qualities over fourteen hundred years ago. Mohammed (محمد - عليه الصلاة و السلام) simply relayed the word of the All-Knowing God (الله - سبحانه و تعالى) which he received from the Archangel Gabriel. It should be clear at this juncture that in this example it is the atheistic arguments that are squarely refuted by science not the Qur'an and in this regard one is not speaking of plausibility but rather certainty backed by modern science.

On mentioning of plausibility, one can also explain the uniquely Islamic issue of Mohammed's (محمد - عليه الصلاة و السلام) Night Journey or Al-Isra' wa Al-Me'raj (الإسراء و المعراج) from a different point of view employing "quantum entanglement" arguments. In this journey, accompanied by the Archangel Gabriel he went to Jerusalem to lead some prophets in the prayer (indicating his elevated status) and then toured the heavens and ultimately came into the physical presence of God (الله - سبحانه و تعالى) where he received the command for the Islamic five daily prayers. This may also solve the age old Muslim scholars' debate about the process taking place by body or soul rendering it immaterial and irrelevant. It must be stressed here that neither Suleiman (سليمان – عليه السلام) nor Mohammed (محمد - عليه الصلاة و السلام) claimed to have carried out their unusual acts using their own powers. The Qur'anic narrative in both cases is crystal clear about who provided the mechanisms. In Suleiman's (سليمان – عليه السلام) case, it was the Jinn and in Mohammed's (محمد - عليه الصلاة و السلام) case, it was the Archangel Gabriel. Islam attributes different nature to Angels and Jinn that are not subject to the laws that govern human interaction with the cosmos. But they all faithfully obey and strictly comply with God's (الله - سبحانه و تعالى) orders nonetheless.

Isara' and Me'raj

The story of Isara' and Me'raj (الإسراء و المعراج) is prominently narrated at the very beginning of Surat Al-Isra' (سورة الإسراء). It is interesting to recognize that the roots of both words are well known and understood in the Arabic language. The word Isra' (الإسراء) means travelling at night which is an act Arabs for millennia carried out and incorporated the expression in their common descriptions. The word Me'raj (المعراج) means changing direction to reach a specific destination which is very well understood by all Arabic speakers. However, the expression never entered the daily description of their movements. The fascinating fact is that the Qur'an distinguishes between the two portions of Mohammad's (محمد - عليه الصلاة و السلام) trip. The first leg; Al-Isra' (الإسراء) from Makkah to Jerusalem is explicitly spelled out while the second; Al-Me'raj (المعراج) which describes the ascendance to the many levels of heaven to personally and physically be in the immediate presence of God (الله - سبحانه و تعالى) is only implicit. The terrestrial leg is something everyone can relate to and the speed with which it was carried out will become easily comprehensible with the passing of time and the invention of the various means of transportation. Since everyone regardless of the timeframe can recognize and scrutinize the possibility of such an act especially when accurate description of landmarks along the route was provided by Mohammad (محمد - عليه الصلاة و السلام), no denial of it was permissible and it was thus explicitly spelled out in the Qur'an. On the other hand, the second leg was beyond the comprehension of individuals of that era when the Qur'an was revealed and as a matter of fact is difficult to explain even fourteen centuries later due to lack of knowledge. One can only speculate that it was movement from the familiar three-dimensional universe to higher dimensions or from one universe to another within a multiverse or employing quantum entanglement, etc. These are mere speculations provided by modern science. Therefore, explicit mention was dropped in order not to overburden human intellect of believers till human knowledge reaches the point of finding out an acceptable non-linguistic explanation if ever there is one. However, since Islam is decidedly holistic what cannot be temporarily comprehensible should be taken as truthful on account of the countless truths and predictions of the entire body of Islam. There is nothing unusual about that attitude since that is exactly how science and human knowledge is acquired and progresses. In a sense, Al-Me'raj (المعراج) is a unique *prediction* advanced by the Qur'an for humans to ponder and try to discover its intricacies if they ever could. Validity of the Qur'an is timeless and human intellect is continuously encouraged and challenged to find out about all the physical truths indicated in it. Islamic tradition is clearly rich with descriptions of encounters between humans (mostly prophets) and angels and Jinn (الجن). However, only two incidents are registered as individual human beings personally and physically be in the immediate presence of God (الله - سبحانه و تعالى). These individuals are Mohammad (محمد - عليه الصلاة و السلام) and Moses (موسى - عليه السلام). There are vivid descriptions of both encounters in both the Qur'an and the Hadith. Mohammad (محمد - عليه الصلاة و السلام) encounter in this state is mainly detailed in Surat Al-Isra' (سورة الإسراء) and Surat Al-Najm (سورة النجم) in addition to numerous narratives recorded in the standard Hadith books such as Sahih Al-Bokhari (صحيح البخارى), Sahih Muslim (صحيح مسلم), Mosnad Imam Ahmad ibn Hanbal (مسند الإمام أحمد), etc. In all these accounts Mohammad (محمد - عليه الصلاة

و السلام) is described as in an absolute state of serenity and peace. On the other hand, Moses' (موسى - عليه السلام) encounter in this state is detailed in Surat Taha (سورة طه), Surat Al-Nisa' (سورة النساء), Surat Al-A'raf (سورة الأعراف), Surat Al-Sho'ra' (سورة الشعراء), Surat Al-Naml (سورة النمل), Surat Al-Qasas (سورة القصص) and Surat Al-Nazia'at (سورة النازعات). In all these accounts, Moses (موسى - عليه السلام) is frightened and desperate due to God's (الله - سبحانه و تعالى) presence. Moses (موسى - عليه السلام) is highly regarded and effusively praised by God (الله - سبحانه و تعالى) in many places in the Qur'an and as such his fright has nothing to do with him as an individual. God (الله - سبحانه و تعالى) showed willingness to reveal Himself to Moses (موسى - عليه السلام) as requested but with an unambiguous warning that he and his surroundings have no possibility of bearing the impact and the consequences which is what Moses (موسى - عليه السلام) experienced right away and was frightened. The difference between Mohammad's (محمد - عليه الصلاة و السلام) and Moses' (موسى - عليه السلام) situations has nothing to do with their persons. It is rather explained in terms of where the encounters took place. Mohammad's (محمد - عليه الصلاة و السلام) encounter was when personally and physically being in the immediate presence of God (الله - سبحانه و تعالى), took place in heaven where the surroundings are compatible with the presence of God (الله - سبحانه و تعالى) while Moses' (موسى - عليه السلام) encounter was when personally and physically being in the immediate presence of God (الله - سبحانه و تعالى), took place on earth where the three dimensional universe cannot accommodate the presence of God (الله - سبحانه و تعالى). As can be easily seen, this approach to interpreting these Qur'anic narratives can be reconciled with any speculation about the various potentialities of multi-dimensional universe or multiverse theories of modern science if one wishes.

Epilogue

It was historically assumed that Christian apprehensions about Islam stemmed from the irreconcilability between the fundamental notions of faith in God (الله – سبحانه و تعالى) being outside the realm of human space and time (Islam) and the requirement that He is physically personified, persecuted and crucified here on earth as a prerequisite for salvation of humans (Christianity). Western Christianity's self-delusion that Islam is simply a Christian heresy satisfied its prejudices and prompted it to chart a course to confront it emotionally, ideologically and physically. However, when the Christian West rid itself of Christianity, it held on fast to all its trepidations about Islam. It has been concluded in the first book of this series; "Why Do They Hate Us So Much?" that the most probable cause for such anxieties was enduring the existential "Ottoman Nightmare". When Western Christianity failed to stand up to assaults of Age of Enlightenment's great thinkers, Islam was ignorantly implicitly dismissed as well without any viable discussion. That served the West well during the disgraceful era of European Colonialism and Imperialism. But no more with Islamic renaissance starting at the beginning of the twentieth century. Currently, another Western self-delusion satisfies its bigotry and irrational hatred of Islam. This is the chimera of "Islamic Terrorism" which does not seem to work despite the untold worldwide misery it has caused. Underlying the fantasy of Muslims' hatred of the West is the supposition of their technical incompetence because of their faith and consequently their jealousy of Western life styles. This book explodes the myth that Islam is a Christian derivation proving that Islam while acknowledging the legitimacy of the messages conveyed by Moses (موسى – عليه السلام) and Jesus Christ (عيسى – عليه السلام), it dissociated itself from the corruptions of human additions that plagued both Judaism and Christianity when Mohammad's (محمد – عليه الصلاة و السلام) prophethood was only days old even before it confronted the obvious immediate threat of unbelievers of Makkah. It therefore substantiates the fact that what atheists advance to dismiss Christianity does not concern Islam in the least. Additionally, it ascertains the historical fact that what is currently called "Western Civilization" is a by-product of the golden age of the "Islamic Civilization". Islam's intellectual heritage is briefly discussed to illustrate that claim. Having firmly established these facts, this undertaking proceeded to construct its main goal of independently showing the affinity of Islam and humanity's unending curiosity about itself, its surroundings, its universe and its purpose displayed in its continuous acquisition of knowledge. Since science has been both humanity's (self-appointed by atheists as its true representatives) tool to acquire knowledge and its formidable means to bring down Christianity, it was natural to follow the progress of scientific pursuits till their state-of-the-art status in the twenty first century. This obviously done as qualitatively as possible without obscuring the very fundamental conceptions of science. The theme that was developed in the other two previous books of this series ascertaining the "Purposefulness" of the unfolding of the on-going creation process is given an elegant historical proof in the elucidation of the "Grand Cosmological Divine Plan". This plan explains in details the correspondence between progress of humanity's knowledge according to its collective brain power and the inevitability of decline of the "Islamic Civilization" exactly when it happened, the rise of the "Western Civilization" at the appropriate time and the perceived current renaissance of the Islamic thought process. The core argument of the book is the compatibility of Islam's sacred narrative and whatever science can irrefutably establish. This is done through subjecting this narrative to the scrutiny of science. It is concluded that there is nothing science finds out that could not be shown compatible with an interpretation of sacred statements of Islam. In the meantime, a call is urged for a "Paradigm Shift" in the usage of the words of the Arabic

Language from the mundane daily applications to the more global meanings according to the rule of that fascinating language. Through it all, nothing was found to create antagonism between Islam and the West other than ignorance and disinformation. It is hoped that this endeavor contributes something to better understanding and appreciation of Islam among the average Western people.

www.ingramcontent.com/pod-product-compliance
Lightning Source LLC
Chambersburg PA
CBHW020625220526
45464CB00001B/32